OF SYMBOLS

$\lvert x \rvert$	the absolute value of x
$\overleftrightarrow{AB}$	line AB
$\overset{\circ}{\longrightarrow}{AB}$	half-line AB
$\overrightarrow{AB}$	ray AB
$\overline{AB}$	line segment AB
$\angle ABC$	angle ABC
$m(\overline{P_r P_s})$	the measure of $\overline{P_r P_s}$
$\approx$	the approximation symbol
$\square\, ABCD$	rectangle $ABCD$
$\mathcal{C}(\square ABCD)$	the area of rectangle $ABCD$
$\triangle ABC$	triangle ABC
$\square\, ABCD$	parallelogram $ABCD$
$P(A)$	the probability of A
$_nP_r$	the number of permutations of n things taken r at a time
$n!$	n factorial
$_nC_r$	the number of combinations of n things taken r at a time
$\overline{X}$	the arithmetic mean
σ	the standard deviation

Introduction to
MATHEMATICS

Introduction to

MATHEMATICS

Bruce E. Meserve

Professor of Mathematics, University of Vermont

Max A. Sobel

Professor of Mathematics, Montclair State College

Prentice-Hall, Inc., Englewood Cliffs, New Jersey

Library of Congress Cataloging in Publication Data

MESERVE, BRUCE ELWYN.
    Introduction to mathematics.

    1. Mathematics—1961– I. Sobel, Max A.,
joint author. II. Title.
QA39.2.M49 1973   510   72–6566
ISBN 0-13-487397-1

Introduction to MATHEMATICS

Bruce E. Meserve / Max A. Sobel

© 1973, 1969, 1964 by PRENTICE-HALL, INC., Englewood Cliffs, New Jersey

10 9 8 7 6 5 4 3 2 1

Printed in the United States of America

PRENTICE-HALL INTERNATIONAL, INC., *London*
PRENTICE-HALL OF AUSTRALIA, PTY. LTD., *Sydney*
PRENTICE-HALL OF CANADA, LTD., *Toronto*
PRENTICE-HALL OF INDIA PRIVATE LIMITED, *New Delhi*
PRENTICE-HALL OF JAPAN, INC., *Tokyo*

CONTENTS

2

An Introduction to **SETS**

3

Concepts of **LOGIC**

4

Systems of NUMERATION   116

5

Mathematical SYSTEMS   158

6

Sets of NUMBERS

7

An Introduction to ALGEBRA

8

An Introduction to GEOMETRY   296

9

An Introduction to PROBABILITY   354

10

An Introduction to STATISTICS

EPILOGUE

ANSWERS TO
ODD-NUMBERED EXERCISES

INDEX

PREFACE

Mathematics can be fun! The authors have embarked with this idea and have gone ahead to introduce a variety of interesting and timely topics without a major emphasis upon the so-called practical applications. This point of view, it is felt, will leave the reader with a better picture of the true meaning and beauty of mathematics as opposed to a traditional approach with a major emphasis on abstract manipulations.

Many users of the prior two editions of this text, teachers and students alike, have made numerous worthwhile suggestions that have been incorporated into this third edition. Some of the key changes feature an early introduction to the basic concepts of sets and logic (Chapters 2 and 3), and the inclusion of a chapter on statistics (Chapter 10). Another distinct new feature is the inclusion of sets of explorations following almost every section throughout the book. These are discovery-type exercises that allow the reader to explore interesting mathematical ideas on his own, an approach that many consider to be among the most rewarding type of activity in mathematics. The entire first chapter, "Explorations With Mathematics", sets the stage for this approach.

Very little mathematical background is required of the reader. It is expected that he will have had some secondary school introduction to algebra and geometry, but no working knowledge of any of the skills normally taught in these subjects is presupposed. Maturity, on the other hand, is expected; and interest in the subject is anticipated.

The topics considered in this book are used throughout many of the activities of citizens in our technological society. Accordingly, this book has been written with a number of different audiences in mind. The subject matter is suitable for the undergraduate college student who has had moderate secondary school training in mathematics, one who is not a mathematics major but who wishes to acquire a basic understanding of the nature of mathematics. Many students seeking a knowledge of basic mathematics are so included. Frequently, prospective elementary-school teachers will be among such students. This book is also appropriate for inservice courses for elementary school and many junior high school teachers. To this end, the emphasis throughout the book is on key concepts and the structure of mathematics, without undue concern over the mechanical procedures.

The famous French mathematician René Descartes concluded his famous *La Géométrie* with the statement: "I hope that posterity will judge me kindly, not only as to the things which I have explained, but also as to those which I have intentionally omitted so as to leave to others the pleasure of discovery." The authors have attempted to provide a great deal of exposition in this text. They have, however, left a great deal for the reader so that he may experience the true beauty of mathematics through discovery.

Bruce E. Meserve
Max A. Sobel

1

Explorations with
MATHEMATICS

Mathematics has numerous practical applications ranging from everyday household usage to the charting of astronauts through outer space. Many people study the subject because of such applications as they relate to their particular fields of interest. Mathematics may also be considered as part of our great cultural heritage, for it has a history that dates back many thousands of years. As such, it should be studied by the average citizen who wishes to be literate in this twentieth century.

Many people study mathematics just for fun! These individuals would rather solve a mathematical puzzle than read a book, watch television, or go to a movie. Admittedly, not everyone does have, or can have, this type of disposition. On the other hand, most of us use mathematical concepts in a variety of ways and have never been given an opportunity to explore some of the more interesting aspects of mathematics.

One of the unique features of this book is the sets of explorations that follow many of the sections. These are collections of exercises for the reader to explore, often leading to some particular discovery or generalization. To set the stage for such developments this first chapter contains a number of interesting explorations in different areas of mathematics. Thus this introductory chapter contains a smorgasbord of items served to whet the reader's appetite for the main course that follows. The material contained in this chapter is not sequential. However, the spirit of this development is significant as an indication of modern approaches to the teaching of mathematics by discovery methods.

1-1

Explorations with Number Patterns

Mathematicians love to search for patterns and generalizations in all branches of their subjects—in arithmetic, in algebra, and in geometry. A search for such patterns may not only be interesting but may also help one develop insight into mathematics as a whole.

Many patterns that often escape notice may be found in the structure of arithmetic. For example, consider these multiples of 9.

$$1 \times 9 = 9$$
$$2 \times 9 = 18$$
$$3 \times 9 = 27$$
$$4 \times 9 = 36$$
$$5 \times 9 = 45$$
$$6 \times 9 = 54$$
$$7 \times 9 = 63$$
$$8 \times 9 = 72$$
$$9 \times 9 = 81$$

What patterns do you notice in the column of multiples on the right? You may note that the sum of the digits in each case is always 9. You should also see that the units digit decreases (9, 8, 7, . . .), whereas the tens digit increases (1, 2, 3, . . .). What lies behind this pattern?

Consider the product

$$5 \times 9 = 45$$

To find 6×9 we need to add 9 to 45. Instead of adding 9, we may add 10 and subtract 1.

$$
\begin{array}{r}
45 \\
+10 \\
\hline
55
\end{array}
\qquad
\begin{array}{r}
55 \\
-1 \\
\hline
54 = 6 \times 9
\end{array}
$$

That is, by adding 1 to the tens digit, 4, of 45, we are really adding 10 to 45. We then subtract 1 from the units digit, 5, of 45 to obtain 54 as our product.

Most people speak of "adding digits," "subtracting from the units digit," and so forth, as we have done. Many teachers are more precise in their terminology and recognize that digits are numerals, that is, symbols for numbers, rather than numbers. Numerals can be written. Only numbers can be added or subtracted. However, unless the more precise terminology is needed to avoid major confusion, we use the commonly accepted phraseology.

The number 9, incidentally, has other fascinating properties. Of special interest is a procedure for multiplying by 9 on one's fingers. For example, to multiply 9 by 3, place both hands together as in the figure, and bend the third finger from the left. The result is read as 27.

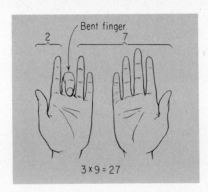

3 x 9 = 27

The next figure shows the procedure for finding the product 7×9. Note that the seventh finger from the left is bent, and the result is read in terms of the tens digit, to the left, and the units digit to the right of the bent finger. (Note that a thumb is considered to be a finger.)

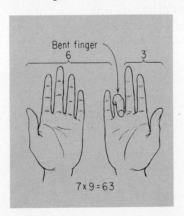

7 x 9 = 63

What number fact is shown in the next figure?

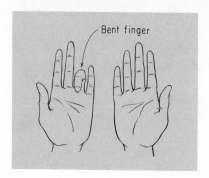

Here is one more pattern related to the number 9. You may, if you wish, verify that each of the following is correct:

$$1 \times 9 + 2 = \quad 11$$
$$12 \times 9 + 3 = \quad 111$$
$$123 \times 9 + 4 = \quad 1{,}111$$
$$1{,}234 \times 9 + 5 = \quad 11{,}111$$
$$12{,}345 \times 9 + 6 = 111{,}111$$

Try to find a correspondence of the number of 1's in the number symbol on the right with one of the numbers used on the left. Now see if you can supply the answers, without computation, to the following:

$$123{,}456 \times 9 + 7 = ?$$
$$1{,}234{,}567 \times 9 + 8 = ?$$

Let us see *why* this pattern works. To do so we shall examine just one of the statements. A similar explanation can be offered for each of the other statements. Consider the statement:

$$12{,}345 \times 9 + 6 = 111{,}111$$

We can express 12,345 as a sum of five numbers as follows:

$$
\begin{array}{r}
11{,}111 \\
1{,}111 \\
111 \\
11 \\
\underline{1} \\
12{,}345
\end{array}
$$

Next we multiply each of the five numbers by 9:

$$11{,}111 \times 9 = 99{,}999$$
$$1{,}111 \times 9 = 9{,}999$$
$$111 \times 9 = 999$$
$$11 \times 9 = 99$$
$$1 \times 9 = 9$$

Finally, we add 6 by adding six ones as in the following array, and find the total sum:

$$99{,}999 + 1 = 100{,}000$$
$$9{,}999 + 1 = 10{,}000$$
$$999 + 1 = 1{,}000$$
$$99 + 1 = 100$$
$$9 + 1 = 10$$
$$1 = \underline{1}$$
$$111{,}111$$

Here is another interesting pattern. After studying the pattern, see if you can add the next four lines to the table.

$$1 \times 1 = 1$$
$$11 \times 11 = 121$$
$$111 \times 111 = 12{,}321$$
$$1{,}111 \times 1{,}111 = 1{,}234{,}321$$
$$11{,}111 \times 11{,}111 = 123{,}454{,}321$$

Do you think that the pattern displayed will continue indefinitely? Compute the product $1{,}111{,}111{,}111 \times 1{,}111{,}111{,}111$ to help you answer this question.

Interesting discoveries can often be made by studying arithmetic patterns. A famous German mathematician by the name of Karl Gauss (1777–1855) is said to have been a precocious child who would often drive his teachers to despair. The story is told that on one occasion his teacher asked him to add a long column of figures, hoping to keep him suitably occupied for some time. Instead, young Gauss recognized a pattern, and gave the answer immediately. He is said to have found the sum of the first 100 counting numbers, as indicated in this array:

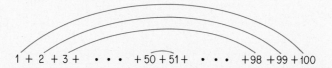

$$1 + 2 + 3 + \cdots + 50 + 51 + \cdots + 98 + 99 + 100$$

He reasoned that there would be 50 pairs of numbers, each with a sum of 101 (consider $100 + 1$, $99 + 2$, $98 + 3$, ..., $50 + 51$). Thus the sum is 50×101, that is, 5050.

Another interesting pattern emerges from the story of the man who was offered a month's employment at the rate of 1¢ for the first day, 2¢ for the second day, then 4¢, 8¢, 16¢, and so forth. He wished to determine what his total wages would be for four weeks of five days each; that is, for 20 working days.

Of course, one could list the 20 daily salaries and add, but this would be a tedious job indeed. Let us, rather, consider the big problem by exploring smaller tasks first. This is frequently a helpful technique in problem solving. Thus we consider total salaries in cents, for five, for six, for seven, and for eight days.

For 5 days	For 6 days	For 7 days	For 8 days
1	1	1	1
2	2	2	2
4	4	4	4
8	8	8	8
16	16	16	16
31	32	32	32
	63	64	64
		127	128
			255

Do you see a pattern emerging? Compare the total for the first five days with the salary for the sixth day; compare the total for six days with the salary for the seventh day. Notice that the total for five days is one cent less than the salary for the sixth day; the total for six days is one cent less than the salary for the seventh day; and so forth. Thus his total salary for ten days will be one cent less than the salary for the eleventh day. That is,

$$1 + 2 + 4 + 8 + 16 + 32 + 64 + 128 + 256 + 512$$
$$= 2(512) - 1, \text{ or } 1023$$

His salary for the eleventh day would be 1024 cents; in 10 days he will earn a *total* of 1023 cents, or $10.23.

To use this approach to answer the original question will require a good deal more work, but will still be easier than the addition of 20 amounts. By a doubling process we first need to find the salary for the twentieth day. The total salary for all 20 days can then be found by doubling this amount (to find the salary for the twenty-first day), and subtracting one.

To generalize, we note that the doubling process starting with one gives powers of two:

$$2^0 = 1, \quad 2^1 = 2, \quad 2^2 = 4, \quad 2^3 = 8, \quad 2^4 = 16, \quad 2^5 = 32, \quad \ldots$$

Then, by our discovery of a pattern, we may say

$$2^0 + 2^1 + 2^2 + 2^3 + \cdots + 2^n = 2^{n+1} - 1$$

When $n = 5$, we have

$$2^0 + 2^1 + 2^2 + 2^3 + 2^4 + 2^5 = 1 + 2 + 4 + 8 + 16 + 32$$
$$= 2^6 - 1 = 63$$

To find the total salary for 20 working days, we need to find the sum:

$$2^0 \quad + \quad 2^1 \quad + \quad 2^2 \quad + \cdots + \quad 2^{19}$$

1st day   2nd day   3rd day   20 th day

From the preceding discussion we know that this sum is equal to $2^{20} - 1$. We can compute 2^{20} by a doubling process, or can use a shortcut such as the following:

$$2^{10} = 2 \times 2^9 = 2 \times 512 = 1024$$

Alternatively, we may say

$$2^{10} = 2^5 \times 2^5 = 32 \times 32 = 1024$$

In a similar manner we compute

$$2^{20} = 2^{10} \times 2^{10} = 1024 \times 1024 = 1,048,576$$

and thus $2^{20} - 1 = 1,048,575$.

In 20 days, our worker would earn a total of \$10,485.75.

In the following exercises the reader will have an opportunity to make other discoveries of his own through careful explorations of patterns.

exercises

1. Verify that the process for finger multiplication shown in this section will work for each of the multiples of nine from 1×9 through 9×9.

2. Follow the procedure outlined in this section and show that

$$1234 \times 9 + 5 = 11,111$$

3. Study the following pattern and use it to express the squares of 6, 7, 8, and 9 in the same manner.

$1^2 = 1$
$2^2 = 1 + 2 + 1$
$3^2 = 1 + 2 + 3 + 2 + 1$
$4^2 = 1 + 2 + 3 + 4 + 3 + 2 + 1$
$5^2 = 1 + 2 + 3 + 4 + 5 + 4 + 3 + 2 + 1$

4. Study the entries that follow and use the pattern that is exhibited to complete the last four rows.

$1 + 3 = 4$ or 2^2
$1 + 3 + 5 = 9$ or 3^2
$1 + 3 + 5 + 7 = 16$ or 4^2
$1 + 3 + 5 + 7 + 9 = ?$
$1 + 3 + 5 + 7 + 9 + 11 = ?$
$1 + 3 + 5 + 7 + 9 + 11 + 13 = ?$
$1 + 3 + 5 + \cdots + (2n - 1) = ?$

5. An addition problem can be checked by a process called "casting out nines." To do this, you first find the sum of the digits of each of the addends (that is, numbers that are added), divide by 9, and record the remainder. Digits may be added again and again until a one-digit remainder is obtained. The sum of these remainders is then divided by 9 to find a final remainder. This should be equal to the remainder found by considering the sum of the addends (that is, the answer), adding its digits, dividing the sum of these digits by 9, and finding the remainder. Here is an example:

Addends	Sum of Digits	Remainders
4,378	22	4
2,160	9	0
3,872	20	2
1,085	14	5
11,495		11

When the sum of the remainders is divided by 9, the final remainder is 2. This corresponds to the remainder obtained by dividing the sum of the digits in the answer $(1 + 1 + 4 + 9 + 5 = 20)$ by 9.

Try this procedure for several other examples and verify that it works in each case.

6. Try to discover a procedure for checking multiplication by casting out nines. Verify that this procedure works for several cases.

7. There is a procedure for multiplying a two-digit number by 9 on one's fingers provided that the tens digit is smaller than the ones digit. The accompanying diagram shows how to multiply 28 by 9.

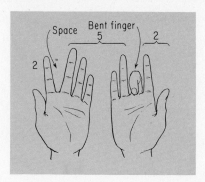

Reading from the left, put a space after the second finger and bend the eighth finger. Read the product in groups of fingers as 252.

Use this procedure to find: **(a)** 9×47; **(b)** 9×39; **(c)** 9×18; **(d)** 9×27. Check each of the answers you have obtained.

8. John offered to work for 1¢ the first day, 2¢ the second day, 4¢ the third day, and so forth, doubling the amount each day. Bill offered to work for $30 a day. Which boy would receive the most money for a job that lasted **(a)** ten days? **(b)** fifteen days? **(c)** sixteen days?

9. Using a method similar to that of Gauss, find:

(a) The sum of the first 80 counting numbers.

(b) The sum of the first 200 counting numbers.

(c) The sum of all the odd numbers from 1 through 49.

(d) The sum of all the odd numbers from 1 through 199.

(e) The sum of all the even numbers from 2 through 400.

10. Use the results obtained in Exercise 9 and try to find a formula for the sum of:

(a) The first n counting numbers, that is,

$$1 + 2 + 3 + \cdots + (n - 1) + n$$

(b) The first n odd numbers, that is,

$$1 + 3 + 5 + \cdots + (2n - 3) + (2n - 1)$$

11. Find the sum:

$$\frac{1}{1 \times 2} + \frac{1}{2 \times 3} + \frac{1}{3 \times 4} + \cdots + \frac{1}{9 \times 10}$$

(a) To help you discover this sum, complete the statements in the array at the top of the next page.

$$\frac{1}{1 \times 2} = \frac{1}{2}$$

$$\frac{1}{1 \times 2} + \frac{1}{2 \times 3} = ?$$

$$\frac{1}{1 \times 2} + \frac{1}{2 \times 3} + \frac{1}{3 \times 4} = ?$$

(b) Compare the answers on the right with the denominator of the last fraction in the given sums. Then attempt a guess at the sum of the original problem.

(c) Use the procedure for your guess to find the following sum; then confirm your answer by addition:

$$\frac{1}{1 \times 2} + \frac{1}{2 \times 3} + \frac{1}{3 \times 4} + \frac{1}{4 \times 5}$$

(d) Find the following sum:

$$\frac{1}{1 \times 2} + \frac{1}{2 \times 3} + \frac{1}{3 \times 4} + \cdots + \frac{1}{98 \times 99} + \frac{1}{99 \times 100}$$

*12. Note the following relationships:

$$\frac{1}{1 \times 2} = \frac{1}{1} - \frac{1}{2}; \quad \frac{1}{2 \times 3} = \frac{1}{2} - \frac{1}{3}; \quad \frac{1}{3 \times 4} = \frac{1}{3} - \frac{1}{4}$$

Use this pattern to confirm your answer in Exercise 11(d).

13. Find the sum:

$$\frac{1}{2} + \frac{1}{2^2} + \frac{1}{2^3} + \frac{1}{2^4} + \cdots + \frac{1}{2^n}$$

To help you discover this sum, complete these partial sums and search for a pattern.

(a) $\dfrac{1}{2} + \dfrac{1}{2^2} = \dfrac{1}{2} + \dfrac{1}{4} = ?$

(b) $\dfrac{1}{2} + \dfrac{1}{2^2} + \dfrac{1}{2^3} = \dfrac{1}{2} + \dfrac{1}{4} + \dfrac{1}{8} = ?$

(c) $\dfrac{1}{2} + \dfrac{1}{2^2} + \dfrac{1}{2^3} + \dfrac{1}{2^4} = \dfrac{1}{2} + \dfrac{1}{4} + \dfrac{1}{8} + \dfrac{1}{16} = ?$

*14. Try to discover a rule that is being used in each case to obtain the answer given. For example, the given information

---

*An asterisk preceding an exercise indicates that the exercise is more difficult or challenging than the others.

$$2, 5 \to 6, \qquad 3, 10 \to 12, \qquad 7, 8 \to 14, \qquad 5, 3 \to 7$$

should lead you to the rule $x, y \to x + y - 1$.

    (a) $3, 4 \to 8;$ $3, 3 \to 7;$ $1, 5 \to 7;$ $2, 8 \to 11.$

    (b) $2, 4 \to 8;$ $3, 5 \to 15;$ $1, 7 \to 7;$ $3, 9 \to 27.$

    (c) $1, 5 \to 1;$ $5, 2 \to 2;$ $3, 9 \to 3;$ $6, 5 \to 5.$

    (d) $4, 8 \to 4;$ $5, 1 \to 5;$ $6, 6 \to 6;$ $8, 2 \to 8.$

    (e) $3, 4 \to 5;$ $4, 4 \to 4;$ $5, 7 \to 0;$ $9, 2 \to 1.$

1-2

Explorations with Geometric Patterns

In the study of geometry we frequently form conclusions on the basis of a small number of examples, together with an exhibited pattern. Consider, for example, the problem of determining the number of triangles that can be formed from a given convex polygon by drawing diagonals from a given vertex P. First we draw several figures and consider the results in tabular form, as follows:

  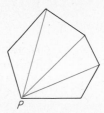

Number of sides of polygon	4	5	6
Number of diagonals from each vertex	1	2	3
Number of triangles formed	2	3	4

From the pattern of entries in the table, it appears that the number of triangles formed is two less than the number of sides of the polygon. Thus we expect that we can form 10 triangles for a dodecagon, a polygon with 12 sides, by drawing diagonals from a given vertex. In general, then, for a polygon with n sides, called an n-gon, we can form $n - 2$ triangles.

This is reasoning by *induction*. We formed a generalization on the basis of several specific examples and an obvious pattern. This procedure does not, however, constitute a proof. In order to *prove* that $n - 2$ triangles can be formed in this way for a polygon with n sides, we must observe that two of the n sides intersect at a common point of the diagonals and that each of the other $n - 2$ sides is used to form a different triangle.

Next let us consider the problem of finding the total number of possible diagonals that may be drawn in a polygon. Again we first draw several figures and consider the results for possible patterns of answers.

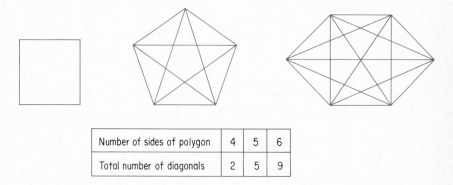

Number of sides of polygon	4	5	6
Total number of diagonals	2	5	9

No pattern appears to be evident. Therefore we shall examine the number of diagonals in one of the figures, the hexagon, and then attempt to generalize our conclusions for a polygon of n sides.

From the vertex A of the hexagon $ABCDEF$ we can draw three diagonals, $\overline{AC}$, $\overline{AD}$, and $\overline{AE}$. We cannot draw diagonals from A to A, from A to B, or from A to F.

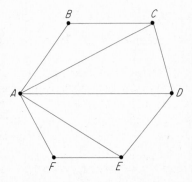

Do you see that three diagonals can be drawn from *each* vertex of the hexagon? This might lead us to conclude, *incorrectly*, that a total of 6×3, that is, 18, diagonals can be drawn. However, the diagonal drawn from A to C is the same line segment as the diagonal that can be drawn from C to A. In other words, if three diagonals are drawn from *each* vertex then each diagonal will be drawn twice. Thus the number of possible diagonals in the hexagon is $\frac{1}{2}(6 \times 3)$, that is, 9.

Let us try to generalize this discovery for an n-gon, a polygon of n sides. Consider a vertex B of a polygon $ABCD \ldots Q$. From B we *cannot* draw diagonals to B, to A, or to C.

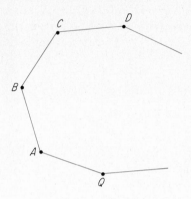

Since an n-gon has n sides and n vertices, we draw only $n - 3$ diagonals from each vertex. (That is, we can draw a diagonal from a given vertex to each vertex except the given one and its two adjacent vertices.) Therefore, we may draw a total of $n \times (n - 3)$ diagonals, not necessarily distinct. As in the case of the hexagon, each diagonal will be drawn twice under this procedure. Therefore, the total number of possible diagonals in an n-gon is given by the formula $\frac{1}{2}n \times (n - 3)$, that is, $\dfrac{n(n - 3)}{2}$.

If we test this formula for the case of the hexagon, $n = 6$, we have $\dfrac{6(6 - 3)}{2} = \dfrac{6(3)}{2} = 9$ as originally stated. Does it appear that $n(n - 3)$ is always divisible by 2 for n greater than 3? Try to explain why this should be so.

It is important to recognize that not all patterns lead to valid generalizations. Patterns offer an opportunity to make reasonable guesses, but these need to be proved before they can be accepted with certainty. Consider, for example, the maximum number of regions into which a circular region can be divided by line segments joining given points on a circle in all possible ways. See the figure at the top of page 14.

Would you agree that a reasonable guess for the number of regions derived from five points is 16? Draw a figure to confirm your conjecture. What is your guess for the maximum number of regions that can be derived from six points? Again, draw a figure to confirm your conjecture. You may be in for a surprise!

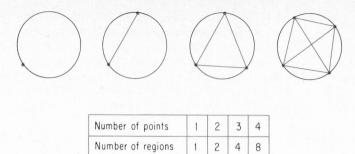

Number of points	1	2	3	4
Number of regions	1	2	4	8

exercises

1. Take a piece of notebook paper and fold it in half. Then fold it in half again and cut off a corner that does not involve an edge of the original piece of paper.

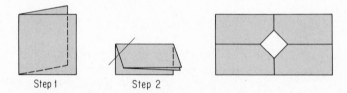

Step 1   Step 2

Your paper, when unfolded, should look like the preceding sketch. That is, with two folds we produced one hole. Repeat the same process but this time make three folds before cutting off an edge. Try to predict the number of holes that will be produced. How many holes will be produced with four folds? With *n* folds?

2. We wish to color each of the pyramids in the accompanying figure so that no two of the faces (sides and base) that have a common edge are of the same color.

 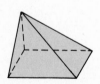 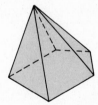

(a) What is the minimum number of colors required for each pyramid?

*(b) What is the relationship between the minimum number of colors required and the number of faces of a pyramid?

3. Consider the following set of figures. In each figure we count the number V of vertices, the number A of arcs, and the number R of regions into which the figure divides the plane. A square, for example, has 4 vertices, 4 arcs, and divides the plane into 2 regions (inside and outside of the square). See if you can discover a relationship between V, R, and A that holds for each case. Confirm your generalization by testing it on several other figures.

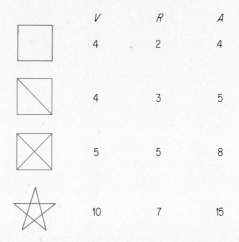

	V	R	A
square	4	2	4
square with diagonal	4	3	5
square with two diagonals	5	5	8
star	10	7	15

*4. What is the maximum number of pieces that can be obtained by slicing an orange with two cuts? With three cuts? With four cuts?

5. Consider the problem of arranging sets of squares that are joined along their edges. In general, these figures are referred to as **polyominoes.** A single square, a **monomino,** can be arranged in only one way. Two squares, a **domino,** can also be arranged in only one way, since the position of the 2×1 rectangle does not affect the arrangement. That is, any two such figures are congruent. There are two distinct arrangements for a **tromino,** that is, three squares.

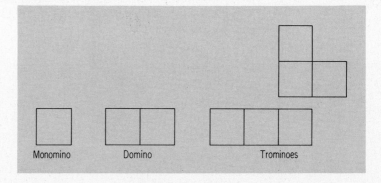

Monomino   Domino   Trominoes

Next consider a **tetromino,** a set of four squares. Here we find that there are five possible arrangements such that no two are congruent.

As in the case of the domino and other arrangements, we exclude any rearrangement that merely consists of a rotation which places the same squares in a congruent position. Here are three of the five possible tetrominoes:

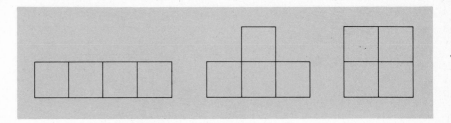

Draw the two remaining tetrominoes.

6. Consider a polyomino formed with five squares, called a **pentomino**. There are 12 such arrangements possible. Try to draw all 12.

7. Try to traverse each of the following figures, tracing each line segment exactly once without removing the point of the pencil from the paper.

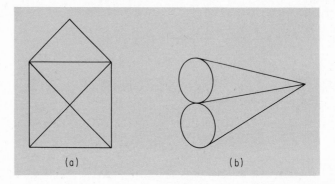

(a)   (b)

8. Twelve matchsticks are arranged to form the figure shown. By removing only two of the matches, form a figure that consists of two squares.

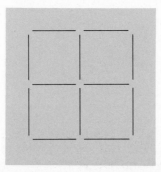

9. Copy the given figure and form a figure consisting of two triangles by removing three of the line segments.

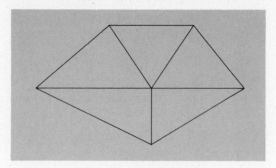

10. Twenty matchsticks are arranged to form the figure shown. By rearranging only three matchsticks, form a new figure that consists of five congruent squares.

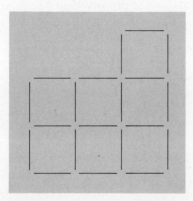

11. A square lake has a tree planted at each of its four corners. Show that the lake can be enlarged to twice its original size without replanting the trees, without having the trees surrounded by water, and without changing the shape of the lake.

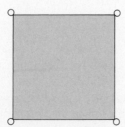

12. About the middle of the nineteenth century a problem related to map making was proposed; mathematicians are still searching for its solution. This problem, known as the **four-color problem**, involves the coloring of maps using at most four colors. When two countries have common boundaries, they must have different colors. When two countries have only single points in common they may use the same color.

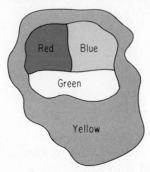

Four colors are required
for this map

(a) Draw a map of five countries that requires only three colors. **(b)** Can you draw a map of five countries that requires only two colors?

1-3
Explorations with Mathematical Recreations

The popularity of mathematics as a means of recreation and pleasure is evidenced by the frequency with which it is found in popular magazines and newspapers. In this section we shall explore several of these recreational aspects of mathematics.

How many of us have ever failed to be impressed by a magic square? The magic square shown here is arranged so that the sum of the numbers in any row, column, or diagonal is always 34.

1	12	7	14
8	13	2	11
10	3	16	5
15	6	9	4

Although there are formal methods to complete such an array, we shall not go into them here. Suffice it to say that such arrangements have fascinated man for many centuries. Indeed, the first known example of a magic square is said to have been found on the back of a tortoise by the Emperor Yu in about 2200 B.C.! This was called the "lo-shu" and appeared as an array of numerals indicated by knots in strings as in the figure. Black knots were used for even numbers and white ones for odd numbers. In modern times this appears as a magic square of third order. The sum along any row, column, or diagonal is 15 as in the next figure.

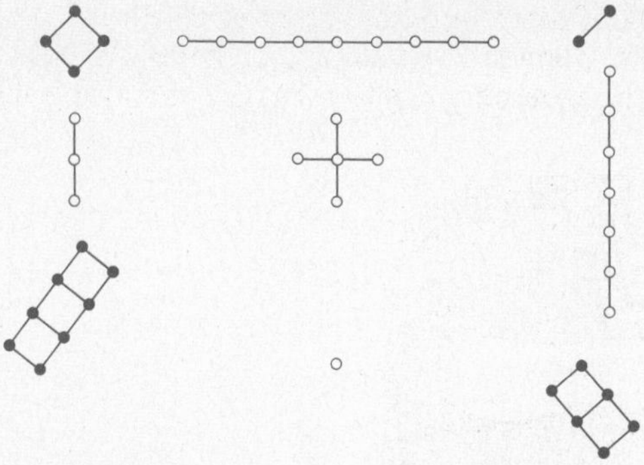

4	9	2
3	5	7
8	1	6

Closely akin to magic squares are square arrays of numbers useful in "mathemagic." Let us "build" a trick together. We begin by forming a square array and placing any six numerals in the surrounding spaces as in the figure. The numbers 3, 4, 1, 7, 2, and 5 are chosen arbitrarily. Next find the sum of each pair of numbers as in a regular addition table.

+	3	4	1
7			
2			
5			

+	3	4	1
7	10	11	8
2	5	6	3
5	8	9	6

Now we are ready to perform the trick. Have someone circle any one of the nine numerals in the box, say 10, and then cross out all the other numerals in the same row and column as 10.

+	3	4	1
7	(10)	11	8
2	5	6	3
5	8	9	6

Next circle one of the remaining numerals, say 3, and repeat the process. Circle the only remaining numeral, 9. The sum of the circled numbers is $10 + 2 + 9 = 22$.

+	3	4	1
7	(10)	11	8
2	5	6	(3)
5	8	(9)	6

The interesting item here is that the sum of the three circled numbers will always be equal to 22, regardless of where you start! Furthermore, note that 22 is the sum of the six numbers outside the square. Try to explain why this trick works, and then build a table with 16 entries.

Another type of mathematical trick that is quite popular is the "think of a number" type. Follow these instructions:

Think of a number.
Add 3 to this number.
Multiply your answer by 2.
Subtract 4 from your answer.
Divide by 2.
Subtract the number with which you started.

If you follow these instructions carefully, your answer will always be 1, regardless of the number with which you start. We can explain why this trick works by using algebraic symbols or by drawing pictures.

Think of a number:	n	☐	(Number of coins in a box)
Add 3:	$n+3$	☐ ○ ○ ○	(Number of original coins plus three)
Multiply by 2:	$2n+6$	☐ ○ ○ ○ ☐ ○ ○ ○	(Two boxes of coins plus six)
Subtract 4:	$2n+2$	☐ ○ ☐ ○	(Two boxes of coins plus two)
Divide by 2:	$n+1$	☐ ○	(One box of coins plus one)
Subtract the original number, n:	$(n+1)-n=1$	○	(One coin is left)

Optical illusions are vivid reminders of the fact that we cannot always trust our eyes. Can you trust yours? Test yourself and see. First guess which of the line segments, *a* or *b,* appears to be the longer in the four parts of the figure. Then use a ruler and check your estimate.

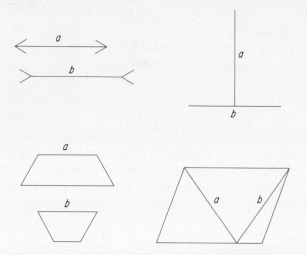

Now see whether you can guess which line segments are parallel, if any, in the three parts of the next figure.

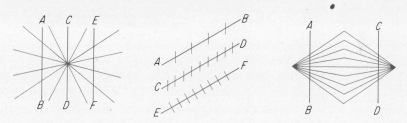

You should have found, in both cases, that looks can be deceiving. In each of the first four drawings, the segments are equal in length; in each of the last three they are parallel!

Here are two other optical illusions that were recently created and that appear most startling at first glance. They serve to show that we cannot always trust our intuition in mathematics, but must attempt to verify (prove) any hunches to be right or wrong.

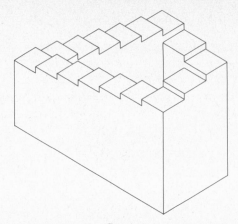

Mathematical fallacies have always intrigued both professional and amateur mathematicians alike. Here is an arithmetic fallacy to puzzle you. You might even consider trying this on your local banker. First you need to deposit $50 in the bank and then make withdrawals in the following manner:

Withdraw $20, leaving a balance of $30.
Withdraw $15, leaving a balance of $15.
Withdraw $ 9, leaving a balance of $ 6.
Withdraw $ 6, leaving a balance of $ 0.

Adding, we have: $50 $51

The total withdrawal is $50, whereas the total of the balances is $51. Can you therefore go to the bank to demand an extra dollar?

Here is a "proof" that $1 = 2$. Even though you may have forgotten the algebra you need to follow this, don't let it stop you; see if you can discover the fallacy.

Let $a = b$. Then $a^2 = b^2 = b \cdot b$

Since $a = b$, we may write $b \cdot b$ as $a \cdot b$. Thus $a^2 = a \cdot b$

Subtract b^2: $a^2 - b^2 = a \cdot b - b^2$

Factor: $(a + b)(a - b) = b(a - b)$

Divide by $a - b$: $$\frac{(a + b)(a - b)}{(a - b)} = \frac{b(a - b)}{(a - b)}$$

Thus $a + b = b$

Since $a = b$, we may write this as $b + b = b$ or $2b = b$

Divide by b: $\dfrac{2b}{b} = \dfrac{b}{b}$

Therefore $2 = 1$

Many people enjoy solving interesting or amusing puzzles. A variety of these, many of which are old-timers, appear in the exercises that follow.

exercises

1. All of the following puzzles have logical answers, but they are not strictly mathematical. See how many you can answer.
    (a) How many two-cent stamps are there in a dozen?
    (b) How many telephone poles are needed in order to reach the moon?
    (c) How far can you walk into a forest?
    (d) Two United States coins total 55¢ in value, yet one of them is not a nickel. Can you explain this?
    (e) How much dirt is there in a hole which is 3 feet wide, 4 feet long, and 2 feet deep?
    (f) There was a blind beggar who had a brother, but this brother had no brothers. What was the relationship between the two?

2. A farmer has to get a fox, a goose, and a bag of corn across a river in a boat which is only large enough for him and one of these three items. Now if he leaves the fox alone with the goose, the fox will eat the goose. If he leaves the goose alone with the corn, the goose will eat the corn. How does he get all items across the river?

3. Three Indians and three missionaries need to cross a river in a boat big enough only for two. The Indians are fine if they are left alone or if they are with the same number or with a larger number of missionaries. They are dangerous if they are left alone in a situation where they outnumber the missionaries. How do they all get across the river without harm?

4. A bottle and cork cost $1.50 together. The bottle costs one dollar more than the cork. How much does each cost?

5. A cat is at the bottom of a 30-foot well. Each day she climbs up 3 feet; each night she slides back 2 feet. How long will it take for the cat to get out of the well?

6. If a cat and a half eats a rat and a half in a day and a half, how many days will it take for 100 cats to eat 100 rats?

7. Ten coins are arranged to form a triangle as shown in the figure at the top of page 24. By rearranging only three of the coins, form a new triangle that points in the opposite direction from the one shown.

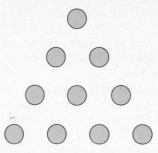

8. Rearrange the eight segments shown so as to form three congruent squares, that is, three squares that are exactly the same size. Each of the four smaller segments is one-half of the length of one of the larger segments.

9. Use six matchsticks, all of the same size, to form four equilateral triangles. (An equilateral triangle has all three sides the same length.)

10. Four matchsticks are arranged to form a "cup" with a coin contained within three of the segments, as shown in the figure. By rearranging only two of the matchsticks, form a new figure that is congruent to the original one (that is, exactly the same in shape but possibly different in position) but with the coin no longer contained within any of the segments.

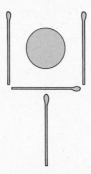

11. A sailor lands on an island inhabited by two types of people. The A's always lie, and the B's always tell the truth. The sailor meets three inhabitants on the beach and asks the first of these: "Are you an A or a B?" The man answers, but the sailor doesn't understand him and asks the second person what he had said. The man replies: "He said that he was a B. He is, and so am I." The third inhabitant then says:

"That's not true. The first man is an *A* and I'm a *B*." Can you tell who was lying and who was telling the truth?

12. A man goes to a well with three cans whose capacities are 3 gallons, 5 gallons, and 8 gallons. Explain how he can obtain exactly 4 gallons of water from the well.

13. Here is a mathematical trick you can try on a friend. Ask someone to place a penny in one of his hands, and a dime in the other. Then tell him to multiply the value of the coin in the right hand by 6, multiply the value of the coin held in the left hand by 3, and add. Ask for the result. If the number given is an even number, you then announce that the penny is in the right hand; if the result is an odd number, then the penny is in the left hand and the dime is in the right hand. Can you figure out why this trick works?

14. Consider a house with six rooms and furniture arranged as in the accompanying figure. We wish to interchange the desk and the bookcase, but in such a way that there is never more than one piece of furniture in a room at a time. The other three pieces of furniture do not need to return to their original places. Can you do this? Try it using coins or other objects to represent the furniture.

Cabinet		Desk
Television set	Sofa	Bookcase

15. Three men enter a hotel and rent a suite of rooms for $30. After they are taken to their rooms the manager discovers he overcharged them; the suite rents for only $25. He thereupon sends a bellhop upstairs with the $5 change. The dishonest bellhop decides to keep $2 and returns only $3 to the men. Now the rooms originally cost $30, but the men had $3 returned to them. This means that they paid only $27 for the room. The bellhop kept $2. $27 + $2 = $29. What happened to the extra dollar?

16. Write the numbers from 1 through 10 using four 4's for each. Here are the first three completed for you:

$$\frac{44}{44} = 1, \quad \frac{4}{4} + \frac{4}{4} = 2, \quad \frac{4 + 4 + 4}{4} = 3$$

17. Arrange two pennies *P* and two dimes *D* as shown at the top of page 26. Try to interchange the coins so that the pennies are at the right and the dimes at the left. You may move only one coin at a time, you may jump over only one coin, and pennies may be moved only to the right whereas dimes may be moved only to the left. No two coins may occupy the same space at the same time. What is the minimum number of moves required to complete the game?

P	P		D	D

18. Repeat Exercise 17 for three pennies and three dimes, using seven blocks. What is the minimum number of moves required to complete the game?

19. What are the next two letters in the sequence:

O, T, T, F, F, S, S, . . . ?

20. What are the next two letters in the sequence:

E, F, F, N, O, S, S, . . . ?

21. Place a half-dollar, a quarter, and a nickel in one position, *A,* as in the figure. Then try to move these coins, one at a time, to position *C*. Coins may also be placed in position *B*. At no time may a larger coin be placed on a smaller coin. This can be accomplished in $2^3 - 1$, that is, 7, moves.

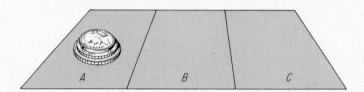

Next add a penny to the pile and try to make the change in $2^4 - 1$, that is, 15, moves.

This is an example of a famous problem called the **Tower of Hanoi.** The ancient Brahman priests were to move a pile of 64 such discs of decreasing size, after which the world would end. This would require $2^{64} - 1$ moves. Try to estimate how long this would take at the rate of one move per second.

22. Here is a game that must be played by two persons. Two players alternate in selecting one of the numbers 1, 2, 3, 4, 5, or 6. After each number is selected, it is added to the sum of those previously selected. For example, if player *A* selects 3 and player *B* selects 5, then the total is 8. If *A* selects 3 again, then the total is 11 and player *B* takes his turn. The object of the game is to be the first one to reach 50. There is a way to win at all times if you are permitted to go first. See if you can discover this method for winning, and then try to play the game with a classmate.

23. Many tricks of magic have their basis in elementary mathematics and may be found in books on mathematical recreations. Here is one example of such a trick.

Have someone place three dice on top of one another while you turn your back. Then instruct him to look at and find the sum of the values shown on the two faces that touch each other for the top and middle dice, the two faces that touch each other for the middle and bottom dice, and the value of the bottom face of the bottom dice. You then turn around and at a glance tell him this sum. The trick is this: You merely subtract the value showing on the top face of the top die from 21. Stack a set of three dice in the manner described and try to figure out why the trick works as it does.

24. You are given a checkerboard and a set of dominoes. The size of each domino is such that it is able to cover two squares on the board. Can you arrange the dominoes in such a way that all of the board is covered with the exception of two squares in opposite corners? (That is, you are to leave uncovered the two squares marked XX in the figure.) Try to explain why you should or should not be able to arrange the dominoes in this way.

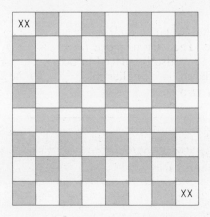

1-4

Explorations with Flow Charts

A flow chart is a method for giving instructions that can be used in a variety of situations. Here are several examples of very simple flow charts. Each one consists of an input, a rule, and an output.

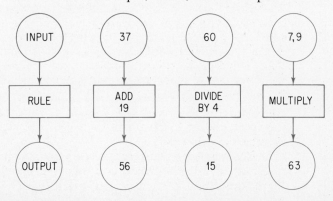

Some flow charts may have a sequence of rules to follow, such as in the next illustration. Note that circular regions are used for inputs and outputs. Rules are usually given within rectangular boxes.

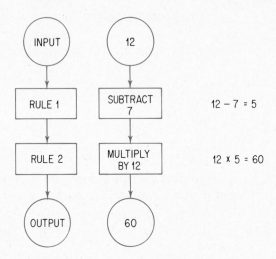

We can also use variables for the input numbers. In the following two flow charts, the two rules have been interchanged. Notice that the two outputs are different. Each of the two flow charts is accompanied by one that illustrates the steps for the replacement $n = 7$.

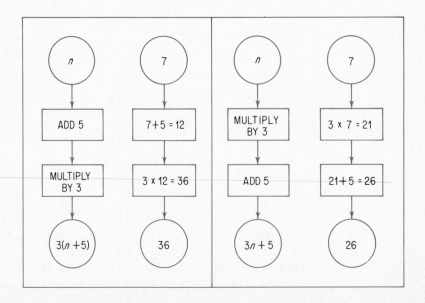

Sometimes a flow chart will contain a decision box that asks a question. In such cases you must answer the question as "YES" or "NO" and follow the arrows out of the box in the indicated direction. Such boxes are usually diamond shaped.

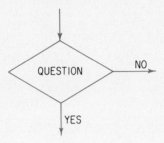

Here is a flow chart that contains a decision box. Notice that you must branch to the right if your answer to the question is negative.

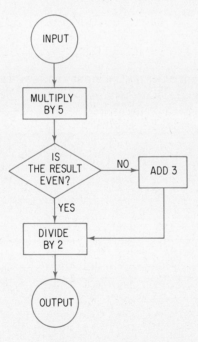

Suppose the input is 3. Then in the first step, $5 \times 3 = 15$. The result is not even, so move to the right in the flow chart and add 3 to obtain 18. In the next step, divide by 2 to obtain the output number, 9. Now find the output for an input of 6. The output should be 15.

Flow charts can also be used to describe the steps in a variety of nonmathematical situations as well. For example, here is a flow chart for cooking a four-minute egg. As with most such charts, INPUT and OUTPUT circles are replaced with START and STOP. Note that this flow chart forces you to loop back at two places if your answer to the questions raised is negative.

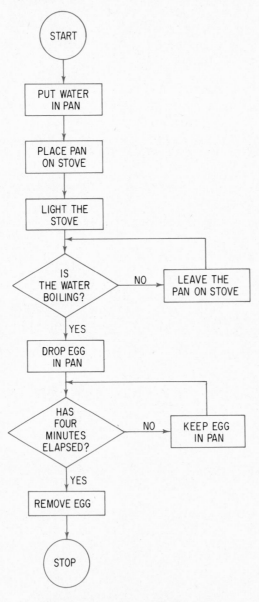

Now try to construct a flow chart to describe some activity in your daily life. More practice with flow charts is given in the exercises that follow.

exercises

For each of the following the rule and the output are given. Find the input.

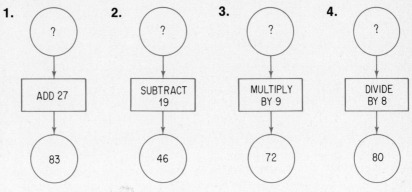

1. ? → ADD 27 → 83

2. ? → SUBTRACT 19 → 46

3. ? → MULTIPLY BY 9 → 72

4. ? → DIVIDE BY 8 → 80

Use the flow chart shown, and the input numbers given in Exercises 5 through 10, to find the outputs.

5. $n = 3$
6. $n = 5$
7. $n = 7$
8. $n = 12$
9. $n = 13$
10. $n = 17$

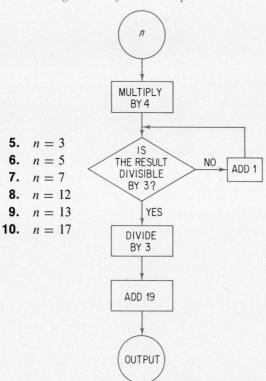

Write an algebraic expression for each output. Then find the output for n = 12.

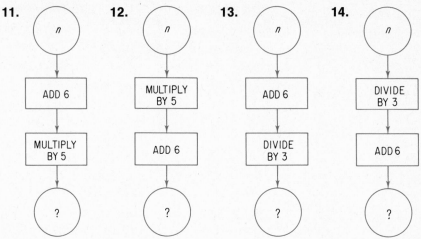

11.   **12.**   **13.**   **14.**

Draw a flow chart to represent each expression.

15. $3(n - 1)$ **16.** $3n - 1$ **17.** $\frac{1}{2}(n + 5)$ **18.** $\frac{1}{2}n + 5$

Flow charts can be used to generate a sequence of terms, as shown in the following two exercises. Each one should give you a set of eight outputs as your answer.

19.   ***20.**

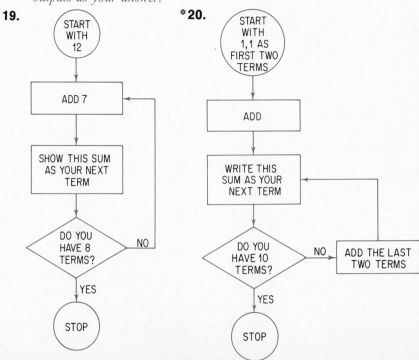

An Introduction to
SETS

The concept of a set pervades all of mathematics and is used in many of our daily activities. Think of the many times that you refer to a pen and pencil set, a set of dishes, a set of encyclopedias, or in some other way identify a collection of items as a set. Statements are important to each of us as the primary basis for written and oral communication. In mathematics we use

Sets of numbers in arithmetic.
Sets of solutions of equations in algebra.
Sets of points in geometry.

Many people consider the use of sets of elements to be a unifying theme throughout all branches of mathematics. The extent to which the concept of a set permeates all of mathematics is illustrated by the fact that set concepts are introduced in the early grades of elementary schools and also provide the basis for graduate mathematics courses at universities.

Before man ever began to count he probably used a matching process to determine the size of a collection. The caveman who had seen five animals roaming in the fields might have shown this by sketching a picture on the wall of his abode.

He was, in fact, using a very sophisticated idea—that of a **one-to-one correspondence.** For each animal he saw in the field he placed a mark on the wall. For each mark his spouse saw, she knew he had seen one animal. They could communicate in this manner without ever knowing about or using the **numeral** 5, which is merely a symbol invented at a later date to represent the size of a collection.

Later, man became more sophisticated and created a set of numbers with which he could count. We call this collection the set C of **counting numbers** and represent them in **set notation** as follows:

$$C = \{1, 2, 3, 4, 5, \ldots\}$$

The three dots indicate that the set continues indefinitely in the pattern indicated, that is, that there are infinitely many elements in the set. The set C is said to be an **infinite set.** A set is usually named by a capital letter, and the **elements** or **members** of the set are indicated within a pair of braces.

Many words in our language indicate sets: *school* of fish, *swarm* of bees, *gaggle* of geese, *herd* of cattle, *squadron* of planes, etc. Try to name at least five other commonly used words that indicate sets. A **set** is a collection of things. Thus we may speak of

The set of letters of the English alphabet.
The set of states of the United States of America.
The set of English names of the days of the week.
The set of counting numbers 1 through 5.

Each of the preceding sets is said to be a **well-defined set.** That is, from the description you can tell whether or not any given element belongs to the set. For example, we know that *r, s,* and *t* are members of the set of letters of the English alphabet, whereas π and 7 clearly are not members.

Can you tell, from the description given, which of the following numbers belong to the set of counting numbers 1 through 5?

3, $2\frac{1}{2}$,   8,   1,   12

You should note that 3 and 1 are members of the set, whereas $2\frac{1}{2}$, 8, and 12 are not. The set of counting numbers 1 through 5 is a well-defined set when the set of counting numbers is defined as the set 1, 2, 3, 4, 5, 6, 7, 8, 9, 10, and so forth.

Not all sets are well-defined sets. See if you can explain why each set in the next list is not a well-defined one.

The set of good tennis players.
The set of interesting numbers.
The set of beautiful movie stars.
The set of well-written books.

In this text we shall be concerned primarily with well-defined sets. Furthermore, we shall agree to name a set by an arbitrarily chosen capital letter and to list the elements of the set within a pair of braces as in the following examples.

EXAMPLE 1

Write in set notation: the set W of English names of the days of the week.

SOLUTION $W = \{$Sunday, Monday, Tuesday, Wednesday, Thursday, Friday, Saturday$\}$

EXAMPLE 2

List the elements in the set T of counting numbers 1 through 5.

SOLUTION $T = \{1, 2, 3, 4, 5\}$

The **membership symbol** $\in$ is often used as in

$2 \in \{1, 2, 3, 4, 5\}$ and $7 \notin \{1, 2, 3, 4, 5\}$

to indicate that elements are or are not members of sets. We read $\in$ as "is a member of" and $\notin$ as "is not a member of" a given set. Where it becomes tedious to list individually all of the members of a set, we frequently use three dots to indicate missing elements.

EXAMPLE 3

List the elements in the set A of letters of the English alphabet.

SOLUTION $A = \{a, b, c, \ldots, z\}$

Notice that when three dots are used to show that some elements are missing, it is necessary to state the elements in some order so that a pattern can be observed and used to identify the missing elements. Any ordered set of elements is a **sequence**. In Example 3 the sequence has a last element and is a **finite set**.

EXAMPLE 4

List the elements in the set G of counting numbers greater than 100.

SOLUTION $G = \{101, 102, 103, \ldots\}$

It is interesting, at times, to seek verbal descriptions for given sets of elements which have already been listed. The next example illustrates this point.

EXAMPLE 5

Write a verbal description for the set $Y = \{1, 3, 5, 7, 9\}$.

SOLUTION There are several correct responses that might be given. Two of these are "The set of odd numbers 1 through 9" and "The set of odd numbers between 0 and 11." Note that the word "between" implies that the first and last numbers (that is, 0 and 11) are not included as members of the given set.

The basic property of a set of elements is the identification of its members. A set is a well-defined set if its members can be identified. The sets $A = \{1, 2, 3\}$ and $B = \{3, 1, 2\}$ are the same since they have the same members. When two sets consist of precisely the same elements, they are said to be **equal sets,** that is, **identical sets.** We write $\{1, 2, 3\} = \{3, 1, 2\}$, since order of listing does not affect membership, $\{1, 1, 2\} = \{1, 2\}$, since listing an element more than once does not affect membership, and in general $A = B$ to show that sets A and B have the same members. We write $A \neq B$ to show that sets A and B do not have the same members.

exercises

Replace the asterisk (*) by $\in$ or $\notin$ to obtain a true statement.

1. $7 * \{1, 3, 5, 7, 9\}$ **2.** $7 * \{2, 4, 6, 8, 0\}$

3. $\triangle * \{., |, \square\}$ **4.** $\triangle * \{/, \wedge, \triangle\}$

Tell whether or not each of the following sets is a well-defined set.

5. The set of U.S. astronauts who have orbited the earth.

6. The set of cities that are state capitals in the United States of America.

7. The set of large states in the United States.

8. The set of states in the United States with good climates.

State whether or not the sets of letters are equal as sets.

9. $\{r, a, t\}, \{t, a, r\}$

10. $\{t, r, a, p\}, \{a, p, a, r, t\}$

11. $\{w, o, l, f\}, \{f, o, l, l, o, w\}$

12. $\{p, r, o, f, e, s, s, o, r\}, \{f, r, a, p, p, e, s\}$

List the elements in each of the following sets, using the notation developed in this section.

13. The set of English names of the months in the year.

14. The set of letters in the English alphabet that precede j.

15. The set of counting numbers between 1 and 10.

16. The set of odd counting numbers less than 25.

Write a verbal description for each of the following sets.

17. $N = \{1, 2, 3, 4, 5, 6\}$

18. $R = \{1, 2, 3, \ldots, 99\}$

19. $S = \{51, 52, 53, \ldots\}$

20. $M = \{5, 10, 15, 20, \ldots\}$

21. $K = \{10, 20, 30, \ldots, 150\}$

22. $T = \{1, 4, 9, 16, 25, 36\}$

*23. $P = \{0, 2, 6, 12, 20, \ldots, 90\}$

*24. $A = \{1, 2, 4, 8, 16, \ldots, 1024\}$

EXPLORATIONS

The order in which the elements of a set are listed is not important in describing a set. Consider, for example, set $A = \{8, 5, 4, 9, 1, 7, 6, 3, 2\}$.

1. Let us consider a verbal description for set A. Since order is not important, one possible answer is "the set of counting numbers from 1 through 9." Try to write at least two other verbal descriptions for set A.

2. Now let us try to prepare a verbal description for set A that also identifies the order in which the elements are listed. One possible solution, quite unsophisticated, is "the set of elements 8, 5, 4, 9, 1, 7, 6, 3, 2 in that order." Try to find another verbal description that identifies the order without actually listing the elements of the set.

2-2
Subsets

The set containing the totality of elements for any particular discussion is called the **universal set** U, and may vary for each discussion. For example, let us agree, for the present, to talk about the set of counting numbers 1 through 9. Then our universal set is

$$U = \{1, 2, 3, 4, 5, 6, 7, 8, 9\}$$

A set A is a **subset** of U if the set A has the property that each element of A is also an element of U. We write $A \subseteq U$ (read "A is included in U"). There are many subsets of U; a few of them are

$A_1 = \{1, 2, 3\}$
$A_2 = \{1, 5, 7, 8, 9\}$
$A_3 = \{2\}$
$A_4 = \{1, 2, 3, 4, 5, 6, 7, 7, 8, 9\}$

Note in particular that the set A_4 contains each of the elements of U and is classified as a subset of U. Any set is said to be a subset of itself.

A set A is said to be a **proper subset** of U if A is a subset of U and there is at least one element of U that is not an element of A. We write $A \subset U$ (read "A is properly included in U"). Intuitively we speak of a proper subset as part of, but not all of, a given set. Each of the subsets A_1, A_2, A_3, and A_4 are subsets of U; the sets A_1, A_2, and A_3 are proper subsets of U; the set A_4 is not a proper subset of U. We may write in symbols:

$$A_1 \subset U, \quad A_2 \subset U, \quad A_3 \subset U, \quad A_4 \subseteq U$$

EXAMPLE 1

List three proper subsets of the set R where $R = \{1, 2, 3, 4\}$.

SOLUTION

Here are three proper subsets:

$$A = \{1, 2\}, \quad B = \{1, 3, 4\}, \quad C = \{1\}$$

Do you see that there are others? Note that the choice of letters to name each of the subsets is completely arbitrary.

The **empty** set or **null** set is the set that contains no elements and is denoted by the symbol $\emptyset$ or by the symbol $\{\ \ \}$. Note that the first symbol given for the empty set is not included within a pair of braces. Some examples of empty sets are

The set of counting numbers between 2 and 3.
The set of states of the United States with borders on both the Atlantic Ocean and the Pacific Ocean.

A set A has been defined to be a subset of a set B if each element of A is also an element of B. If we wish to prove that a set A is not a subset of a set B, we must find an element of A that is not an element of B. For example, if $G = \{1, 3, 5\}$ and $H = \{1, 2, 3, 4, 5\}$, then $G \subseteq H$ but $H \nsubseteq G$, since $2 \in H$ and $2 \notin G$; also $4 \in H$ and $4 \notin G$. Notice that since the empty set has no elements, there are no elements of the empty set that can fail to be elements of a set B. Thus the empty set is a subset of every set B; $\emptyset \subseteq B$ for every set B. This property of the empty set is often emphasized by defining a set A to be a *subset* of a set B if there are no elements of A that are not elements of B.

EXAMPLE 2

List all the possible subsets of the set $\{1, 2\}$.

SOLUTION $\emptyset$, $\{1\}$, $\{2\}$, $\{1,2\}$

Notice in the solution of Example 2 that the subsets of $\{1,2\}$ are the sets obtained by selecting or not selecting each of the elements of the given set. If we select both elements, we obtain $\{1,2\}$; if we select exactly one of the elements, we obtain $\{1\}$ or $\{2\}$; if we do not select either element, we obtain the empty set. Since there are two elements and two choices (take it or leave it) for each element, there are 2×2 subsets, as we obtained.

The subsets obtained in Example 2 may be paired as

$\{1,2\}$ with $\emptyset$, and $\{1\}$ with $\{2\}$

Then the two sets of each pair have no common elements and contain all elements of the given set $\{1,2\}$. Relative to the given set, if we had the set $\{1\}$, we could find the corresponding set $\{2\}$; if we had the set $\{2\}$, we could find the corresponding set $\{1\}$; if we had the set $\{1,2\}$ we could find the set $\emptyset$; and if we had the set $\emptyset$, we could find the set $\{1,2\}$. This process may be described as finding the complement of a set relative to the given set $\{1,2\}$.

For any given universal set U, each set A has a complement A' (also written $\bar{A}$) that consists of the elements of U that are not elements of A and is called the **complement of A relative to U**. For example, if $U = \{1,3,5,7,9\}$ and $A = \{1,3,7\}$, then $A' = \{5,9\}$. If $A = U$, then $A' = \emptyset$; if $A = \emptyset$, then $A' = U$. In other words, the complement of the universal set is the empty set; the complement of the empty set is the universal set.

EXAMPLE 3

List the subsets of $\{1,2,3\}$ and pair each set with its complementary set.

SOLUTION $\{1,2,3\}$ $\{2,3\}$ $\{1,3\}$ $\{1,2\}$
         $\emptyset$ $\{1\}$ $\{2\}$ $\{3\}$

It seems reasonable to expect that all possible subsets should be obtained by considering the elements three at a time, two at a time, one at a time, and zero at a time. As a check that we have found all possible subsets, notice that the original set had three elements; we have two choices (takes it or leave it) for each element; $2 \times 2 \times 2 = 8$, and we found eight subsets.

exercises

1. Describe three empty sets.

Let $U = \{c, a, r, t\}$ and list the elements of A' when A is defined as follows:

2. $\{c, a, r\}$ **3.** $\{r, a, t\}$

4. $\{a, t\}$ **5.** $\varnothing$

Let $U = \{p, r, o, f, e, s, s, o, r\}$ and list the elements of A' when A is defined as follows:

6. $\varnothing$ **7.** $\{r, o, p, e\}$

8. $\{r, o, s, e\}$ **9.** $\{p, r, o, s, e\}$

10. Let U be the set of counting numbers $\{1, 2, 3, \ldots\}$. Let A be the set of even counting numbers and describe the set A'.

Let $U = \{1, 2, 3, 4, 5, 6, 7, 8, 9\}$, $A = \{1, 3, 5, 7, 9\}$, and $B = \{3, 6, 9\}$, and list the elements in each set.

11. A' **12.** B'

13. List the elements in set A if $U = \{1, 2, 3, 4, 5, 6\}$ and $A' = \{1, 4, 6\}$.

14. List the elements in set B if $U = \{2, 4, 6, 8, 10\}$ and $B' = \varnothing$.

List all possible subsets of each set.

15. $\{a\}$ **16.** $\{a, b\}$

17. $\{a, b, c\}$ **18.** $\{a, b, c, d\}$

19. $\{a, b, c, d, e\}$ **20.** $\varnothing$

List all possible proper subsets of each set.

21. $\{a\}$ **22.** $\{a, b\}$

23. $\{a, b, c\}$ **24.** $\{a, b, c, d\}$

25. $\{a, b, c, d, e\}$ **26.** $\varnothing$

EXPLORATIONS

1. Use the results obtained in Exercises 15 through 20, and complete this table.

Number of elements	0	1	2	3	4	5	6
Number of subsets	1						

2. Use the results obtained in Exploration 1 to conjecture a formula for the number N of subsets that can be formed from a set consisting of n elements.

3. Use the results of Exercises 21 through 26 and construct a table to show the number of proper subsets of sets.

Number of elements	0	1	2	3	4	5	6
Number of proper subsets	0						

4. Use the results obtained in Exploration 3 to conjecture a formula for the number N of proper subsets that can be formed from a set consisting of n elements.

2-3
Equivalent Sets

If you have a set of books, you can count them and use a number to tell how many books there are in the set. If you have a set of bookmarks, you can count them and use a number to tell how many bookmarks there are in the set. If your purpose was to be sure that you had exactly one bookmark for each of your books, you could have done this without counting either the books or the bookmarks. Instead you could have placed one bookmark in each book. If this one-to-one matching of books and bookmarks could

be continued until all books and bookmarks were used, you would know that you had the same number of books in the set of books as you had bookmarks in the set of bookmarks. We would then say that there is a one-to-one correspondence between the books and the bookmarks.

Two sets, $X = \{x_1, x_2, \ldots\}$ and $Y = \{y_1, y_2, \ldots\}$, are said to be in **one-to-one correspondence** if we can find a pairing of the x's and y's such that each x corresponds to one and only one y and each y corresponds to one and only one x. Consider the sets $R = \{a, b, c\}$ and $S = \{\$, +, \%\}$. We can match the element $a \in R$ with any one of the elements in S (three choices); we can then match $b \in R$ with any one of the other two elements of S (two choices), and then match $c \in R$ with the remaining element of S (one possibility). Thus a one-to-one correspondence of the sets R and S can be shown in $3 \times 2 \times 1$ (that is, 6) ways:

$$\{a, \quad b, \quad c\} \qquad \{a, \quad b, \quad c\}$$
$$\{\$, \quad +, \quad \%\} \qquad \{\$, \quad \%, \quad +\}$$

$$\{a, \quad b, \quad c\} \qquad \{a, \quad b, \quad c\}$$
$$\{+, \quad \%, \quad \$\} \qquad \{+, \quad \$, \quad \%\}$$

$$\{a, \quad b, \quad c\} \qquad \{a, \quad b, \quad c\}$$
$$\{\%, \quad \$, \quad +\} \qquad \{\%, \quad +, \quad \$\}$$

Two sets A and B that can be placed in a one-to-one correspondence are said to be **equivalent sets** (written $A \leftrightarrow B$). Any two equivalent sets have the same number of elements, that is, the same **cardinality.**

You may use the concept of a one-to-one correspondence to show whether any two sets of elements have the same number of members. You also use this concept of a one-to-one correspondence when you count the elements of a set. To count, you use the set of counting numbers

$$\{1, 2, 3, 4, 5, 6, \ldots\}$$

As in §2-1, the dots indicate that the pattern continues indefinitely. If you were to count the books in the set shown in the previous figure, you would use the set of numbers $\{1, 2, 3, 4, 5\}$. You would form a one-to-one correspondence of the books with the numbers. If you were to count the bookmarks, you would use the same set of numbers. See page 44.

Two sets that may be placed in one-to-one correspondence with the same set of numbers may be placed in one-to-one correspondence with each other and thus have the same number of elements.

Any set A of elements that may be placed in one-to-one correspondence with the set of elements $\{1, 2, 3, 4, 5\}$ is said to have five elements; we write $n(A) = 5$ to show that the number of elements in the set A is 5. Any set B of elements that may be placed in one-to-one correspondence with the set of elements $\{1, 2, 3, 4, \ldots, k - 1, k\}$ is said to have k elements; $n(B) = k$. Notice this use of the set of counting numbers and one-to-one

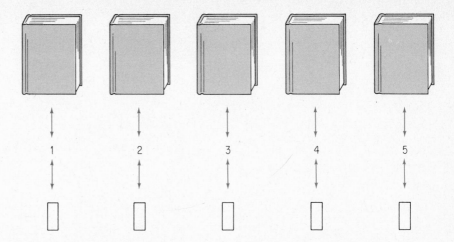

correspondences to determine how many elements there are in a set; that is, to determine the **cardinal number** of a set. When the counting numbers are taken in order, the last number used in the one-to-one correspondence is the cardinal number of the set. We define the cardinality of the empty set to be zero; that is, $n(\varnothing) = 0$.

We have used one-to-one correspondence to determine the cardinal number of a set of numbers. Specifically, we asserted that the last counting number used would be the cardinal number of the set. However, is there always a "last" counting number used? Consider each of the following sets:

$A = \{1, 2, 3, 4, 5, \ldots, n, \ldots\}$
$B = \{2, 4, 6, 8, 10, \ldots, 2n, \ldots\}$
$C = \{1, 3, 5, 7, 9, \ldots, 2n - 1, \ldots\}$
$D = \{2, 2^2, 2^3, 2^4, 2^5, \ldots, 2^n, \ldots\}$
$E = \left\{1, \dfrac{1}{2}, \dfrac{1}{3}, \dfrac{1}{4}, \dfrac{1}{5}, \ldots, \dfrac{1}{n}, \ldots\right\}$
$F = \{10, 10^2, 10^3, 10^4, \ldots, 10^n, \ldots\}$

In each case the set fails to have a last element; consequently, the set does not have a counting number as its cardinal number.

EXAMPLE

Consider the sets A and B:

$A = \{1, 2, 3, 4, 5, \ldots, n, \ldots\}$
$B = \{2, 4, 6, 8, 10, \ldots, 2n, \ldots\}$

Explain: (a) $B \subseteq A$; (b) $B \neq A$; (c) $B \subset A$; (d) B is equivalent to A.

SOLUTION   (a) Each element of B is also an element of A; that is, B is a subset of A, $B \subseteq A$.

(b) There is at least one element of A (for example, 1) that is not an element of B; thus B and A are different sets, $B \neq A$.

(c) The set B is a subset of A [see part (a)] that does not contain all elements of A [see part (b)]; that is, B is a proper subset of A, $B \subset A$.

(d) The one-to-one correspondence of n to $2n$ between the elements of A and the elements of B establishes the equivalence of A and B as shown in this array.

$$A: \{1, \quad 2, \quad 3, \quad 4, \quad \ldots, \quad n, \ldots\}$$
$$\updownarrow \quad \updownarrow \quad \updownarrow \quad \updownarrow \qquad\quad \updownarrow$$
$$B: \{2, \quad 4, \quad 6, \quad 8, \quad \ldots, \quad 2n, \ldots\}$$

Notice that we have shown that the set A is equivalent to one of its proper subsets.

The empty set and any set that has a counting number as its cardinal number are called *finite sets*. A set that is equivalent to one of its proper subsets is an *infinite set*. That is, an infinite set is one that can be placed in a one-to-one correspondence with a proper subset of itself. Any infinite set fails to have a last element and fails to have a counting number as its cardinal number.

The counting numbers serve as the cardinal numbers for finite sets. However, no counting number can serve in such a capacity for an infinite set. Therefore, we need a new kind of number for cardinal numbers of infinite sets. The cardinal number of an infinite set is called a **transfinite cardinal number.** However, as in the case of the books and bookmarks, we do not need to find the cardinal numbers of two sets in order to show that they have the same number of elements. We need only demonstrate that a one-to-one correspondence exists between the sets. In the example, the sets A and B have the same number of elements, $n(A) = n(B)$, because there is a one-to-one correspondence of n to $2n$ between the elements of the sets. In other words, there are just as many even counting numbers as there are counting numbers.

Each of the sets identified as A, B, C, D, E, and F may be placed in one-to-one correspondence with the set A. Each of these sets is an infinite set and each has the same transfinite cardinal number as the set of counting numbers.

The properties of infinite sets confused mathematicians for centuries. Indeed, it is only within the last century that infinite sets have been reasonably well understood. At the turn of the twentieth century, a German mathematician named Georg Cantor developed an entire theory of infinite sets of numbers. He assigned the symbol $\aleph_0$ (read "aleph-null") to represent the transfinite cardinal number of the set of counting numbers. It is correct then to say that there are $\aleph_0$ counting numbers, just as you might say that there are 7 days of the week or 10 fingers on your hands. Furthermore, any set that can be matched in a one-to-one correspondence with the set of counting numbers is also of size $\aleph_0$. We shall consider a few properties of transfinite cardinal numbers and some of their apparent paradoxes in the next section.

exercises

Show all possible one-to-one correspondences of the two sets.

1. $\{1, 2\}$ and $\{p, q\}$
2. $\{1, 2, 3\}$ and $\{x, y, z\}$
3. $\{1, 2, 3, 4\}$ and $\{r, e, s, t\}$

Describe a set with each given cardinal number.

4. 5   5. 6
6. 7   7. 0

Find the cardinal number of each of the sets in Exercises 8 through 13.

8. $\{\triangle, >, =, \square\}$ 9. $\{3, 5, 7\}$
10. $\{x\}$ 11. $\{11, 12, \ldots, 18\}$
12. $\{6\}$ 13. $\varnothing$

14. Give a set that may be placed in one-to-one correspondence with the set of counting numbers.

15. Give a set that may be placed in one-to-one correspondence with the set of even counting numbers.

16. Consider the set $A = \{a, b, c, d, e\}$ and find a set that is (**a**) equivalent and identical (equal) to A; (**b**) equivalent but not equal to A.

17. Explain whether (**a**) any two equal sets are necessarily equivalent; (**b**) any two equivalent sets are necessarily equal.

Show a one-to-one correspondence between the sets of each pair.

18. The set of counting numbers 4 through 8 and the set of vowels in the English language.

19. The set of vowels in the English language and the set of even counting numbers less than or equal to 10.

20. The set of counting numbers greater than 100 and the set of all counting numbers.

*21. The set of positive odd integers and the set of positive integral multiples of 5.

EXPLORATIONS

The concept of infinity as used in mathematics is a very difficult one for most people to acquire. It is helpful to contrast the idea of very large sets with that of infinite sets.

For each of the following finite situations, first make an educated guess. Then find some way to approximate the correct answer.

1. To the nearest day, how long would it take to count to one billion at the rate of one number per second?

2. Estimate how many pennies it would take to make a stack one inch high. Approximately how high would a stack of one million pennies be?

3. One million one-dollar bills are placed end-to-end along the ground. To the nearest ten miles, how long would this strip of bills run?

4. To the nearest million, estimate the number of seconds that elapse in a century.

5. How long would it take to spend $1,000,000 at the rate of $1.00 per minute?

One of the largest numbers ever named is a googol, which has been defined as 1 followed by 100 zeros:

10000000000000000000000
00000000000000000000
00000000000000000000
00000000000000000000
00000000000000000000

This number is larger than what is considered to be the total number of protons or electrons in the universe! A googol can be expressed, using exponents, as 10^{100}.

Even larger than a googol is a googolplex, defined as 1 followed by a googol of zeros. One famous mathematician claimed that there would not even be room between the earth and the moon to write all the zeros in a googolplex!

*6. How many zeros are there in the number represented as a googol times a googol? Express this number using exponents. Is this number smaller or larger than a googolplex?

2-4
Sets of Numbers

The set of numbers that we use in counting

$$\{1, 2, 3, 4, 5, \ldots\}$$

is called the set of *counting numbers* and has been considered in several earlier sections. The counting numbers provide the basis for the development of other sets of numbers and also for consideration of various uses of numbers. Chapter 6 will be devoted to a detailed study of various sets of numbers.

Counting numbers and zero may be used as *cardinal numbers* of finite sets of elements. Counting numbers may also be used to assign an order to the elements of a finite set. For example, you expect page 15 of this book to follow page 14. Numbers used to assign an order to the elements of a set are often called ordinal numbers.

Numbers are also used in other ways. For example, neither cardinality nor order is significant for the number of your driver's license, your telephone number, or your social security number. Such numbers are used solely for identification.

The use of numbers as cardinal numbers was considered in the previous section. At that time the symbol $\aleph_0$ was introduced to represent the transfinite cardinal number of the set of counting numbers. The discussion of transfinite numbers gives rise to some very interesting apparent paradoxes. One of the most famous of these is the story of the infinite house. This is a house that contains an infinite number of rooms, numbered 1, 2, 3, 4, 5, and so on. Each room is occupied by a single tenant. That is, there is a one-to-one correspondence between rooms and occupants. There are $\aleph_0$ rooms, and $\aleph_0$ occupants. One day a stranger arrived at the house and asked to be admitted. The caretaker was an amateur mathematician and was able to accommodate this visitor in the following manner. He asked the occupant of room 1 to move to room 2, the occupant of room 2 to move to room 3, the occupant of room 3 to move to room 4, and in general the occupant of room n to move to room $n + 1$. Now everyone had a room and room number 1 was available for the visitor! In other words, we have demonstrated this interesting fact:

$$\aleph_0 + 1 = \aleph_0$$

Several days later an infinite number of visitors arrived at the house, all demanding individual rooms! Again the caretaker was able to accommodate them. He merely asked each tenant to move to a room number that was double his current room number. That is, the occupant of room 1 moved to room 2, the occupant of room 2 moved to room 4, room 3 to room 6, and room n to room $2n$. After this move was made, the new arrivals were placed in rooms 1, 3, 5, and so forth. Here we have an example of another interesting fact in the language of transfinite arithmetic:

$$\aleph_0 + \aleph_0 = \aleph_0$$

We can demonstrate these interesting arithmetic properties of $\aleph_0$ through the use of sets of numbers. To do so, consider the following sets:

$A = \{1, 2, 3, 4, \ldots, n, \ldots\}$
$B = \{2, 4, 6, 8, \ldots, 2n, \ldots\}$
$C = \{1, 3, 5, 7, \ldots, 2n - 1, \ldots\}$

1. $\aleph_0 + 1 = \aleph_0$. The set A has the transfinite cardinal number $\aleph_0$; the set A with one element 0 added has cardinal number $\aleph_0 + 1$. The equivalence of these two sets may be shown as follows:

$$\{0, \quad 1, \quad 2, \quad 3, \quad 4, \ldots, \quad n, \ldots\}$$
$$\uparrow \quad \uparrow \quad \uparrow \quad \uparrow \quad \uparrow \quad \quad \uparrow$$
$$\{1, \quad 2, \quad 3, \quad 4, \quad 5, \ldots, \quad n+1, \ldots\}$$

2. $\aleph_0 + \aleph_0 = \aleph_0$. The sets B and C each have the transfinite cardinal number $\aleph_0$; the union of these sets has the cardinal number $\aleph_0 + \aleph_0$. However, sets B and C together form set A, which has cardinal number $\aleph_0$.

There are still many unanswered questions about transfinite numbers. For example, it can be shown that $\aleph_0^{\aleph_0}$ leads to another transfinite cardinal number that is larger than $\aleph_0$ and is often called $\aleph_1$. The problem of determining whether or not there is a cardinal number between $\aleph_0$ and $\aleph_1$ has occupied mathematicians up to recent times and has still not been completely resolved. Indeed, you may be reading about further discoveries on this topic at any time, inasmuch as a number of mathematicians are currently at work on this famous problem.

exercises

In each of the following tell whether the number is used as a cardinal number, an ordinal number, or merely for identification.

1. This is the first problem on the list.
2. There are 20 volumes in the set of encyclopedias.
3. Mathematics is discussed in the 12th volume.

4. My license plate number is EZT 133.
5. Dorothy is in the eighth row.
6. There are 35 students in the class.
7. I am listening to 104 on the FM dial.
8. It takes 9 men to field a baseball team.
9. This is the third time he is taking the course.
10. The card catalog lists the book as 510.7.

Show a one-to-one correspondence between the sets of each pair.

11. The set of counting numbers and the set of odd counting numbers.
12. The set of counting numbers and the set of counting numbers greater than 25.
13. The set of even counting numbers and the set of odd counting numbers.
14. The set of even counting numbers and the set of even counting numbers greater than 100.
15. The set of odd counting numbers greater than 100 and the set of even counting numbers greater than 100.
16. The set of odd counting numbers greater than 50 and the set of odd counting numbers greater than 150.

Classify each set as finite or infinite:

17. $A = \{1, 2, 3, \ldots\}$
18. $B = \{2, 4, 6, \ldots, 1000\}$
19. $C = \{2, 4, 6, \ldots, 1{,}000{,}000\}$
20. $D = \{1, 3, 5, \ldots, 2n - 1, \ldots\}$
21. The number of inhabitants in the world
22. The number of words in this book.
23. The number of counting numbers greater than 1000.
*24. The number of counting numbers less than a googol.

Use a one-to-one correspondence with the set of counting numbers to show each of the following facts.

25. $\aleph_0 + 3 = \aleph_0$ 26. $\aleph_0 + 5 = \aleph_0$
27. $\aleph_0 - 1 = \aleph_0$ 28. $\aleph_0 - 3 = \aleph_0$

EXPLORATIONS

Consider the fact that a line segment contains an infinite number of points. From this one may demonstrate the fact that two line segments of unequal length nevertheless contain the same number of points. Thus consider segments *AB* and *CD* where lines *CA* and *DB* meet at a point *P*.

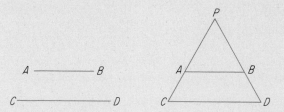

We can show that there is a one-to-one correspondence between the points of segments *AB* and *CD*. Given any point *M* on *CD*, draw line segment *PM*. This segment crosses *AB* at a point *M′* that corresponds to *M*. Also given any point *N* on segment *AB*, draw line segment *PN* and extend it until it meets *CD* at a point *N′* that corresponds to *N*. Thus there is a one-to-one correspondence between the points of segments *AB* and *CD*, and there are just as many points on segment *AB* as there are on segment *CD*.

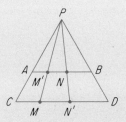

1. Demonstrate a one-to-one correspondence between the points on the circle and the points on the square in the figure shown.

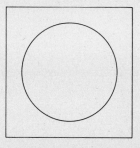

2. Examine several high school mathematics textbooks and describe the manner in which the concept of infinity is treated at this level of instruction.

2-5

Intersection and Union

Let us consider two sets A and B, defined as follows:

$A = \{1, 2, 3, 4, 5, 6, 7\}$
$B = \{2, 5, 7, 8, 9\}$

From these two sets let us form another set C, whose members are those elements that appear in each of the two given sets:

$C = \{2, 5, 7\}$

The set C consists of the elements that the sets A and B have in common and is called the **intersection** of the sets A and B. Formally, we say that the intersection of two sets A and B (written $A \cap B$) is the set of elements that are members of both of the given sets.

EXAMPLE 1

If $A = \{a, r, e\}$ and $B = \{c, a, t\}$, find $A \cap B$.

SOLUTION $A \cap B = \{a\}$; that is, a is the only letter that appears in each of the two given sets.

EXAMPLE 2

If $X = \{1, 2, 3, 4, 5\}$ and $Y = \{4, 5, 6, 7\}$, find $X \cap Y$.

SOLUTION $X \cap Y = \{4, 5\}$

EXAMPLE 3

If $X = \{1, 3, 5, \ldots\}$ and $Y = \{2, 4, 6, \ldots\}$, find $X \cap Y$.

SOLUTION   Sets X and Y have no elements in common and are said to be **disjoint sets;** their intersection is the null set. Thus we may write $X \cap Y = \varnothing$.

From the two given sets A and B of this section let us next form another set D, whose members are those elements that are elements of at least one of the two given sets:

$D = \{1, 2, 3, 4, 5, 6, 7, 8, 9\}$

The set D is called the **union** of sets A and B. Formally, we say that the union of two sets A and B (written $A \cup B$) is the set of elements that are members of at least one of the given sets.

EXAMPLE 4

Let A represent the set of names of boys on a particular committee, $A = \{$Bill, Bruce, Max$\}$. Let B represent the set of names of boys on another committee, $B = \{$Bruce, John, Max$\}$. Find $A \cup B$.

SOLUTION $A \cup B = \{$Bill, Bruce, John, Max$\}$. Here the union of the two sets is the set of names of the boys who are on at least one of the two committees. Note that the names Bruce and Max appear only once in the set $A \cup B$ even though these names are listed for both committees.

EXAMPLE 5

If $A = \{1, 2, 3, 4, 5\}$ and $B = \{3, 5, 7, 9\}$, find $A \cup B$.

SOLUTION $A \cup B = \{1, 2, 3, 4, 5, 7, 9\}$

EXAMPLE 6

If $U = \{1, 2, 3, \ldots\}$, $A = \{1, 2, 3, 4, 5\}$, and $B = \{3, 4, 5, 6, 7\}$, list the elements in $A' \cap B'$.

SOLUTION $A' = \{6, 7, 8, 9, 10, \ldots\}$, $B' = \{1, 2, 8, 9, 10, \ldots\}$, and $A' \cap B' = \{8, 9, 10, \ldots\}$.

exercises

For each of the following sets list the elements in **(a)** $A \cup B$; **(b)** $A \cap B$.
1. $A = \{1, 2, 3\}$, $B = \{1, 3, 5, 7\}$.
2. $A = \{3, 4, 5\}$, $B = \{4, 5, 6, 7\}$.
3. $A = \{1, 3, 5, 7\}$, $B = \{7, 9\}$.
4. $A = \{2, 4, 6, 8\}$, $B = \{4, 6, 7, 8\}$.
5. $A = \{1, 5, 10\}$, $B = \{3, 7, 8\}$.

6. $A = \{1, 3, 5, \ldots\}$, $B = \{2, 4, 6, \ldots\}$.

7. $A = \varnothing$, $B = \{1, 2, 3, \ldots\}$.

8. $A = \{1, 2, 3, \ldots\}$, $B = \{1, 3, 5, \ldots\}$.

For each of the given universal sets list the elements in **(a)** A'; **(b)** B'; **(c)** $A' \cup B'$; **(d)** $A' \cap B'$.

9. $U = \{1, 2, 3, 4, 5\}$, $A = \{1, 2\}$, $B = \{1, 3, 5\}$.

10. $U = \{1, 2, 3, \ldots, 10\}$, $A = \{1, 3, 5, 7, 9\}$, $B = \{2, 4, 6, 8, 10\}$.

11. $U = \{1, 2, 3, \ldots\}$, $A = \{1, 3, 5, \ldots\}$, $B = \{2, 4, 6, \ldots\}$.

12. $U = \{1, 2, 3, 4, 5, 6, 7\}$, $A = \varnothing$, $B = \{1, 2, 3, 4, 5, 6, 7\}$.

13. $U = \{1, 2, 3\}$, $A = \{1\}$, $B = \{3\}$.

14. $U = \{1, 2, 3, \ldots, 10\}$, $A = \{1, 3, 5, 7, 9\}$, $B = \{1, 2, 3, 4, 5\}$.

For Exercises 15 through 20, let $U = \{1, 2, 3, 4, 5, 6, 7, 8, 9, 10\}$, $A = \{1, 3, 4, 5, 7, 9\}$, *and* $B = \{2, 4, 5, 9, 10\}$, *and list the elements in each set.*

15. $A' \cap B'$ **16.** $(A \cup B)'$

17. $A' \cup B$ **18.** $(A' \cap B)'$

19. $A \cap B$ **20.** $(A \cup B')'$

21. If $A \subseteq B$, describe **(a)** $A \cap B$; **(b)** $A \cup B$; **(c)** $A \cap B'$; **(d)** $A' \cup B$.

22. If $A \subseteq B$ and $B \subseteq A$, describe **(a)** $A \cup B$; **(b)** $A \cap B$; **(c)** $A \cap B'$; **(d)** $A' \cup B$.

For Exercises 23 through 28, let $U = \{1, 2, 3, \ldots\}$, $A = \{1, 3, 5, \ldots\}$, $B = \{2, 4, 6, \ldots\}$, *and list the elements in each set.*

23. $A \cup B$ **24.** $A' \cup B$

25. $A \cap B$ **26.** $(A \cap B)'$

27. $A \cap B'$ **28.** $(A \cup B)'$

EXPLORATIONS

The flow chart on page 55 can be used to determine whether an element n is a member of the intersection of two sets, A and B.

1. Use the flow chart to find $A \cap B$ where $A = \{k, l, m\}$ and $B = \{m, n\}$.

2. Construct a flow chart that can be used to determine whether an element n is a member of the union of two sets, A and B.

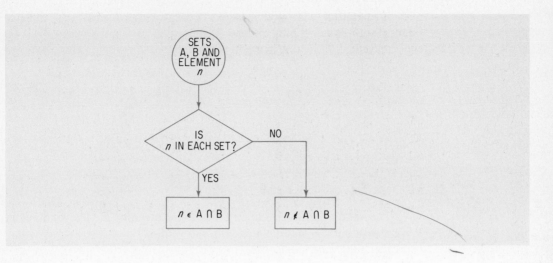

2-6
Sets of Points

Relationships among sets are often represented by sets of points. For example, we may represent the universal set U as a set of points on a line, a set A by the shaded subset of U, and the set A' by the remaining points of U.

The sets of points that are most frequently used to represent sets of elements are rectangular regions and circular regions, as identified in the following figure.

Rectangular region   Circular region

Rectangular regions and circular regions may be formally introduced but are used here simply as the sets of points that first grade children would color in coloring the points of a rectangle (or circle) and its interior points. Since there is no confusion about which points are meant, we shall not concern ourselves with formal definitions.

We shall use rectangular regions to represent universal sets and frequently shade the regions under consideration. As in the case of A', we use a dashed line for the part of the boundary of a region that does not belong to the region.

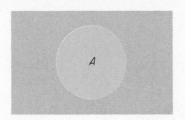

The shading is often omitted when the meaning is clear without it. With or without the shading, the figures are called **Euler diagrams.** We may use Euler diagrams to show the intersection of two sets.

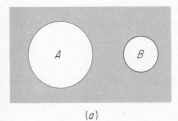

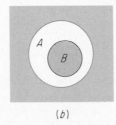

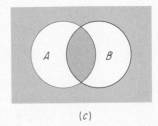

(a)   (b)   (c)

Note above that in Figure (a), $A \cap B$ is the empty set; in Figure (b), $A \cap B = B$. We may also use Euler diagrams to show the union of two sets. Note below that in Figure (b), $A \cup B = A$.

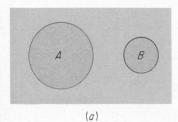

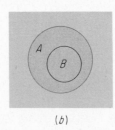

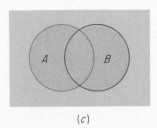

(a)   (b)   (c)

The figures for intersection and union may also be used to illustrate the following properties of any two sets A and B:

$$(A \cap B) \subseteq A, \quad (A \cap B) \subseteq B$$
$$A \subseteq (A \cup B), \quad B \subseteq (A \cup B)$$

We consider only well-defined sets (§2-1), and thus each element of the universal set U is a member of exactly one of the sets A and A'. When two sets A and B are considered, an element of U must belong to exactly one of these four sets:

$A \cap B$,   i.e., both A and B
$A \cap B'$,   i.e., A and not B
$A' \cap B$,   i.e., B and not A
$A' \cap B'$,   i.e., neither A nor B

An Euler diagram in which each of these four regions is represented is often called a **Venn diagram.**

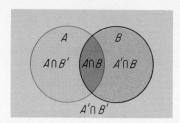

Venn diagrams may be used to show that two sets refer to the same set of points—that is, are equal. In the examples that follow note that Venn diagrams are used as a visual aid for what otherwise would involve very abstract thinking.

EXAMPLE 1

Show by means of a Venn diagram that $(A \cup B)' = A' \cap B'$.

SOLUTION

We make separate Venn diagrams for $(A \cup B)'$ and $A' \cap B'$. The diagram for $(A \cup B)'$ is made by shading the region for A horizontally, shading the region for B vertically, identifying the region for $A \cup B$ as consisting of the points in regions that are shaded in any way (horizontally, vertically, or both horizontally and vertically), and identifying the region for $(A \cup B)'$ as consisting of the points in the region without horizontal or vertical shading.

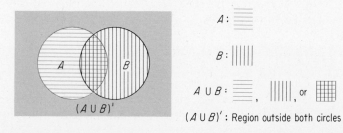

$(A \cup B)'$: Region outside both circles

The diagram for $A' \cap B'$ is made by shading the region for A' horizontally, shading the region for B' vertically, and identifying the region for $A' \cap B'$ as consisting of all points in regions that are shaded both horizontally and vertically.

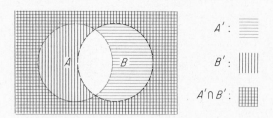

The solution is completed by observing that the region for $(A \cup B)'$ in the first Venn diagram is identical with the region for $A' \cap B'$ in the second Venn diagram; the two sets have the same elements and therefore are equal.

We may also use Venn diagrams for three sets. In this case there are eight regions that must be included, and the figure is usually drawn as follows:

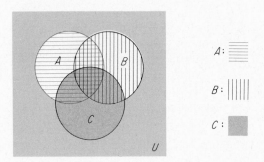

Venn diagrams for three sets are considered in these two examples and in Exercises 19 through 32. Venn diagrams for four sets are considered in the Explorations.

EXAMPLE 2

Show that $A \cap (B \cup C) = (A \cap B) \cup (A \cap C)$.

SOLUTION

Set A is shaded with vertical lines; $B \cup C$ is shaded with horizontal lines. The intersection of these sets, $A \cap (B \cup C)$, is the subset of U that has both vertical and horizontal shading.

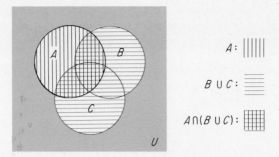

The set $A \cap B$ is shaded with horizontal lines; $A \cap C$ is shaded with vertical lines. The union of these sets is the subset of U that is shaded with lines in either or in both directions.

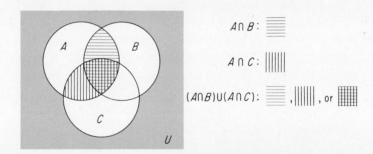

Note that the final results in the two diagrams are the same, thus showing the equivalence of the set $A \cap (B \cup C)$ and the set $(A \cap B) \cup (A \cap C)$.

EXAMPLE 3

Find (a) $n(A \cap B \cap C)$; (b) $n(A \cap B' \cap C)$.

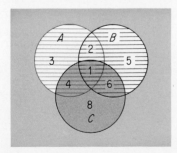

SOLUTION   (a) There is one element in the intersection of all three sets. Thus $n(A \cap B \cap C) = 1$. (b) There are four elements that are in both sets A and C, but not in set B. Thus $n(A \cap B' \cap C) = 4$.

EXAMPLE 4

In a group of 35 students, 15 are studying algebra, 22 are studying geometry, 14 are studying trigonometry, 11 are studying both algebra and geometry, 8 are studying geometry and trigonometry, 5 are studying algebra and trigonometry, and 3 are studying all three subjects. How many of these students are not taking any of these subjects? How many are taking only geometry?

SOLUTION

This problem can easily be solved by means of a Venn diagram with three circles to represent the set of students in each of the listed subject-matter areas. It is helpful to start with the information that there are 3 students taking all three subjects. We write the number 3 in the region that is the intersection of all three circles; $n(A \cap G \cap T) = 3$. Then we work backward; since 5 are taking algebra and trigonometry, there must be 2 in the region representing algebra and trigonometry but *not* geometry; $n(A \cap G' \cap T) = 2$. Continuing in this manner, we enter the given data in the figure.

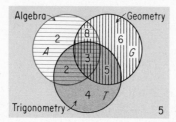

Since the total of the numbers in the various areas is 30, there must be 5 students not in any of the classes listed in the various regions. Also, reading directly from the figure, we find that there are 6 students taking geometry only.

The terms Euler diagram, Euler circle, and Venn diagram are used synonymously in some books. Their use in this book is the historical one and should provide a basis for comprehension of any other special uses encountered.

exercises

Consider the diagrams on page 61 and find each number.

1. **(a)** $n(A \cap B)$ **(b)** $n(A)$
   (c) $n(B \cap A')$ **(d)** $n(B \cup A)$

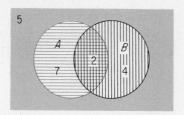

2. **(a)** $n(P \cup Q)$ **(b)** $n(P' \cap Q')$
    (c) $n(P' \cup Q)$ **(d)** $n(P \cup Q')$

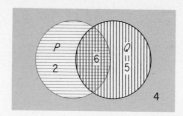

Use an Euler diagram to represent each of the following:

3. $A \subset B$.

4. A and B are disjoint sets.

5. $A \cup B$ when $A \cap B = \varnothing$.

6. $A \subseteq B$ and $B \subseteq A$.

7. $(A \cup B)'$ when A and B are disjoint sets.

8. $(A \cup B \cup C)'$ when A, B, and C are disjoint sets.

Represent each set by a Venn diagram.

9. $A' \cup B$ **10.** $A' \cap B$

11. $A \cap B'$ **12.** $A \cup B'$

Show each relation by Venn diagrams.

13. $A \cup B = B \cup A$

14. $A \cap B = B \cap A$

15. $(A \cap B) \subseteq A$

16. $A \subseteq (A \cup B)$

17. $(A \cap B)' = A' \cup B'$

18. $A \cup B' = (A' \cap B)'$

Consider the diagrams on page 62 and find each number.

19. **(a)** $n(A \cap B \cap C)$ **(b)** $n(A \cap B \cap C')$
    (c) $n(A \cap B' \cap C')$ **(d)** $n(A)$
    (e) $n(A \cup B)$ **(f)** $n(B \cup C)$

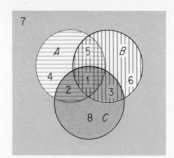

20. **(a)** $n(R' \cap S \cap T)$ **(b)** $n(R')$
    (c) $n(R' \cup S)$ **(d)** $n(S' \cup T')$
    (e) $n(R' \cup S' \cup T')$ **(f)** $n(R \cup S' \cup T)$

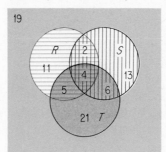

21. **(a)** $n(A \cup B)$ **(b)** $n(B \cap C)$
    (c) $n(A \cap B')$ **(d)** $n(A \cup B \cup C)$
    (e) $n(A \cup B' \cup C')$ **(f)** $n(A \cap B' \cap C')$

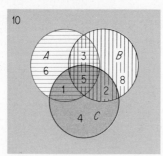

22. **(a)** $n(X \cup Y)$ **(b)** $n(X \cup Z)$
    (c) $n(X')$ **(d)** $n(X \cup Y')$
    (e) $n(X' \cap Y \cap Z)$ **(f)** $n(X \cap Y' \cap Z')$

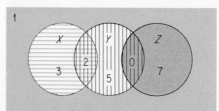

Use Venn diagrams to solve each problem.

23. In a survey of 50 students, the following data were collected. There were 19 taking biology, 20 taking chemistry, 19 taking physics, 7 taking physics and chemistry, 8 taking biology and chemistry, 9 taking biology and physics, 5 taking all three subjects. How many of the group were not taking any of the three subjects? How many were taking only chemistry? How many were taking physics and chemistry, but not biology?

24. A survey was taken of 30 students enrolled in three different clubs, A, B, and C. Show that the following data that were collected are inconsistent: 18 in A, 10 in B, 9 in C, 3 in B and C, 6 in A and B, 9 in A and C, 2 in A, B, and C.

Represent each of the following by a Venn diagram.

25. **(a)** $A \cap B \cap C$ **(b)** $A \cap B \cap C'$
     (c) $A \cap B' \cap C$ **(d)** $A \cap B' \cap C'$

26. **(a)** $A' \cap B \cap C$ **(b)** $A' \cap B \cap C'$
     (c) $A' \cap B' \cap C$ **(d)** $A' \cap B' \cap C'$

Show by Venn diagrams.

27. $A \cup (B \cup C) = (A \cup B) \cup C$
28. $A \cap (B \cap C) = (A \cap B) \cap C$

Identify each of the shaded regions by means of set notation.

29.
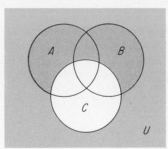

30.
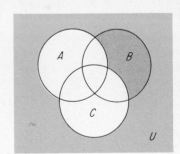

31.
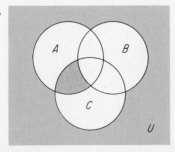

32.
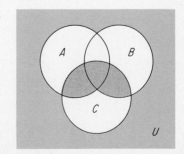

EXPLORATIONS

1. Venn diagrams may be drawn for four sets, *A, B, C,* and *D.* The region *D* must intersect each of the eight regions of the Venn diagram for *A, B,* and *C.* The new Venn diagram has sixteen regions as numbered in the figure.

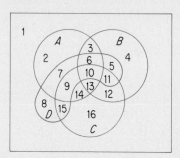

Identify each of the sixteen regions by its number and as an intersection of four of the sets *A, B, C, D, A', B', C', D'.*

Use numbers from the figure for Exploration 1 to identify each region. For example, the region A ∩ C consists of the regions 9, 10, 13, and 14.

2. *A*

3. *D*

4. *B'*

5. $(A \cup B) \cap D$

6. $A' \cap (B \cup D)$

7. $(A \cup C) \cap (B' \cup D')$

chapter test

1. State whether or not the sets $\{b, a, t\}$ and $\{l, a, b\}$ are **(a)** equal; **(b)** equivalent.

2. Let $U = \{m, e, m, o, r, a, b, l, e\}$ and list the elements of A' when A is defined as **(a)** $\{m, o, r, e\}$; **(b)** $\{l, a, b, o, r\}$.

3. Show all possible one-to-one correspondences of the two sets $\{t, o, p\}$ and $\{r, a, t\}$.

4. List all possible subsets of $\{h, a, t\}$.

Let $U = \{0, 1, 2, 3, 4, 5, 6, 7, 8, 9\}$, $A = \{2, 4, 6\}$, and $B = \{0, 3, 6, 9\}$. Then list the elements in each set.

5. $A' \cap B$

6. $(A \cup B)'$

Use a Venn diagram to represent each set.

7. $A \cap B'$

8. $A' \cap B \cap C$

Consider the given diagram and find each number.

9. $n(A \cap B \cap C')$

10. $n(A \cup B \cup C')$

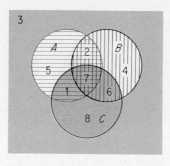

3

Concepts of
LOGIC

You undoubtedly make numerous conscious and subconscious decisions. Many of these decisions are based upon reasoning. Your satisfactions from your decisions often depend upon the validity of your reasoning and thus upon logical concepts. In this sense an understanding of logical concepts can improve your personal opportunities to find the kind of life that you want and to enjoy the kind of life that you find.

Reasoning is based upon the handling of facts, usually in the form of statements. You may think of reasoning as an "algebra" of statements. In this chapter truth values of statements are considered in an effort to recognize relations among statements. Then some of these relations are selected to provide the basis for the discussion of valid arguments, logical reasoning, and proofs. In the process you will learn how to recognize several different forms of the same statement in the English language, and to express statements in forms that can be easily understood. Many people feel that the study of logical concepts is worthwhile solely for the purpose of increasing their facility with the English language.

A sentence that is either *true* or *false* is often called a statement. Each of the following is an example of a simple statement:

Today is Friday in Chicago.
This is a sunny day.

Notice that a command such as "Stand up and be counted" is neither true nor false and is not considered to be a statement as we are using the word here.

A compound statement is formed by combining two or more simple statements. An example is the following:

Today is Friday in Chicago and this is a sunny day.

In this illustration the two simple statements are combined by the connective "and." Other connectives could also have been used. Consider the same simple statements using the connective "or":

Today is Friday in Chicago or this is a sunny day.

We shall consider such compound statements and determine the conditions under which they are true or false. In doing this we use letters or variables to represent statements and symbols to represent connectives. For example, we may use p and q to represent these simple statements:

p: Today is Friday in Chicago.
q: This is a sunny day.

The following connectives are commonly used:

$\wedge$: and
$\vee$: or
$\sim$: not

We may use p and q as previously defined and write these statements in symbolic form together with their translations in words:

$p \wedge q$: Today is Friday in Chicago and this is a sunny day.
$p \vee q$: Today is Friday in Chicago or this is a sunny day.
$\sim p$: Today is not Friday in Chicago.

EXAMPLE 1

Translate $p \wedge (\sim q)$, where p and q are as given in this section.

SOLUTION   Today is Friday in Chicago and this is not a sunny day.

EXAMPLE 2

Write, in symbolic form: Today is not Friday in Chicago or this is not a sunny day.

SOLUTION $(\sim p) \vee (\sim q)$

exercises

1. Use p: Jim is tall; q: Bill is short. Think of "short" as "not tall," and write each of these statements in symbolic form.
  (a) Jim is short and Bill is tall.
  (b) Neither Jim nor Bill is tall.
  (c) Jim is not tall and Bill is short.
  (d) It is not true that Jim and Bill are both tall.
  (e) Either Jim or Bill is tall.

2. Assume that Bill and Jim are both tall. Which of the statements in Exercise 1 are true?

3. Use p: Joan is happy; q: Mary is sad. Think of "sad" as "not happy," and write each of these statements in symbolic form:
  (a) Joan and Mary are both happy.
  (b) Either Joan is happy or Mary is happy.
  (c) Neither Joan nor Mary is happy.
  (d) It is not true that Joan and Mary are both sad.
  (e) It is not true that neither Joan nor Mary is happy.

4. Assume that Joan and Mary are both happy. Which of the statements in Exercise 3 are true?

5. Use p: I like this book; q: I like mathematics. Give each of these statements in words.
  (a) $p \wedge q$ **(b)** $\sim q$
  (c) $\sim p$ **(d)** $(\sim p) \wedge (\sim q)$

6. Continue as in Exercise 5 with the statements:
  (a) $(\sim p) \wedge q$ **(b)** $p \vee q$
  (c) $\sim(p \wedge q)$ **(d)** $\sim[\sim p) \wedge q]$

7.   Assume that you like this book and that you like mathematics. Which of the statements in Exercises 5 and 6 are true for you?

8.   Assume that you like this book but that you do not like mathematics. Which of the statements in Exercises 5 and 6 are true for you?

*9.   Assume that two given statements p and q are both true and indicate whether you would expect each of the following statements to be true.

   (a) $p \wedge q$ **(b)** $p \vee q$
   (c) $p \vee (\sim q)$ **(d)** $(\sim p) \vee q$

*10.   Repeat Exercise 9 under the assumption that p is true and q may be true or false.

EXPLORATIONS

Logical statements are often represented by electric circuits. For a single statement q think of an electric light cord from a wall outlet to an electric light bulb and with a switch in the middle of the cord.

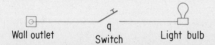

If the switch is closed the bulb is on; if the switch is open the bulb is off.

For two statements p and q we use two switches. The following circuits are particularly useful and common.

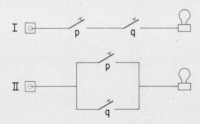

   1.   Modify circuit I as needed and indicate whether the light is on or off for each of these situations.

(a) p closed, q closed

(b) p closed, q open

(c) p open, q closed

(d) p open, q open

2. Explain why it seems reasonable to call circuit I a $p \wedge q$ circuit.

3. Repeat Exploration 1 for circuit II.

4. Explain why it seems reasonable to call circuit II a $p \vee q$ circuit.

The statement $(p \vee q) \wedge r$ may be represented by the following circuit where each circuit is left open until instructions are received for that switch.

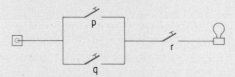

Sketch a circuit for each of these statements.

5. $(p \wedge q) \wedge r$

6. $(p \wedge q) \vee r$

7. $p \wedge (q \vee r)$

3-2
Truth Values of Statements

We first consider the **negation** $(\sim p)$ of a statement p. The symbol was introduced in the last section. As we would expect from the meaning of the word "negation," if p is true, then $\sim p$ is false; if p is false, then $\sim p$ is true. The truth values of the statement $\sim p$ are given in the accompanying **truth table,** where T stands for "true" and F stands for "false."

p	$\sim p$
T	F
F	T

We often use Euler diagrams to represent situations in which a statement p is true.

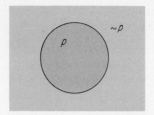

Next consider the simple statements:

p: It is snowing.
q: This is the month of December.

The truth of any compound statement, such as $p \wedge q$, is determined by the truth of each of the simple statements. Since each of the statements p and q may be either true or false, there are four distinct possibilities:

p true, q true
p true, q false
p false, q true
p false, q false

For each of these possibilities we wish to determine the truth value of the statement $p \wedge q$. When we say that $p \wedge q$ is true, we mean that both p and q are true. We may define the truth values of $p \wedge q$ by means of the given truth table.

p	q	$p \wedge q$
T	T	T
T	F	F
F	T	F
F	F	F

The statement $p \wedge q$ is called the **conjunction** of p and q. The truth values of $p \wedge q$ are independent of the *meanings* assigned to the variables. For example, if you are told that each of two simple statements is true, it follows immediately that the conjunction of these statements is also true; in any other circumstance, it follows immediately that the conjunction of the statements is false.

Venn diagrams may be used to represent the truth values of two statements p and q. Note that in the following figure we have four possible regions indicated. These regions correspond to the four possibilities for truth and falsity for two general statements (also to the four rows of the truth table) and may be labeled by conjunctions of statements. Notice the similarity of the meanings of $\wedge$ for statements and $\cap$ for sets.

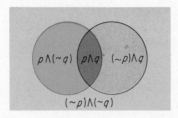

Note that the use of general (or abstract) statements p and q enables us to discuss all four possible pairs of truth values of the statements. For any two specific statements one or more of these four cases may not arise.

We next consider the compound statement $p \vee q$, called the **disjunction** of p and q. We translate "$\vee$" as "or" even though we shall use the word "or" more precisely than in the ordinary English language. As we shall use it here, "p or q" means that at least one of the statements p, q is true; that is, either p is true, or q is true, or both p and q are true. In other words, the meaning of $\vee$ for statements corresponds to the meaning of $\cup$ for sets.

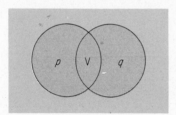

The truth values of $p \vee q$ are defined by the accompanying truth table. From the table we deduce the fact that the compound statement $p \vee q$ is true unless both p and q are false.

p	q	$p \vee q$
T	T	T
T	F	T
F	T	T
F	F	F

We can use truth tables to summarize the truth values of various compound statements. To illustrate this procedure we shall construct a truth table for the statement $p \wedge (\sim q)$.

First set up a table with the appropriate headings as follows:

p	q	p	$\wedge$	$(\sim q)$
T	T			
T	F			
F	T			
F	F			

Now complete the column headed "p" by using the truth values that appear under p in the first column. In the column headed "$\sim q$" write the negation of the values given under q in the second column. (Why?) Our table now appears as follows:

p	q	p	$\wedge$	$(\sim q)$
T	T	T		F
T	F	T		T
F	T	F		F
F	F	F		T

Finally we find the conjunctions of the values given in the third and fifth columns. The completed table appears as follows, with the order in which the columns were considered indicated by (a), (b), and (c), and the final results in bold print:

p	q	p	$\wedge$	$(\sim q)$
T	T	T	**F**	F
T	F	T	**T**	T
F	T	F	**F**	F
F	F	F	**F**	T

$\qquad\qquad(a)\quad(c)\quad(b)$

We can summarize the above table by saying that the statement $p \wedge (\sim q)$ is true only in the case when p is true and q is false. Thus the statement "Today is Wednesday and it is not snowing" is a true statement on a hot Wednesday in July. (We must assume, of course, that it will not snow on a hot day in July.)

exercises

Construct truth tables for each statement.

1. $(\sim p) \wedge q$ **2.** $(\sim p) \vee q$

3. $(\sim p) \vee (\sim q)$ **4.** $(\sim p) \wedge (\sim q)$

5. $\sim(p \wedge q)$ **6.** $p \vee (\sim q)$

In Exercises 7 through 10 copy and complete each truth table.

7.

p	q	$\sim$	$\big[$	p	$\vee$	$(\sim q)\big]$
T	T					
T	F					
F	T					
F	F					

              (d) (a) (c) (b)

8.

p	q	$\sim$	$\big[(\sim p)$	$\vee$	$q \big]$
T	T				
T	F				
F	T				
F	F				

              (d) (a) (c) (b)

9.

p	q	$\sim$	$\big[(\sim p)$	$\wedge$	$(\sim q)\big]$
T	T				
T	F				
F	T				
F	F				

              (d) (a) (c) (b)

10.

p	q	$\sim$	$\big[(\sim p)$	$\vee$	$(\sim q)\big]$
T	T				
T	F				
F	T				
F	F				

              (d) (a) (c) (b)

11. Define $p \underline{\vee} q$ to mean "p or q but not both" and construct a truth table for this connective.

12. Construct a truth table for $p \,|\, q$, which we define to be true when p and q are not both true and to be false otherwise.

13. In Exercise 12 express $p \,|\, q$ in terms of other connectives that we have previously defined.

14. Use p: I like this book; q: I like mathematics. Tell the conditions under which each of the statements in Exercises 1 through 6 is true.

EXPLORATIONS

We may use the presence of an electric current to indicate that p is true and the absence of an electric current to indicate that p is false.

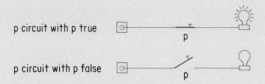

Since we are using circuits solely to improve our understanding of logical statements, we may be a bit fancy in our development of a "negator." To indicate $\sim p$, the negation of p, we need to supply current when p is false and to cut off current when p is true. We may use an electric magnet and these symbols.

An electric magnet with its coil.

A magnet with no current in its coil
does not attract the other part of the switch.

A magnet with current in its coil attracts
the other part of the switch.

Thus, an electric magnet exerts a force if and only if there is an electric current in its coil of wire. When there is no current to the magnet, the circuit is extended with a new flow of current.

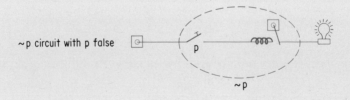

When there is a current to the magnet, it cuts off (breaks) the new circuit.

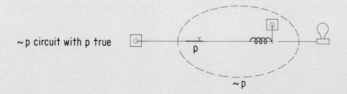

Whenever we use a negator we shall draw a dashed line around it and the switch or switches that are being negated. Then

$\sim(p \vee \sim q)$ for p false and q false may be represented by the following circuit.

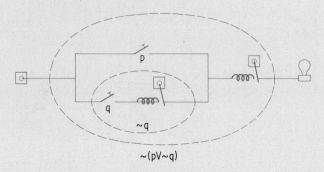

$\sim(pV{\sim}q)$

Note that, as indicated in the figure, if q is false, then $\sim q$ is true, $(p \vee \sim q)$ is true, and $\sim(p \vee \sim q)$ is false.

Assume that p is true and q is false. Then sketch a circuit for each statement.

1. $(\sim p) \vee q$

2. $p \vee \sim q$

3. $\sim(p \wedge q)$

The statement $(p \wedge q) \vee [\sim(p \vee q)]$ may be represented by the following circuit, leaving all circuits open until further instructions are received and assuming that all switches for any one statement (such as p) are open or shut together—that is, are linked together. For this representation, each negator must be placed so that the switches for the expression to be negated are directly supplied with current. For example, we think of $p \wedge (\sim q)$ as $(\sim q) \wedge p$ in sketching a circuit.

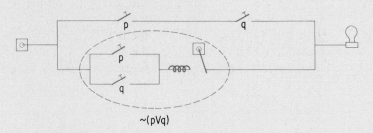

$\sim(pVq)$

See Exercises 11 and 12 and sketch circuits for each statement.

4. $p \veebar q$

5. $p \mid q$

Many of the statements that we make in everyday conversation are based upon a condition. For example, consider the following statements.

If the sun shines, then I will cut the grass.
If I have no homework, then I will go bowling.
If I bribe the instructor, then I will pass the course.

Each of these statements is expressed in the **if-then** form:

If p, then q.

Any if-then statement can be expressed in symbols as

$$p \rightarrow q$$

which is read either as "if p, then q" or as "p implies q." The symbol "$\rightarrow$" is called the **implication symbol**; it is a connective used to form a compound statement called a **conditional statement.** Then the statement p is the **premise** and q is the **conclusion** of the conditional statement $p \rightarrow q$.

Our first task is to consider the various possibilities for p and q in order to define $p \rightarrow q$ for each of these cases. One way to do this would be to present a completed truth table and to accept this as our definition of $p \rightarrow q$. Let us, however, attempt to justify the entries in such a table. Consider again the conditional statement:

If the sun shines, then I will cut the grass.

Now if the sun shines and I do cut the grass, then the statement is obviously true. On the other hand, the statement is false if the sun shines and I do not cut the grass. Assume now that the sun is not shining; then, whether I cut the grass or not, the original statement is true in that I only declared my intentions under the condition that the sun shines. We can summarize these assertions by means of the given truth table.

p	q	$p \rightarrow q$
T	T	T
T	F	F
F	T	T
F	F	T

Consider the statement:

If it rains, then I will give you a ride home.

Have I lied to you:

1. If it rains and I give you a ride home?
2. If it rains and I do not give you a ride home?
3. If it does not rain and I give you a ride home?
4. If it does not rain and I do not give you a ride home?

According to the accepted meanings of the words used, you have a right to feel that I lied to you only if it rains and I do not give you a ride home. In other words, the conditional statement is true unless the premise is true and the conclusion is false.

Many people feel uneasy about some of the truth values of $p \to q$. Actually truth values must be assigned to the conditional statement for each of the four possible combinations of truth values of p and q. Thus the only debatable question is

Under what conditions is a statement of the form
$p \to q$ considered to be a true statement?

In other words, we are concerned with the use of words in the English language. If we are to communicate with each other, we must accept definitions of the meanings of the words that we use. The accepted meaning of $p \to q$ is given in the truth table. Other meanings for the symbols could have been used, but since this meaning has been accepted we should use some other symbol if we wish to ascribe another meaning to a compound statement in terms of p and q.

EXAMPLE 1

Give the truth value of each statement.
(a) If $5 + 7 = 12$, then $6 + 7 = 13$.
(b) If $5 \times 7 = 35$, then $6 \times 7 = 36$.
(c) If $5 + 7 = 35$, then $6 + 7 = 13$.
(d) If $5 + 7 = 35$, then $6 \times 7 = 36$.

SOLUTION

Think of each statement in the form $p \to q$.
(a) For p true and q true, the statement $p \to q$ is *true*.
(b) For p true and q false, the statement $p \to q$ is *false*.
(c) For p false and q true, the statement $p \to q$ is *true*.
(d) For p false and q false, the statement $p \to q$ is *true*.

According to the truth table for $p \rightarrow q$, any statement of the form "if p, then q" is false only when p is true and q is false. On the other hand, if p is false, then the statement $p \rightarrow q$ is accepted as true regardless of the truth value of q. Thus each of the following statements is true by this definition:

> If $2 + 3 = 7$, then George Washington is now the President of the United States.
> If $2 + 3 = 7$, then the moon is made of green cheese.
> If $2 + 3 = 7$, then July follows June.

If you have difficulty accepting any of these statements as true, then you should review the definition of the truth values of if-then statements. Remember also that there need be no relationship between p and q in an if-then statement, although we tend to use such statements in this way in everyday life.

EXAMPLE 2

Find all possible replacements for x such that the following statement is true.

If $3 \times 4 = 34$, then $x - 3 = 5$.

SOLUTION

Think of the statement in the form $p \rightarrow q$ for

p: $3 \times 4 = 34$
q: $x - 3 = 5$

Since p is false, the given statement $p \rightarrow q$ is true for all values of x. That is, regardless of the value assigned to x, the statement

If $3 \times 4 = 34$, then $x - 3 = 5$

is always true.

exercises

In Exercises 1 through 4, consider the statements:

p: You will study hard.
q: You will get an A.

Then translate each of the following symbolic statements into an English sentence.

1. $p \rightarrow q$ **2.** $q \rightarrow p$
3. $(\sim p) \rightarrow (\sim q)$ **4.** $(\sim q) \rightarrow (\sim p)$
5. Repeat Exercises 1 through 4 for the statements:

p: The triangle is equilateral.
q: The triangle is isosceles.

6. Give the truth value of each statement.
  (a) If $2 \times 3 = 5$, then $2 + 3 = 6$.
  (b) If $2 \times 3 = 5$, then $2 + 3 = 5$.
  (c) If $2 + 3 = 5$, then $2 \times 3 = 5$.
7. Give the truth value of each statement.
  (a) If $5 \times 6 = 56$, then $5 + 6 = 11$.
  (b) If $5 \times 6 = 42$, then $5 + 6 = 10$.
  (c) If $5 \times 6 = 42$, then $5 + 5 = 10$.
8. Assume that $a \times b = c$, $b \times c = d$, and $c \neq d$. Then give the truth value of each statement.
  (a) If $a \times b = c$, then $b \times c = d$.
  (b) If $a \times b = d$, then $b \times c = c$.
  (c) If $a \times b = d$, then $b \times c = d$.
  (d) If $a \times b = c$, then $b \times c = c$.
9. Show by means of truth tables that $(\sim p) \vee q$ has the same truth values as $p \rightarrow q$.
10. Show that $\sim [p \wedge (\sim q)]$ has the same truth values as $p \rightarrow q$.

Construct truth tables for:

11. $(p \rightarrow q) \wedge (q \rightarrow p)$
12. $q \rightarrow [(\sim p) \vee q]$

Copy and complete each truth table.

13.

p	q	$[(\sim p)$	$\wedge$	$q]$	$\longrightarrow$	$(p \vee q)$
T	T					
T	F					
F	T					
F	F					

(a) (c) (b) (e) (d)

14.

p	q	$[(\sim p)$	$\vee$	$q]$	$\longrightarrow$	$(p \vee q)$
T	T					
T	F					
F	T					
F	F					

(a) (c) (b) (e) (d)

Find all possible replacements for x for which each sentence is a true sentence.

15. If $2 + 3 = 5$, then $x + 1 = 8$.

16. If $2 + 3 = 6$, then $x + 1 = 8$.

*17. If $x + 1 = 8$, then $2 + 3 = 5$.

*18. If $x + 1 = 8$, then $2 + 3 = 6$.

*19. If $3 - x = 1$, then $2 \times 5 = 13$.

*20. If $3 - x = 1$, then $2 \times 5 = 10$.

EXPLORATIONS

1. The $p \vee q$ circuit provides a basis for defining addition of two elements

1   presence of current at light bulb or through a switch
0   absence of current at light bulb or through a switch

Use sketches of a $p \vee q$ circuit as necessary and complete the following *addition table*.

		q	
+		0	1
p	0		
	1		

2. Use sketches of a $p \wedge q$ circuit as necessary and complete the following *multiplication table*.

		q	
x		0	1
p	0		
	1		

The algebra of the two elements 0 and 1 with addition and multiplication defined as in Explorations 1 and 2 is the **binary Boolean algebra.** It is named after the English mathematician George Boole (1815–1864). We compare this algebra of the numbers 0 and 1 with the algebra of sets $\emptyset$ and U, where as usual $\emptyset$ is the empty set and U is the universal set.

3. Interpret $+$ as $\cup$ and complete the following addition table.

$+$	$\emptyset$	U
$\emptyset$		
U		

4. Interpret $\times$ as $\cap$ and complete the following multiplication table.

$\times$	$\emptyset$	U
$\emptyset$		
U		

5. Two sets of elements, each with relations that may be interpreted as $+$ and $\times$, are said to be **isomorphic sets** if
(i) There is a one-to-one correspondence of the elements of the sets such that
(ii) each sum corresponds to the sum of corresponding elements, and
(iii) each product corresponds to the product of corresponding elements.
Show that the set of elements of the binary Boolean algebra considered in Explorations 1 and 2 and the sets of the algebra of sets considered in Explorations 3 and 4 are isomorphic.

3-4
Equivalent Statements

Two statements are **equivalent statements** if they are either both true or both false. If p and q are equivalent statements, we write

$$p \leftrightarrow q, \quad \text{read as } \text{``}p \text{ is equivalent to } q\text{''}$$

Thus the symbol $\leftrightarrow$ serves as another connective that may be used for any two statements.

EXAMPLE 1

Show by means of a truth table that

$$(p \rightarrow q) \leftrightarrow [(\sim q) \rightarrow (\sim p)]$$

SOLUTION

p	q	$[p \rightarrow q]$	$\longleftrightarrow$	$[(\sim q)$	$\longrightarrow$	$(\sim p)]$
T	T	T	T	F	T	F
T	F	F	T	T	F	F
F	T	T	T	F	T	T
F	F	T	T	T	T	T
		(a)	(e)	(b)	(d)	(c)

The desired equivalence is always true, as indicated in column (e). By comparing columns (a) and (d) you will see that both statements have the same truth values.

The following pairs of equivalent statements are considered in Exercises 1 through 4.

1. $\sim(p \wedge q)$ and $(\sim p) \vee (\sim q)$
2. $\sim(p \vee q)$ and $(\sim p) \wedge (\sim q)$
3. $p \rightarrow q$ and $q \vee (\sim p)$
4. $(\sim p) \rightarrow (\sim q)$ and $q \rightarrow p$

The first two of these pairs of equivalent statements correspond to properties of sets and their complements

$$(A \cap B)' = A' \cup B'$$
$$(A \cup B)' = A' \cap B'$$

and are called *De Morgan's laws for statements*. The third pair of equivalent statements enables us to replace any statement of implication by a disjunction.

There are many other forms in which a conditional statement may appear. We shall explore several of these here. Given any simple statement of implication, $p \rightarrow q$, three other related statements may be identified.

Statement:	$p \to q$	If p, then q.
Converse:	$q \to p$	If q, then p.
Inverse:	$(\sim p) \to (\sim q)$	If not p, then not q.
Contrapositive:	$(\sim q) \to (\sim p)$	If not q, then not p.

The statement proved in Example 1 may now be expressed in the form:

Any statement of implication is equivalent to its contrapositive.

Here are five examples of statements of implication with their converses, inverses, and contrapositives.

1. *Statement:*　If it is snowing, I leave my car in the garage.
    Converse:　If I leave my car in the garage, it is snowing.
    Inverse:　If it is not snowing, I do not leave my car in the garage.
    Contrapositive:　If I do not leave my car in the garage, it is not snowing.

2. *Statement:*　If $\triangle ABC \cong \triangle XYZ$, then $\triangle ABC \sim \triangle XYZ$.
    Converse:　If $\triangle ABC \sim \triangle XYZ$, then $\triangle ABC \cong \triangle XYZ$.
    Inverse:　If $\triangle ABC$ is not congruent to $\triangle XYZ$, then $\triangle ABC$ is not similar to $\triangle XYZ$.
    Contrapositive:　If $\triangle ABC$ is not similar to $\triangle XYZ$, then $\triangle ABC$ is not congruent to $\triangle XYZ$.

3. *Statement:*　If x is negative, then $x \neq 0$.
    Converse:　If $x \neq 0$, then x is negative.
    Inverse:　If x is not negative, then $x = 0$.
    Contrapositive:　If $x = 0$, then x is not negative.

4. *Statement:*　If $x + 2 = 5$, then $x = 3$.
    Converse:　If $x = 3$, then $x + 2 = 5$.
    Inverse:　If $x + 2 \neq 5$, then $x \neq 3$.
    Contrapositive:　If $x \neq 3$, then $x + 2 \neq 5$.

5. *Statement:*　$p \to (\sim q)$.
    Converse:　$(\sim q) \to p$.
    Inverse:　$(\sim p) \to \sim (\sim q)$, which can be simplified as $(\sim p) \to q$.
    Contrapositive:　$\sim (\sim q) \to (\sim p)$, or simply $q \to (\sim p)$.

The following truth tables for these variants of a conditional statement $p \to q$ summarize this discussion and specify the truth values for each statement. From the table you should see again that the contrapositive of a statement is equivalent to the statement. Also note that the converse and inverse of any simple statement of implication are equivalent.

		Statement	Converse	Inverse	Contrapositive
p	q	$p \rightarrow q$	$q \rightarrow p$	$(\sim p) \rightarrow (\sim q)$	$(\sim q) \rightarrow (\sim p)$
T	T	T	T	T	T
T	F	F	T	T	F
F	T	T	F	F	T
F	F	T	T	T	T

EXAMPLE 2

Write the contrapositive of the following statement:

If I work hard, then I will pass the course.

SOLUTION

If I do not pass the course, then I have not worked hard.

This statement is equivalent to the original one. If you fail the course, then you have not worked hard, because if you had worked hard then you would have passed the course.

exercises

Show by means of truth tables that the following statements are equivalent.

1. $\sim(p \wedge q)$ and $(\sim p) \vee (\sim q)$.
2. $\sim(p \vee q)$ and $(\sim p) \wedge (\sim q)$.
3. $p \rightarrow q$ and $q \vee (\sim p)$.
4. $(\sim p) \rightarrow (\sim q)$ and $q \rightarrow p$.

Write the converse, inverse, and contrapositive of each statement.

5. If we can afford it, then we buy a new car.
6. If we play pingpong, then you win the game.
7. If two sides and the included angle of one triangle are congruent to two sides and the included angle of another triangle, then the triangles are congruent.
8. If $x > 2$, then $x \neq 0$.
9. If $x(x - 1) = 0$, then $x = 1$.
10. $(\sim p) \rightarrow q$.

Exercises 11, 12, 13, and 14, refer to the statements given in Exercises 7, 8, and 9. Tell whether you accept as always true:

11. The given statement.

12. The converse of the given statement.

13. The inverse of the given statement.

14. The contrapositive of the given statement.

EXPLORATIONS

Express each of the following statements in symbols and explain why it is an accepted property of statements.

1. Any negation of a conjunction of statements may be expressed as the disjunction of the negations of the given statements.

2. Any negation of a disjunction of statements may be expressed as the conjunction of the negations of the statements.

Note: The statements in Explorations 1 and 2 are often called De Morgan's laws for statements.

3. Any implication of the form $p \rightarrow q$ may be expressed as the disjunction of $\sim p$ and q.

4. Any equivalence may be expressed as a conjunction of disjunctions.

Note: The results summarized in Explorations 1 through 4 are extensively used by logicians and mathematicians who are concerned with the identification of statements as true or false.

3-5
Forms of Statements

Any two equivalent statements may be considered as different forms of the same statement. The recognition of different forms of a statement is particularly useful when statements of implication are under consideration. Frequently the words *necessary* and *sufficient* are used in conditional statements. For example, consider the statement:

Working hard is a sufficient condition for passing the course.

Let us use p to mean "work hard" and q to represent "pass the course." We need to decide whether the given statement means "if p, then q" or "if q, then p." The word sufficient can be interpreted to mean that working hard is adequate or enough—but not necessary—for passing. That is, there may be other ways to pass the course, but working hard will do it. Thus we interpret the statement to mean:

If you work hard, then you will pass the course.

The symbolic statement $p \rightarrow q$ may thus be used for each of these statements:

If p, then q.
p if a sufficient condition for q.

Next consider the statement

Working hard is a necessary condition for passing the course.

Here you are told that working hard is necessary or essential in order to pass. That is, regardless of what else you do, you had better work hard if you wish to pass. However there is no assurance that working hard alone will do the trick. It is necessary, but not sufficient. (You may also have to get good grades.) Therefore, we interpret the statement to mean

If you pass the course, then you have worked hard.

The symbolic statement $q \rightarrow p$ may thus be used for each of these statements:

If q, then p.
p is a necessary condition for q.

Still another form to consider is the statement "q, only if p." In terms of the example used in this section we may write this as

You will pass the course only if you work hard.

Note that this does *not* say that working hard will insure a passing grade. It does mean that if you have passed, then you have worked hard. That is, "q, only if p" is equivalent to the statement "if q, then p."
We can also interpret this in another way. This statement "q, only if p" means "if not p, then not q." The contrapositive of this last statement, however, is "if q, then p." In terms of our illustration this means that if you do not work hard, then you will not pass. Therefore, if you pass, then you have worked hard.

To summarize our discussion to date, each of the following statements represents an equivalent form of writing the statement $p \to q$:

If p, then q.
q, if p.
p implies q.
q is implied by p.
p is a sufficient condition for q.
q is a necessary condition for p.
p, only if q.

The last form is probably the hardest to use. It should appear reasonable when considered as a restatement of the fact that if $p \to q$, then it is impossible to have p without having q. Many people find it helpful to remember these facts:

The premise is sufficient.
The conclusion is necessary.
If you can identify one part of a conditional statement as necessary, then the other part is sufficient.
If you can identify one part of a conditional statement as sufficient, then the other is necessary.
The premise is the "if part."
The conclusion is the "only-if part."
If you can identify one part of a conditional statement as the premise (if part), then the other part is the conclusion (only-if part).
If you can identify one part of a conditional statement as the conclusion (only-if part), then the other part is the premise (if part).

The seven ways of asserting that a statement p implies a statement q illustrate the difficulty of understanding the English language. We shall endeavor to reduce the confusion by considering statements of implication in the form

$$p \to q$$

when stated in symbols and, when stated in words, in the form

If p, then q.

EXAMPLE 1

Write each statement in if-then form.
(a)   Right angles are congruent.
(b) $x > 5$, only if $x \geq 0$.

SOLUTION   (a) If two angles are right angles, then they are congruent.

(b) If $x > 5$, then $x \geq 0$.

EXAMPLE 2

Translate into symbolic form, using

p: I will work hard.
q: I will get an A.

(a) I will get an A only if I work hard.
(b) Working hard will be a sufficient condition for me to get an A.
(c) If I work hard, then I will get an A, and if I get an A, then
     I will have worked hard.

SOLUTION   (a) $q \rightarrow p$;   (b) $p \rightarrow q$;   (c) $(p \rightarrow q) \wedge (q \rightarrow p)$.

The statement $(p \rightarrow q) \wedge (q \rightarrow p)$ in Example 2(c) is one way of stating that p and q are equivalent, $p \leftrightarrow q$. Any statement of implication is sometimes called a conditional statement; any statement of the equivalence of two statements is then called a **biconditional statement**. The symbol $\leftrightarrow$ is referred to as the **biconditional symbol** or the **equivalence symbol**.

Any biconditional statement

$$p \leftrightarrow q$$

is a statement that p is a sufficient condition for q and also p is a necessary condition for q. We may condense this by saying that p is a **necessary and sufficient condition** for q. The biconditional statement may be stated in any one of these forms:

p if and only if q; that is, p **iff** q.
p implies and is implied by q.
p is a necessary and sufficient condition for q.

EXAMPLE 3

Complete a truth table for the statement:

$(p \rightarrow q) \wedge (q \rightarrow p)$

SOLUTION

p	q	$(p \rightarrow q)$	$\wedge$	$(q \rightarrow p)$
T	T	T	T	T
T	F	F	F	T
F	T	T	F	F
F	F	T	T	T

(a) (c) (b)

In Example 3 we constructed a truth table for $(p \rightarrow q) \wedge (q \rightarrow p)$. However, we have previously agreed that this conjunction of statements is equivalent to $p \leftrightarrow q$. This enables us to construct a truth table for $p \leftrightarrow q$.

p	q	$p \leftrightarrow q$
T	T	T
T	F	F
F	T	F
F	F	T

From the truth table we see that $p \leftrightarrow q$ is true when p and q are both true or are both false. Thus each of these statements is true:

$2 \times 2 = 4$ if and only if $7 - 5 = 2$.   (Both parts are true.)
$2 \times 2 = 5$ if and only if $7 - 5 = 3$.   (Both parts are false.)

Each of the following statements is false because exactly one part of each statement is false:

$2 \times 2 = 4$ if and only if $7 - 5 = 3$.
$2 \times 2 = 5$ if and only if $7 - 5 = 2$.

As we have previously noted, we use the symbol "$\leftrightarrow$" to show that two statements are equivalent. Consider these statements:

p: I cut the grass this afternoon.
q: The sun is shining.

Then the statement "$p \leftrightarrow q$," that is, p if and only if q, is true in these two cases:

1.   I cut the grass and the sun is shining.
2.   I do not cut the grass and the sun is not shining.

In all other cases, the statement "$p \leftrightarrow q$" is false. Briefly, two statements "p" and "q" are equivalent if each implies the other; in other words, we may write "$p \leftrightarrow q$" if it is true that $p \to q$ and also that $q \to p$.

EXAMPLE 4

Under what conditions is the following statement true?

I will get an A if and only if I work hard.

SOLUTION
The statement is true in each of two cases:

(a) You get an A and you work hard.
(b) You do not get an A and you do not work hard.

exercises

Write each statement in if-then form.

1. All ducks are birds.
2. Vertical angles are congruent.
3. Complements of the same angle are congruent.
4. Supplements of congruent angles are congruent.
5. Any two parallel lines are coplanar.
6. All triangles are polygons.
7. All circles are round.
8. All mathematics books are dull.
9. All teachers are boring.
10. All p are q.
11. You will like this book only if you like mathematics.
12. A necessary condition for liking this book is that you like mathematics.
13. To like this book it is sufficient that you like mathematics.
14. A sufficient condition for liking this book is that you like mathematics.
15. Liking this book is a necessary condition for liking mathematics.

Write each statement in symbolic form using:

p: The sun shines.
q: I cut the grass.

16. If the sun shines, then I cut the grass.
17. I will cut the grass only if the sun shines.

18. If the sun does not shine, then I do not cut the grass.
19. The sun's shining is a necessary condition for me to cut the grass.
20. I cut the grass if and only if the sun shines.
21. If the sun shines, then I do not cut the grass.
22. For me to cut the grass it is sufficient that the sun shine.
23. A necessary and sufficient condition for me to cut the grass is that the sun shine.

Write each statement in symbolic form using:

p: I miss my breakfast.
q: I get up late.

24. I miss my breakfast if and only if I get up late.
25. A necessary condition for me to miss my breakfast is that I get up late.
26. For me to miss my breakfast it is sufficient that I get up late.
27. A necessary and sufficient condition for me to miss my breakfast is that I get up late.
28. For me not to miss my breakfast it is necessary that I do not get up late.

Express each statement in if-then form and classify as true or false:
29. $11 - 3 > 8$ if $9 + 3 < 10$.
30. A necessary condition for 2×2 to be equal to 5 is that $8 - 5 = 3$.
31. For 7×4 to be equal to 25 it is sufficient that $5 + 3 = 8$.
32. $7 \times 6 = 40$ only if $8 \times 5 \neq 40$.
33. $7 \times 6 = 42$ only if $8 \times 5 \neq 40$.

Decide relative to each given assertion whether a young man who makes the assertion, receives the money, and fails to marry the young lady can be sued for breach of promise.
34. I will marry your daughter only if you give me $10,000.
35. A sufficient condition for me to marry your daughter is that you give me $10,000.

EXPLORATIONS

We frequently state things in a variety of ways. In Explorations 1, 2, and 3, restate each sentence in at least four equivalent ways.

1. Learning your part is necessary for you to be in the play.

2. It is sufficient to come to rehearsals to sing in the chorus.

3. Doing the daily work regularly will enable you to pass the course.

4. Difficulties often arise with the words "only if." For example, the statement "I will get an A, only if I work hard" means "If I get an A, then I have worked hard." However, in everyday language most people tend to interpret the first statement (incorrectly) as "If I work hard, then I will get an A." Find several other common examples of such confusions involving "only if."

5. A statement such as "You may leave early if you finish the quiz" logically means "If you finish the quiz, then you may leave early." However many tend to interpret this permissive use of *if* as meaning *only if.* Thus they would then incorrectly think of the given statement as meaning "If you leave early, then you have finished the quiz."

The use of an if-then statement to give permission frequently has the special meaning of "don't . . . unless . . . ." With this usage, the statement

You may leave early if you finish the quiz

is interpreted as

Don't leave early unless you have finished the quiz.

Prepare a set of three "if-then" statements that give permission, and rewrite each in the form "don't . . . unless . . . ."

3-6
The Nature of Proof

How do you "prove" a statement to a friend? Undoubtedly there are several ways, including these three:

1. Find the statement in an encyclopedia or other reference that he will accept without further proof.
2. Prove to him that the statement is a necessary consequence of some statement that he accepts.
3. Prove to him that the statement cannot be false.

In mathematics there are also several ways of "proving" statements. In essence each proof is based upon:

1. Statements that are accepted as true (assumed).
2. Sequences of statements (arguments) such that each statement is either assumed or is a *logical consequence* of the preceding statements, and the statement to be proved is included in the sequence.
3. Proofs that statements cannot be false.

Thus the "key" to our understanding the nature of proof is our understanding of "logical consequences." We shall find in §3-7 that each logical consequence of a statement depends not only upon the given statement but also upon one or more statements that cannot fail to be true. Any statement such as $p \lor (\sim p)$ that is always true is called a **tautology**.

EXAMPLE

Determine whether or not the statement

$$[(p \to q) \land p] \to q$$

is a tautology.

SOLUTION

p	q	$[(p \longrightarrow q)$	$\land$	$p]$	$\longrightarrow$	q
T	T	T	T	T	T	T
T	F	F	F	T	T	F
F	T	T	F	F	T	T
F	F	T	F	F	T	F
		(a)	(c) (b)		(e)	(d)

The truth value of the given statement is T under all possible situations, as shown in column (e); therefore the statement is a tautology.

exercises

Determine whether each statement is a tautology.

1. $p \lor (\sim p)$
2. $\sim[p \land (\sim p)]$
3. $[(p \to q) \land (\sim q)] \to (\sim p)$
4. $[(p \to q) \land q] \to p$
5. $[(p \to q) \land (\sim p)] \to (\sim q)$

6. $(p \wedge q) \leftrightarrow (q \wedge p)$

7. $[(p \vee q) \wedge (\sim p)] \rightarrow q$

8. $(p \vee q) \rightarrow (q \vee p)$

9. $[(p \rightarrow q) \wedge (q \rightarrow r)] \rightarrow (p \rightarrow r)$

EXPLORATIONS

There exist situations in which statements are neither true nor false and our usual rules of logic cannot be used.

1. Select a plain 3×5 card or similar piece of paper. On one side write

The statement on the other side of this card is true.

Then on the other side of the card write

The statement on the other side of this card is false.

Discuss the possible sets of truth values for these two statements.

2. There is reported to be a town in which the barber is a man who shaves all men who do not shave themselves. Who shaves the barber?

3. Discuss the truth values of the statement

The sentence you are reading is false.

There exist situations in which a knowledge of tautologies can provide the basis for asking appropriate questions to solve very perplexing problems.

Suppose that you are a prisoner standing alone with the executioner in the execution chamber. You must choose one of two chairs and sit in it. One chair is an electrified chair that kills anyone who sits in it. The other chair is harmless.

You are allowed to ask the executioner just one question that he may answer by "yes" or "no." Furthermore, you know that the executioner either always tells the truth or always lies but you do not know whether he tells the truth or lies.

What question could you as the condemned prisoner ask and determine without any doubt which chair is the safe chair? Try to answer this question of life and death before reading further.

This form of a relatively common problem situation was presented by Julian C. Stanley on page 74 of the January 1957 issue of *School Science and Mathematics* under the title "Clever question beats the heat." Mr. Stanley later suggests the question "If I were to ask you, 'Is this the electrified chair?' would you say 'Yes'?"

Explain how the executioner would provide you with the desired information in each of these situations, that is, in all possible situations.

4. You identify the electrified chair and the executioner always tells the truth.

5. You identify the electrified chair and the executioner always lies.

6. You identify the harmless chair and the executioner always tells the truth.

7. You identify the harmless chair and the executioner always lies.

Suppose that you approach a fork in the road while traveling to the river in a strange country. What single question could you ask to find your way in each of these situations?

8. You encounter a native who either always tells the truth or always lies but you do not know which.

9. You encounter two natives, one of whom always tells the truth and one of whom always lies, but you do not know which is which.

3-7
Valid Arguments

We shall consider proofs in terms of the patterns (arguments) formed by the statements used. The assumed statements are identified as given statements and called premises of the argument. The statements to be proved are called conclusions of the argument.

EXAMPLE 1

Determine whether the following argument is valid:
Given: If Mary is a junior, she is taking algebra.
Given: Mary is a junior.
Conclusion: Mary is taking algebra.

SOLUTION   Use

p: Mary is a junior
q: Mary is taking algebra

and think of the argument as:

Given: $p \rightarrow q$.
Given: p.
Conclusion: q.

The argument is valid if and only if the statement

$$[(p \rightarrow q) \wedge p] \rightarrow q$$

is a tautology. This statement is a tautology (§3-6, Example). Thus the argument is valid.

The argument used in Example 1 illustrates one of the basic rules of inference, the **law of detachment,** also called **modus ponens:**

If a statement of the form "if p, then q" is assumed to be true, and if p is known to be true, then q must be true.

EXAMPLE 2

Determine whether or not the following argument is valid:
Given: If $3x - 1 = 8$, then $3x = 9$.
Given: $3x - 1 = 8$.
Conclusion: $3x = 9$.

SOLUTION   For

p: $3x - 1 = 8$
q: $3x = 9$

the argument is valid by *modus ponens.* It has the form of the tautology

$$[(p \rightarrow q) \wedge p] \rightarrow q$$

EXAMPLE 3

Determine whether the following argument is valid:
Given: If you worked hard, then you passed the course.
Given: You passed the course.
Conclusion: You worked hard.

SOLUTION   For

p: You worked hard
q: You passed the course

the argument has the form

$$[(p \rightarrow q) \wedge q] \rightarrow p$$

This statement is not a tautology (§3-6, Exercise 4). Thus the argument is not valid.

Notice that anyone who uses the argument in Example 3 to convince his friends that he has worked hard is hoping that his friends will think that the converse of any true conditional statement is also true; that is,

If $p \rightarrow q$, then $q \rightarrow p$.

We know from our comparison of truth values of $p \rightarrow q$ and $q \rightarrow p$ that a statement does not necessarily imply its converse. As another example of this type of reasoning consider the advertisement:

If you want to be healthy, eat KORNIES.

The advertiser hopes that the consumer will fallaciously assume the converse statement:

If you eat KORNIES, then you will be healthy.

The argument is not valid and is called a **fallacy**.

EXAMPLE 4

Determine whether the following argument is valid:
Given: If you worked hard, then you passed the course.

Given: You did not pass the course.
Conclusion: You did not work hard.

SOLUTION For

> *p*: You worked hard
> *q*: You passed the course

the argument has the form

$$[(p \rightarrow q) \land (\sim q)] \rightarrow (\sim p)$$

This statement is a tautology (§3-6, Exercise 3); the argument is valid.

Notice that the argument in Example 4 is based upon the equivalence of any given conditional statement $p \rightarrow q$ and its contrapositive statement $(\sim q) \rightarrow (\sim p)$ as we observed in §3-4.

EXAMPLE 5

Determine whether the following argument is valid:
Given: If you worked hard, then you passed the course.
Given: You did not work hard.
Conclusion: You did not pass the course.

SOLUTION For

> *p*: You worked hard
> *q*: You passed the course

the argument has the form

$$[(p \rightarrow q) \land (\sim p)] \rightarrow (\sim q)$$

This statement is not a tautology (§3-6, Exercise 5), and the argument is not valid. It is another example of a fallacy.

Notice that anyone who uses the argument in Example 5 expects his listeners to assume that the inverse $(\sim p) \rightarrow (\sim q)$ of any true statement $p \rightarrow q$ must be true. As in the case of a conditional statement and its converse, we know that a statement does not necessarily imply its inverse. As another example of this type of reasoning, consider the advertisement:

If you brush your teeth with SCRUB, then you will have no cavities.

The advertiser would like you to assume, fallaciously, the inverse statement:

If you do not brush your teeth with SCRUB, then you will have cavities.

The final form of valid reasoning that we shall consider here is based upon the tautology.

$$[(p \rightarrow q) \wedge (q \rightarrow r)] \rightarrow (p \rightarrow r)$$

(§3-6, Exercise 9). Consider these specific examples:

1. If you like this book, then you like mathematics.
   If you like mathematics, then you are intelligent.
   Therefore, if you like this book, then you are intelligent.

2. If a polygon is a square, then it has 3 sides.
   If a polygon has 3 sides, then it has 4 angles.
   Therefore, if a polygon is a square, then it has 4 angles.

3. If a polygon is a square, then it has 4 sides.
   If a polygon has 4 sides, then it has 3 angles.
   Therefore, if a polygon is a square, then it has 3 angles.

In the second example, both premises are false, the conclusion is true, and the argument is valid. In the third example, the first premise is true, the second premise is false, the conclusion is false, yet the argument is again valid, since it is based upon a tautology. Thus the truth of the conclusion does not enter in any way into a discussion of the validity of an argument. The *validity* of an argument depends solely upon its form; that is, whether or not it is based upon a tautology. An argument is valid if the conjunction of the premises implies the conclusion. The *truth* of a conclusion of a valid argument depends completely upon the truth of the premises.

It is important to emphasize the fact that validity has nothing to do with the question of whether the conclusion is true or false. The conclusion may be false, yet the argument may be valid if the chain of reasoning is correct. On the other hand, the conclusion may be true, yet the argument may not be valid if the reasoning is incorrect.

In each of the exercises that follow, bear in mind that the validity of each argument depends upon its form rather than the truth of the sentences used.

exercises

Determine whether each argument is valid or not valid.

1. *Given:* If Elliot is a freshman, then Elliot takes mathematics.
   Given: Elliot is a freshman.
   Conclusion: Elliot takes mathematics.

2. *Given:* If you like dogs, then you will live to be 120 years old.
   Given: You like dogs.
   Conclusion: You will live to be 120 years old.

3. If the Yanks win the game, then they win the pennant.
   They do not win the pennant.
   Therefore, they did not win the game.

4. If you like mathematics, then you like this book.
   You do not like mathematics.
   Therefore, you do not like this book.

5. If you work hard, then you are a success.
   You are not a success.
   Therefore, you do not work hard.

6. If you are reading this book, then you like mathematics.
   You like mathematics.
   Therefore, you are reading this book.

7. If you are reading this book, then you like mathematics.
   You are not reading this book.
   Therefore, you do not like mathematics.

8. If you work hard, then you will pass the course.
   If you pass the course, then your teacher will praise you.
   Therefore, if you work hard, then your teacher will praise you.

9. If you like this book, then you like mathematics.
   If you like mathematics, then you are intelligent.
   Therefore, if you are intelligent, then you like this book.

10. If you are a blonde, then you are lucky.
    If you are lucky, then you will be rich.
    Therefore, if you do not become rich, then you are not a blonde.

Supply a conclusion so that each of the following arguments will be valid

11. If you drink milk, then you will be healthy.
    You are not healthy.
    Therefore, ...

12. If you eat a lot, then you will gain weight.
    You eat a lot.
    Therefore, ...

13. If you like to fish, then you enjoy swimming.
    If you enjoy swimming, then you are a mathematician.
    Therefore, ...

14. If you do not work hard, then you will not get an A.
    If you do not get an A, then you will have to repeat the course.
    Therefore, ...

15. If you like this book, then you are not lazy.
If you are not lazy, then you will become a mathematician.
Therefore, ...

EXPLORATIONS

The following explorations are based upon sets of premises written by Charles Lutwidge Dodgson, who used the name of Lewis Carroll as author of *Alice's Adventures in Wonderland* and *Through the Looking Glass.* For additional such explorations, see some of Dodgson's other books.

Supply a conclusion so that each argument will be valid.

1. Babies are illogical.
Nobody is despised who can manage a crocodile.
Illogical persons are despised.
Therefore, ...

2. No ducks waltz.
No officers ever decline to waltz.
All my poultry are ducks.
Therefore, ...

3. No terriers wander among the signs of the zodiac.
Nothing that does not wander among the signs of the zodiac
is a comet.
Nothing but a terrier has a curly tail.
Therefore, ...

3-8

Euler Diagrams

In the eighteenth century the Swiss mathematician Leonhard Euler used diagrams to present a visual approach to the study of the validity of arguments. The diagrams used in this section are an aid to reasoning but are not essential to the reasoning process. We use a region, usually a circular region P to represent the situations in which a statement p is true. Then p is false in all situations represented by points of P'. Similarly, we use a region Q for a statement q, and so forth.

First consider the statement $p \rightarrow q$ (if p, then q). This statement is equivalent to saying "all p are q," that is, p is a subset of q; if you have p, then you must have q. We show this by drawing circles to represent p and q with p drawn as a subset of q. Consider the statements:

> p: It is a lemon.
> q: It is a piece of fruit.

Then the figure below may be used as a visualization of the statement "All lemons are pieces of fruit." Statements p and q are called consistent statements since they can both be true of the same object.

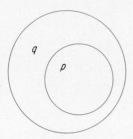

Next consider the statements:

> p: It is a lemon.
> s: It is an orange.

Statements p and s are called contrary statements since they cannot both be true of the same object. If we wish to draw a diagram to represent the statement "No lemon is an orange," then we may draw two circles whose intersection is the empty set:

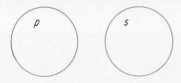

Note that this same diagram can be used to represent either of the following equivalent statements:

> If it is a lemon, then it is not an orange.
> If it is an orange, then it is not a lemon.

We may also represent statements such as the following by means of Euler diagrams:

> p: It is a lemon.
> $\sim p$:   It is not a lemon.

If we draw a circle to represent the statement p, then $\sim p$ may be represented by the set of points outside of p; that is, by the complement of p.

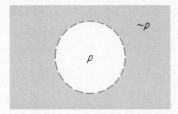

Statements such as p and $\sim p$ are called **contradictory statements.** Exactly one of the statements must be true; exactly one of the statements must be false. If either statement is true, then the other is false. If either statement is false, then the other is true.

Let us now return to the question of validity and explore a visual approach to the argument

$$[(p \rightarrow q) \wedge (q \rightarrow r)] \rightarrow (p \rightarrow r).$$

We have seen that the statement "if p, then q" is equivalent to saying "all p's are q's" and may be drawn with p as a subset of q. Similarly, the statement "if q, then r" can be written in the form "all q's are r's"; that is, q is a subset of r. To represent these two statements we construct circles in such a way that the set of points representing p is a subset of the set of points representing q, and the set of points representing q is a subset of the set of points representing r, as in the diagram.

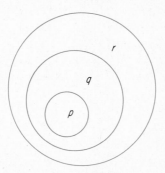

We have followed instructions carefully for our two premises. From the diagram it is clear that the set of points representing p must be a subset of the set of points representing r. That is, we conclude that "all p's are r's" or "if p, then r," which is what we set out to demonstrate as the conclusion of our argument.

From the final diagram it should also be clear that each of the following conclusions would *not* be valid:

All r's are q's. $r \rightarrow q$
All q's are p's. $q \rightarrow p$

Can you see that each of the following conclusions follows logically from the diagram drawn?

If not r, then not p. $(\sim r) \rightarrow (\sim p)$
It not q, then not p. $(\sim q) \rightarrow (\sim p)$

Euler diagrams may be used to test the validity of an argument. We draw one or more Euler diagrams, making use of the premises and avoiding the introduction of additional premises. The structure of the diagram indicates whether or not the argument is valid. As in §3-7, we disregard the truth values of the premises and conclusions.

EXAMPLE 1

Test the validity of this argument:

All undergraduates are sophomores.
All sophomores are attractive.
Therefore, all undergraduates are attractive.

SOLUTION   A diagram of this argument follows:

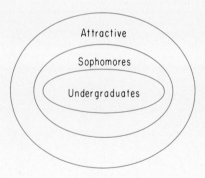

The argument is valid.

Notice in Example 1 that you may or may not agree with the conclusion or with the premises. The important thing is that you are *forced* to draw the diagram in this manner. The conclusion is an inescapable consequence of the given premises. You should not allow the everyday meaning of words to alter your thinking. Thus the preceding argument can be stated abstractly as in the following three statements.

All *u*'s are *s*'s.
All *s*'s are *a*'s.
Therefore, all *u*'s are *a*'s.

Here we rely on logic alone rather than any preconceived notions about the meaning of such words as "undergraduate," "sophomore," and "attractive."

EXAMPLE 2

Draw a diagram and test the validity of this argument:

All freshmen are clever.
All attractive people are clever.
Therefore, all freshmen are attractive.

SOLUTION   The first premise tells us that the set of freshmen is a subset of the set of clever individuals. This is drawn as follows:

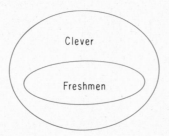

Next we need to draw a circle to represent the set of attractive people as a subset of the set of clever individuals. However, there are several possibilities here:

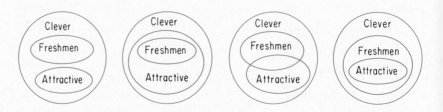

Each of the figures drawn represents a distinct possibility, but only one shows that all freshmen are attractive. Therefore, since we are not *forced* to arrive at this conclusion, the argument is said to be *not valid*. Each of the figures, on the other hand, forces you to arrive at the following valid conclusions:

Some clever people are attractive.
Some clever people are freshmen.

Consider next the statement:

Some freshmen are attractive.

This statement means that there exists *at least one* freshman who is attractive. It does not, however, preclude the possibility that all freshmen are attractive. Now consider the following argument whose validity we wish to check:

Some freshmen are attractive.
All girls are attractive.
Therefore, some freshmen are girls.

To test the validity of this argument we draw Euler circles for the two premises and then see whether we are forced to accept the conclusion. Since some freshmen are attractive, we draw the following but recognize that the set of freshmen could also be drawn as a proper subset of the set of attractive people.

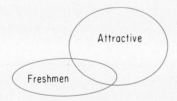

Next we draw a circle to represent the set of girls as a proper subset of the set of attractive people. Here are several possible ways to do this:

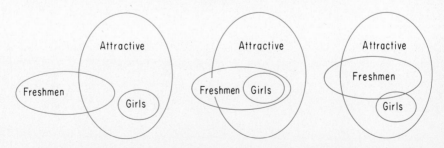

Notice that in two of the figures we find that some freshmen are girls. But we are not forced to adopt this conclusion, as seen in the first of the three figures drawn. Accordingly, we conclude that the argument is not valid.

EXAMPLE 3

Test the validity of the following argument:

All *a*'s are *b*'s.
Some *a*'s are *c*'s.
Therefore, some *c*'s are *b*'s.

SOLUTION

Note that since some *a*'s are *c*'s, then there are also some *c*'s that are *a*'s. However, since all *a*'s are *b*'s, it follows that some *c*'s must also be *b*'s, as in these diagrams:

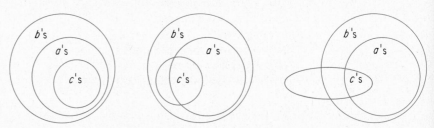

The argument is valid.

exercises

Use each of the following diagrams and write at least five valid conclusions.

1.   **2.**

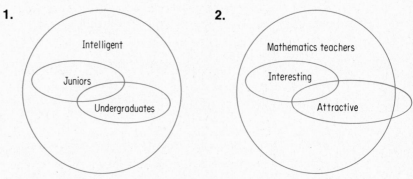

Draw a diagram and test the validity of each argument.

3. All students love mathematics.
Harry is a student.
Therefore, Harry loves mathematics.

4. All juniors are brilliant.
All brilliant people love mathematics.
Therefore, if you are a junior, then you are brilliant.

5. All girls are beautiful.
All beautiful people like this book.
Therefore, if you like this book, then you are a girl.

6. All mathematics teachers are dull.
Some Ph.D.'s are dull.
Therefore:

   (a) Some mathematics teachers are Ph.D.'s.
   (b) Some dull people have Ph.D.'s.

7. All juniors are clever.
Some juniors are males.
Therefore:

   (a) Some males are clever.
   (b) Some males are juniors.
   (c) Some clever people are males.
   (d) All males are juniors.

8. All boys are handsome.
Some boys are athletes.
Therefore, some athletes are handsome.

9. All mathematics teachers are interesting.
All attractive individuals are interesting.
Some mathematics teachers are kind.
Therefore:

   (a) Some interesting people are kind.
   (b) Some mathematics teachers are attractive.
   (c) All mathematics teachers are attractive.
   (d) All mathematics teachers are kind.
   (e) Some kind individuals are attractive.
   (f) No mathematics teachers are attractive.
   (g) No attractive individuals are interesting.

10. All x's are y's.
Some y's are z's.
Therefore:

   (a) Some x's are z's.   **(b)** Some z's are y's.

11. All a's are b's.
All b's are c's.
Some d's are not c's.
Therefore:

   (a) All a's are c's.   **(b)** Some d's are not b's.
   (c) Some d's are not a's.   **(d)** Some d's are c's.

12. All booms are mooms.
Some booms are tooms.
Some zooms are booms.
Therefore:

   (a) Some mooms are booms.

(b)   Some zooms are mooms.
(c)   Some mooms are tooms.
(d)   No toom is a zoom.
(e)   Some zooms are moom-tooms.

EXPLORATIONS

1.   Find a truth table for the connective NOR where "*p* NOR *q*" means "neither *p* nor *q*."

2.   Find a truth table for $\sim(p$ NOR $q)$.

Find one or more statements with each of the given sets of truth values.

p	*q*	3.	4.	5.	6.	7.	8.	9.	10.
T	T	F	T	T	T	T	F	F	F
T	F	T	F	T	T	F	T	F	F
F	T	T	T	F	T	F	F	T	F
F	F	T	T	F	F	F	F	F	T

p	*q*	11.	12.	13.	14.	15.	16.
T	T	T	T	T	F	F	F
T	F	T	T	F	F	F	T
F	T	F	T	F	F	T	T
F	F	T	T	T	F	T	F

3-9
Quantifiers

Some statements are intended to be *universally* true for the people or things under consideration:

All sophomores are attractive.
All squares are rectangles.

Other statements are intended to assure us that there *exists* at least one case in which the statement is true.

>   Some teachers are intelligent.
>   Some parallelograms are squares.

These last two statements can also be written as:

>   There exists at least one teacher who is intelligent.
>   There exists at least one parallelogram that is a square.

The words "all" and "some" serve as quantifiers. "All" may be used to assert that a statement is universally true and is called a **universal quantifier**. "Some" may be used to assert that there exists a case for which the statement is true and is called an **existential quantifier**.

In mathematics, quantifiers are particularly helpful to express sentences so that they can be identified as true or false. For example,

$$x + 2 = 5$$

is true if $x = 3$ and is false otherwise. Thus we don't know whether $x + 2 = 5$ is true or false unless we know about the values of x. We know that the sentence is not true for all whole numbers; that is, it is false to say

$$x + 2 = 5 \text{ for all whole numbers } x.$$

We know that the sentence is true for at least one whole number. There exists a whole number x such that $x + 2 = 5$. We may write this in symbols for whole numbers x in this way:

$$\exists_x: \quad x + 2 = 5$$

We use $\exists_x$ as the symbol for the existential quantifier "there exists an x such that."

Now consider this sentence:

$$x + 2 = 2 + x \quad \text{for all real numbers.}$$

By the commutative property of addition we know that this is always true. We write this in symbols for real numbers x in this way:

$$\forall_x: \quad x + 2 = 2 + x$$

We use $\forall_x$ as the symbol for the universal quantifier "for all x."

Careful consideration of quantifiers is particularly important because of the common popular confusion of "all are not" and "not all are." The following statements are equivalent:

Not all students are sophomores.
Some students are not sophomores.

Similarly, these statements are equivalent:

All my students are not dishonest.
No one of my students is dishonest.
All of my students are honest.

Notice that to negate (deny) a universally quantified statement it is only necessary to find one case for which the statement is false, that is, one counterexample. Thus to prove that not all students are sophomores it is sufficient to find one student who is not a sophomore. However, to negate (deny) an existentially quantified statement it is necessary to prove that the statement is false for *all* students. Thus to prove that no one of my students is dishonest it is necessary to prove that all of my students are honest (not dishonest).

EXAMPLE 1

State in words the negation of each statement.

(a)   There is a rectangle that is not a square.
(b)   All triangles are right triangles.

SOLUTION   (a)   All rectangles are square.
                    (b)   There exists a triangle that is not a right triangle.

EXAMPLE 2

State in words the negation of the statement "Some books are worth reading."

SOLUTION   It is often helpful to consider the denial of a statement by various rephrasings of the statement. Thus we can go from the given statement to "There is at least one book that is worth reading." The denial of this statement is then "There aren't any books worth reading," which can finally be stated in the form "All books are not worth reading."

exercises

In Exercises 1 through 4 assign a value to the variable to make each sentence true.

1. $x - 2 = 7$ **2.** $2x = 12$

3. $2x + 3 = 15$ **4.** $x^2 = 9$

5. In Exercises 1 through 4 assign a value to the variable to make each sentence false.

6. In Exercises 1 through 4 use a quantifier to make each sentence true.

7. In Exercises 1 through 4 use a quantifier to make each sentence false.

State in words the negation of each statement.

8. All apples are pieces of fruit.

9. All numbers are whole numbers.

10. Any whole number is positive.

11. Any fraction represents a rational number.

12. There exist rational numbers that are not integers.

13. There exist real numbers that are not rational numbers.

14. Some complex numbers are real numbers.

15. Some rational numbers are integers.

EXPLORATIONS

1. Discuss the validity of the arguments used in the following situation.

A man approached the girl at the check-out counter of a local market and asked the price of a box of blueberries.

Clerk: "65 cents a box."

Customer: "What! They're selling them for 55 cents a box across the street."

Clerk: "Why don't you buy them there?"

Customer: "Because they're all sold out."

Clerk: "Oh! If we were all sold out, our price would be 45 cents a box."

2. Try to find other descriptions of situations, such as the above, which involve logical concepts.

chapter test

1. Use p: John studies hard; q: Jim is lazy. Think of "is lazy" as "does not study hard" and write each of these statements in symbolic form:
   (a) Neither John nor Jim is lazy.
   (b) It is not true that John and Jim are both lazy.

2. Copy and complete the truth table.

p	q	$\sim$	$[(\sim p)$	$\lor$	$q]$
T	T				
T	F				
F	T				
F	F				
		(d)	(a)	(c)	(b)

3. Give the truth value of each statement:
   (a) If $6 + 2 = 10$, then $7 \times 2 = 11$.
   (b) If $7 \times 2 = 10$, then $6 \times 2 = 12$.

4. Write the converse, inverse, and contrapositive of the statement:

   If $x < 3$, then $x \neq 3$.

5. Write each statement in if-then form:
   (a) Right angles are congruent.
   (b) To like to swim it is necessary that you like water.

6. Write each statement in symbolic form using:

   p: I wear a coat.
   q: It is snowing.

   (a) For me to wear a coat it is necessary that it be snowing.
   (b) I wear a coat only if it is snowing.

7. Write each statement in symbolic form:
   (a) p is sufficient for q.
   (b) p is necessary and sufficient for $(\sim q)$.

8. Identify as true or false:
   (a) $3 \times 5 = 35$ only if $6 \times 5 = 70$.
   (b) For $2 \times x = 6$ it is sufficient that $2 \times 5 = 6$.

9. Test the validity of each argument:
   (a) All dogs love their masters.
   If an animal loves its master, it should be well cared for.
   Therefore, all dogs should be well cared for.
   (b) If x is a positive integer, then $x^2 > 0$.
   $x^2 \not> 0$.
   Therefore, x is not a positive number.

10. Use a truth table to determine whether or not the statement

$$[(p \to q) \wedge (\sim q)] \to (\sim p)$$

is a tautology.

4

Systems of
NUMERATION

We habitually take for granted the use of our system of numeration, as well as our computational procedures. However, these represent the creative work of man through the ages. We can gain a better appreciation of our system of numeration and methods of computation by examining other systems.

Our number concepts are based upon sets. We have considered equivalent sets as sets that have the same number of elements. We can also match sets to determine whether one set has more or less elements than another. Early man spent many years finding ways of indicating how many elements there were in a set. Undoubtedly just as you and I might hold up three fingers to indicate a set of three elements, early man used matching sets. For example, he may have used sets of pebbles to keep track of the number of sheep that he had.

There is evidence that members of the Bugilai tribe in British New Guinea indicated numbers by pointing in order to that number of parts of the body. The pointing is reported to have been done by the right index finger to these parts of the body in order:

left-hand little finger	1	tarangesa
left-hand ring finger	2	meta kina
left-hand middle finger	3	guigimeta
left-hand index finger	4	topea
left-hand thumb	5	manda
left wrist	6	gaben
left elbow	7	trankgimbe
left shoulder	8	podei
left breast	9	ngama
right breast	10	dala

Note that no words or numerals were needed for this sort of communication. Later, when number words were developed in the Bugilai dialect, they had original meanings from the parts of body, as indicated in the second and third columns of the array. Other tribes developed similar procedures and some tribes could designate numbers up to 31 using parts of the body. One tribe is reported to have been able to represent all numbers up to 100.

Parts of the body could be used to point out matching sets for small sets of elements. As soon as man started thinking of a part of the body as representing a set, such as a hand representing a set of five elements, he was starting to develop numerals. The use of a set of parts of the body to match one-to-one a specified set was a use of numbers without numerals. Only the matching was involved. There was a set of five fingers but man had not reached the stage of thinking of the set of five fingers as a single symbol, a hand, to represent five.

Throughout this chapter we shall be exploring ways of representing numbers by symbols, that is, systems of numeration. We first examine the system of numeration used by the ancient Egyptians as found in their hieroglyphics. Then we move to contemporary times including the study of binary notation, the mathematical basis for the modern computer. Throughout this exploration the reader should note the liberation in mankind's thinking that has taken place from early times to the present day, a liberation that matches his creative developments in other areas as well.

4-1

Egyptian Numeration

Can you write a dog? You can write the word "dog" or the name of a particular dog. But you cannot write a dog.

Can you write a number? You can write the name of a particular number or a symbol for a number. But you cannot write a number. Here are a few symbols for the number three:

$|||$, 3, $6/2$, $\frac{3}{1}$, $\sqrt{9}$

A name or symbol for a number is a **numeral.** We shall make this distinction between a number and a numeral only where it proves helpful in understanding some concept.

Consider the number that represents the number of items in the following collection of marks.

X X X X X

X X X X X

We would write this number as 10 in our system of notation. The ancient Egyptians used the symbol ∩. The ancient Babylonians used the symbol ❮. The Romans used X. All these symbols—and there are others—are merely numerals; that is, they are different ways of representing (providing a name for) the same number.

Let us explore one system, the one used by the ancient Egyptians, in greater detail. They used a new symbol for each power of ten. Here are some of their symbols, a description of the physical objects they are supposed to represent, and the number they represent as expressed in our notation:

|	Vertical staff	1
∩	Heel-bone	10
𓍢	Scroll	100
𓆼	Lotus flower	1000
𓂭	Pointing finger	10,000

This Egyptian system is said to have a **base** of ten, but has no place value. The "base" ten is due to the use of powers of ten. We call our system of numeration a **decimal system** to emphasize our use of powers of ten. The absence of a place value means that the position of the symbol does not affect the number represented. For example, in our decimal system of numeration, 23 and 32 represent different numbers. In the Egyptian system ∩| and |∩ are different ways of writing eleven, that is, different names or numerals for the same number. (The former notation is the one normally found in their hieroglyphics.) Here are some other comparisons of decimal and ancient Egyptian number symbols.

25	∩∩ | | | | |
142	𓍢 ∩∩∩∩ | |
12,321	𓂭 𓆼𓆼 𓍢𓍢𓍢 ∩∩ |

Computation in the ancient Egyptian system is possible, although tedious. For example, we use these steps to add 27 and 35.

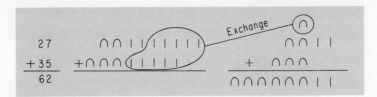

Observe that, in this Egyptian system, an indicated collection of ten ones was replaced by a symbol for ten before the final computation took place. In our decimal system we mentally perform a similar exchange of ten ones for a ten when we express $(7 + 5)$ as one ten and two ones. We exchange kinds of units in a similar manner in subtraction.

The ancient Egyptians multiplied by a process of doubling. This process is based on the fact that any number may be represented as a sum of powers of two. For example, $19 = 1 + 2 + 16$. To find the product 19×25 we first double 25.

   ① x 25 = ㉕

   ② x 25 = ㊿

   4 x 25 = 100

   8 x 25 = 200

  ⑯ x 25 = ④⓪⓪

Then we find the product 19×25 by adding the multiples of 25 that correspond to 1, 2, and 16.

$$19 = 1 + 2 + 16$$
$$19 \times 25 = (1 + 2 + 16) \times 25$$
$$= 25 + 50 + 400 = 475$$

EXAMPLE 1

Use the Egyptian method of multiplication to find the product 23×41.

SOLUTION

$1 \times 41 = 41$		
$2 \times 41 = 82$		$23 = 1 + 2 + 4 + 16$
$4 \times 41 = 164$		$23 \times 41 = (1 + 2 + 4 + 16) \times 41$
$8 \times 41 = 328$		$= 41 + 82 + 164 + 656 = 943$
$16 \times 41 = 656$		

Later the Egyptians adopted a more refined and automatic procedure for multiplication known as **duplation and mediation** that involves doubling one factor and halving the other. For example, to find the product 19×25 we may successively halve 19, discarding remainders at each step, and successively double 25.

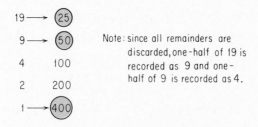

19 ⟶	25
9 ⟶	50
4	100
2	200
1 ⟶	400

Note: since all remainders are discarded, one-half of 19 is recorded as 9 and one-half of 9 is recorded as 4.

This process is complete when a 1 appears in the column of numbers that are being halved. Opposite each number in this column of halves there is a corresponding number in the column of numbers being doubled. The product 19×25 is found as the sum of the numbers that are opposite the *odd* numbers in the column of halves.

$$19 \times 25 = 25 + 50 + 400 = 475$$

Note that this process automatically selects the addends to be used in determining the product; one need not search for the appropriate powers of two to be used.

EXAMPLE 2

Use the process of duplation and mediation to find the product 23×41.

SOLUTION

$$23 \longrightarrow \boxed{41}$$
$$11 \longrightarrow \boxed{82}$$
$$5 \longrightarrow \boxed{164} \qquad 23 \times 41 = 41 + 82 + 164 + 656 = 943$$
$$2 \qquad 328$$
$$1 \longrightarrow \boxed{656}$$

exercises

Write in ancient Egyptian notation.

1. 25   **2.** 138

3. 1426   **4.** 40

5. 12,407   **6.** 5723

Write in decimal notation.

7. ∩∩ I I   **9.** ⌀ ⌀ I I   **11.** ⌀ ⌀ ⌀ ⌀ ∩∩ I I I I

8. ⌀ ⌀ ∩ I   **10.** ⌀ ⌀ ⌀ ⌀ ∩ I   **12.** ⌀ ⌀ ⌀ ∩ I I

Write in ancient Egyptian notation and perform the indicated operation in that system.

13.   42   **14.**   153
     +21   +62

15.   238   **16.**   431
    +135   −213

17. 1243   **18.**   507
  −137   − 124

Use the Egyptian method of doubling to find these products.

19. 17×45 **20.** 15×35

21. 27×31 **22.** 31×19

Use the Egyptian method of duplation and mediation to find these products.

23. 19×33 **24.** 24×51

25. 21×52 **26.** 37×80

27. The Egyptians did not have a symbol for zero. Why do we need such a symbol in our system of numeration? Why did they find it unnecessary to invent such a symbol in order to have symbols for other numbers?

EXPLORATIONS

Roman numerals have been widely used in the past and are not unusual in MCMLXXIII, that is, 1973. These explorations are intended to help you recognize that it is relatively easy to represent whole numbers as Roman numerals. Some computations with Roman numerals are as easy as with our ordinary numerals, but other computations with Roman numerals are very awkward.

1. Use Roman numerals and count from I to XXIX.

2. Use Roman numerals and count by fives from V to L.

3. Describe the use of addition and subtraction in the representation of numbers by Roman numerals, such as XXIX.

4. There are a few instances of "cancellation" in our ordinary numerals such that correct answers are obtained. For example,

$$\frac{1\cancel{6}}{\cancel{6}4} = \frac{1}{4}$$

Find at least five instances where such cancellation may be properly used in Roman numerals and at least one instance where it may not be used.

5. Multiply (a) IX by X; (b) XI by V.

6. Do you find one addition problem easier than the other? If so, which one is the easier one?

$$\begin{array}{r} 25 \\ +12 \\ \hline \end{array} \qquad \begin{array}{r} \text{XXV} \\ +\text{XII} \\ \hline \end{array}$$

7. Do you find one subtraction problem easier than the other? If so, which one is the easier one?

$$\begin{array}{r} 25 \\ -12 \\ \hline \end{array} \qquad \begin{array}{r} \text{XXV} \\ -\text{XII} \\ \hline \end{array}$$

8. Do you find one multiplication problem easier than the other? If so, which one is the easier one?

$$\begin{array}{c} 26 \\ \times 5 \\ \hline \end{array} \qquad \begin{array}{c} \text{XXVI} \\ \times \text{V} \\ \hline \end{array}$$

9. Do you find one division problem easier than the other? If so, which one is the easier one?

$$4\overline{)32} \qquad \text{IV}\overline{)\text{XXXII}}$$

10. Describe some of the probable reasons all of the major trading communities and nations used other notations or devices such as an abacus instead of Roman numerals for their computations.

4-2
Other Methods of Computation

There exist numerous examples of the ways in which ancient man performed his computations. Some of these may prove of interest to the reader. The method of multiplication that appeared in one of the first published arithmetic texts in Italy, the *Treviso Arithmetic* (1478), is an interesting one. Let us use it to find the product of 457 and 382. (We shall refer to this process here as "galley" multiplication, although it was called "Gelosia" multiplication in the original text.)

First prepare a "galley" with three rows and three columns, and draw the diagonals, as in the figure. Our choice for the number of rows and columns is based on the fact that we are to multiply two three-digit numerals.

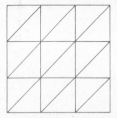

Place the digits 4, 5, and 7 in order from left to right at the top of the columns. Place the digits 3, 8, and 2 in order from top to bottom at the right of the rows. Then each product of a digit of 457 and a digit of 382 is called a **partial product** and is placed at the intersection of the column and row of the digits. The diagonal separates the digits of the partial product (tens digit above units digit). For example, $3 \times 7 = 21$, and this partial product is placed in the upper right-hand corner of the "galley"; $5 \times 8 = 40$, and this partial product is placed in the center of the galley; $4 \times 2 = 8$, and this partial product is entered as 08 in the lower left-hand corner of the galley. See if you can justify each of the entries in the completed array shown.

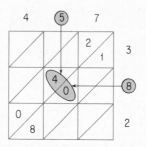 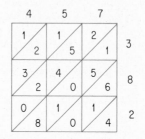

After all partial products have been entered in the galley, we add along diagonals, starting in the lower right-hand corner and carrying to the next diagonal sum where necessary. The next diagram indicates this pattern.

The completed problem appears in the form shown below. We read the final answer, as indicated by the arrow in the figure, as 174,574. Note that we read the digits in the opposite order to that in which they were obtained.

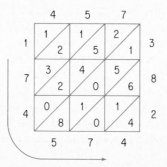

EXAMPLE

Use galley multiplication and multiply 372 by 47.

SOLUTION

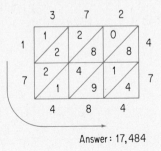

Answer: 17,484

This procedure works because we are really listing all partial products before we add. Compare the following two computations.

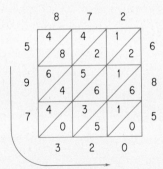

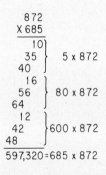

Note that the numerals along the diagonals correspond to those in the columns at the right.

The English mathematician John Napier made use of this system as he developed what proved to be one of the forerunners of the modern computing machines. His device is referred to as **Napier's rods**, or **Napier's bones**, named after the material on which he had numerals printed. Napier (1550–1617) is often spoken of as the inventor of logarithms.

To make a set of these rods we need to prepare a collection of strips of paper, or other material, with multiples of each of the digits listed. Study the set of rods shown in Figure (A) on page 126.

Note, for example, that the rod headed by the numeral 9 lists the multiples of 9:   9, 18, 27, 36, 45, 54, 63, 72, and 81

We can use these rods to multiply two numbers. To multiply 7 × 483, place the rods headed by numerals 4, 8, and 3 alongside the index, as shown in Figure (B) on page 126.

Index	0	1	2	3	4	5	6	7	8	9
1	0/0	0/1	0/2	0/3	0/4	0/5	0/6	0/7	0/8	0/9
2	0/0	0/2	0/4	0/6	0/8	1/0	1/2	1/4	1/6	1/8
3	0/0	0/3	0/6	0/9	1/2	1/5	1/8	2/1	2/4	2/7
4	0/0	0/4	0/8	1/2	1/6	2/0	2/4	2/8	3/2	3/6
5	0/0	0/5	1/0	1/5	2/0	2/5	3/0	3/5	4/0	4/5
6	0/0	0/6	1/2	1/8	2/4	3/0	3/6	4/2	4/8	5/4
7	0/0	0/7	1/4	2/1	2/8	3/5	4/2	4/9	5/6	6/3
8	0/0	0/8	1/6	2/4	3/2	4/0	4/8	5/6	6/4	7/2
9	0/0	0/9	1/8	2/7	3/6	4/5	5/4	6/3	7/2	8/1

(A)

Index	4	8	3
1	0/4	0/8	0/
2	0/8	1/6	0/
3	1/2	2/4	0/
4	1/6	3/2	1/
5	2/0	4/0	1/
6	2/4	4/8	1/
7	2/8	5/6	2/
8	3/2	6/4	2/
9	3/6	7/2	2/

(B)

Consider the row of numerals alongside 7 on the index.

Add along the diagonals, as in "galley multiplication."

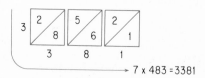

7 x 483 = 3381

The same arrangement of rods may be used to read immediately the product of 483 and any other one-digit number. With practice, one can develop skill in using these rods for rapid computation. For example, 8 × 483 = 3864 as follows:

 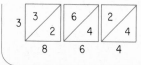

Combining the two previous results, we can find the product 87×483:

$$
\begin{array}{r}
483 \\
\times 87 \\
\hline
\end{array}
\qquad
\begin{array}{r}
7 \times 483 = 3{,}381 \\
80 \times 483 = 38{,}640 \\
\hline
87 \times 483 = 42{,}021
\end{array}
$$

Note that we are able to read only products with a one-digit multiplier directly from the rods.

exercises

Multiply, using the "galley" method.

1. $\begin{array}{r} 492 \\ \times 37 \end{array}$ 2. $\begin{array}{r} 568 \\ \times 429 \end{array}$

3. $\begin{array}{r} 432 \\ \times 276 \end{array}$ 4. $\begin{array}{r} 4876 \\ \times 27 \end{array}$

5. $\begin{array}{r} 7025 \\ \times 398 \end{array}$ 6. $\begin{array}{r} 5081 \\ \times 2376 \end{array}$

Construct a set of Napier's rods and use them to find each of the following products.

7. $\begin{array}{r} 256 \\ \times 8 \end{array}$ 8. $\begin{array}{r} 427 \\ \times 9 \end{array}$

9. $\begin{array}{r} 387 \\ \times 5 \end{array}$ 10. $\begin{array}{r} 592 \\ \times 7 \end{array}$

11. $\begin{array}{r} 7256 \\ \times 8 \end{array}$ 12. $\begin{array}{r} 427 \\ \times 36 \end{array}$

EXPLORATIONS

About 1650 B.C., an Egyptian scribe named Ahmes copied an earlier manuscript which he described as "the entrance into knowledge of all existing things and all obscure secrets." Ahmes' copy is often called the Rhind Papyrus. It contains 85 problems. Here is one of those problems.

A number and its one-fourth added together become 15. What is the number?

Variables were not known 3000 years ago. Algebra did not exist. Problems were solved by arithmetic with a procedure that we now call the *method of false position*. This method can be used for some problems but is not useful for many others. For the preceding problem, we note that we need to take one-fourth of the number, so we try 4.

$$4 + \tfrac{1}{4}(4) = 4 + 1 = 5$$

If we try 4, we get 5. But we need 15, that is, 3×5. Therefore the answer is 3×4, that is, 12.

Try the method of false position for each of these problems from the Rhind Papyrus.

1. A number and its one-fifth added together become 21. What is the number?

2. A number, its one-third, and its one-quarter added together become 2. What is the number?

3. If a number and its two-thirds are added together and from the sum one-third of the sum is subtracted, then 10 remains. What is the number?

Next try to determine the conditions under which the method of false position can be used.

4. Make up a problem that can be solved by the method of false position and explain why the method may be used.

5. Make up a problem that cannot be easily solved by the method of false position and explain why the method of false position is not appropriate for this problem.

4-3
Decimal Notation

We now turn our attention to the decimal system of numeration. It is a decimal system in that it is based on powers or groups of tens. Furthermore, it has place value in that the value of any digit used depends upon the position which it occupies. Thus, the two numerals 4 in 484 have quite different values.

To illustrate this latter concept we will write this number, 484, in what is known as expanded notation:

484 = 4 hundreds + 8 tens + 4 ones
   = (4 × 100) + (8 × 10) + (4 × 1)

Note that one of the numerals 4 represents 4 hundreds, whereas the other 4 represents 4 units, that is, 4 ones.

It is convenient to use exponents when one is writing a number in expanded notation. An **exponent** is a number that tells how many times another number, called the **base,** is used as a factor in a product. For example, in the expression 7^2, the numeral 2 is the exponent and 7 is the base. Note that $7^2 = 7 \times 7 = 49$; also $7^3 = 7 \times 7 \times 7 = 343$.

Using exponents to write 484 in expanded notation, we have

484 = (4 × 100) + (8 × 10) + (4 × 1)
   = $(4 \times 10^2) + (8 \times 10) + (4 \times 1)$

Zero plays an important role in place value notation. Indeed, it is the use of the symbol 0 that allows us to write numbers as large as we wish in decimal notation, using only the ten digits. In Egyptian numeration, ∩ represents 10, ⌒ represents 100, ⌠ represents 1000, etc. Each new power of 10 requires a new symbol, whereas in our decimal system the symbol 0 allows us to write 10, 100, 1000, etc., using only the digits 0 and 1.

EXAMPLE 1

Write 2306 in expanded notation.

SOLUTION

$$2306 = (2 \times 10^3) + (3 \times 10^2) + (0 \times 10) + (6 \times 1)$$

We may use the exponent 1 to indicate that a number is to be used as a factor only once; thus $10^1 = 10$. We define $10^0 = 1$. Using these exponents, we have

$$2306 = (2 \times 10^3) + (3 \times 10^2) + (0 \times 10^1) + (6 \times 10^0)$$

We often think in terms of expanded notation when we read decimal numerals:

92	ninety-two (nine tens with two)
492	four hundred ninety-two
1492	one thousand four hundred ninety-two

Ninety-two was interpreted nine tens *with* two instead of nine tens *and* two because we shall later use "and" to indicate the location of the decimal point.

Each decimal numeral can also be read as a sequence of digits:

92   nine two
492   four nine two
1492   one four nine two

There are still other ways, as illustrated by the familiar example,

1492   fourteen ninety-two

For large numbers, the expanded notation is used for reading the thousands, millions, billions, and so forth. For example, note that we read

2,107,456,072

as two billion one hundred seven million four hundred fifty-six thousand seventy-two.

EXAMPLE 2

Write in decimal notation:

one billion fifty-six million seven thousand ninety

SOLUTION   1,056,007,090.

exercises

Write in expanded decimal notation.

1. 235   **2.** 407
3. 1492   **4.** 1776
5. 17,259   **6.** 250,167
7. One thousand seventy-five.
8. Two hundred thousand six hundred seven.
9. Ten million six hundred fifty thousand two.
10. Five hundred million fifty thousand five.

Read each decimal numeral.

11. 1720   **12.** 1072
13. 107,020   **14.** 1,007,020
15. 17,000,002   **16.** 107,002,000

Write in decimal notation.

17. Twenty five thousand seventeen.
18. Two million six hundred twelve.

19. Five million six hundred thousand.
20. Seven billion fifty thousand one.
21. $(3 \times 10^3) + (2 \times 10^2) + (5 \times 10^1) + (3 \times 10^0)$
22. $(2 \times 10^5) + (4 \times 10^4) + (1 \times 10^3) + (3 \times 10^2)$
$$+ (5 \times 10^1) + (7 \times 10^0)$$
23. $(4 \times 10^5) + (3 \times 10^3) + (2 \times 10^2) + (5 \times 10^0)$
24. $(7 \times 10^6) + (8 \times 10^5) + (5 \times 10^4)$
25. $(9 \times 10^6) + (7 \times 10^0)$

EXPLORATIONS

Any whole number N less than 10,000 can be expressed in the form

$$N = 1000T + 100h + 10t + u$$

where $\{T, h, t, u\} \subseteq \{0, 1, 2, 3, 4, 5, 6, 7, 8, 9\}$. Then

$$\frac{N}{2} = \frac{1000T}{2} + \frac{100h}{2} + \frac{10t}{2} + \frac{u}{2} = 500T + 50h + 5t + \frac{u}{2}$$

Thus $N/2$ is a whole number if and only if $u/2$ is a whole number; that is, N is divisible by 2 if and only if u is divisible by 2.

$$\frac{N}{3} = \frac{(999 + 1)T}{3} + \frac{(99 + 1)h}{3} + \frac{(9 + 1)t}{3} + \frac{u}{3}$$

$$= 333T + 33h + 3t + \frac{T + h + t + u}{3}$$

Thus $N/3$ is a whole number if and only if $(T + h + t + u)/3$ is a whole number; that is, N is divisible by 3 if and only if the sum of its decimal digits is divisible by 3.

$$\frac{N}{4} = \frac{1000T}{4} + \frac{100h}{4} + \frac{10t}{4} + \frac{u}{4} = 250T + 25h + \frac{10t + u}{4}$$

Thus $N/4$ is a whole number if and only if $(10t + u)/4$ is a whole number; that is, N is divisible by 4 if and only if the number represented by its tens and units digits is divisible by 4.

According to these rules 1972 is divisible by 2 since its units digit 2 is divisible by 2; 1972 is not divisible by 3 since the sum of its digits is 19 and 19 is not divisible by 3, 1972 is divisible by 4 since 72 is divisible by 4. In the case of divisibility by 3 the process may be continued as follows: 1972 is not divisible by 3 since the sum of its digits is 19, the sum of the digits of 19 is 10, the sum of the digits of 10 is 1, and 1 is not divisible by 3.

Other explanations (proofs) and examples of divisibility rules are considered in the following explorations.

For numbers expressed in decimal notation, make a flow chart for testing the divisibility of numbers by each given number.

1. 2   2. 5
3. 3   4. 9
5. 4   6. 6

Use the procedures described by the flow charts for Explorations 1 through 6, and test each number for divisibility by (a) 2; (b) 3; (c) 4; (d) 5; (e) 6; (f) 9.

7. 5280   8. 1728
9. 16,527   10. 27,540
11. 19,278   12. 37,600

4-4

Other Systems of Numeration

In our decimal system objects are grouped and counted in tens and powers of ten. For example, the diagram shows how one might group and count 134 items.

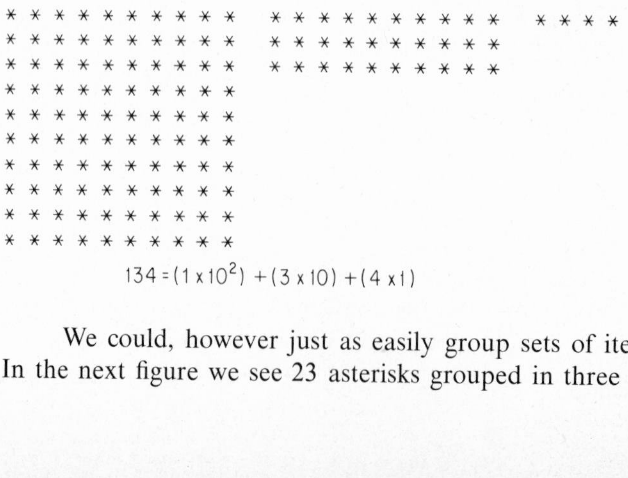

```
* * * * * * * * * *    * * * * * * * * * *    * * * *
* * * * * * * * * *    * * * * * * * * * *
* * * * * * * * * *    * * * * * * * * * *
* * * * * * * * * *
* * * * * * * * * *
* * * * * * * * * *
* * * * * * * * * *
* * * * * * * * * *
* * * * * * * * * *
* * * * * * * * * *
```

$$134 = (1 \times 10^2) + (3 \times 10) + (4 \times 1)$$

We could, however just as easily group sets of items in other ways. In the next figure we see 23 asterisks grouped in three different ways.

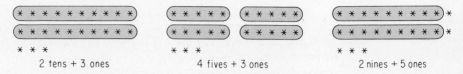

2 tens + 3 ones   4 fives + 3 ones   2 nines + 5 ones

If we use a subscript to indicate our manner of grouping, we may write many different numerals (names) for the number of items in the same collection:

$$23_{\text{ten}} \qquad = \qquad 43_{\text{five}} \qquad = \qquad 25_{\text{nine}}$$
$$(2 \ tens + 3 \ \text{ones}) \quad (4 \ fives + 3 \ \text{ones}) \quad (2 \ nines + 5 \ \text{ones})$$

43_{five} is read "four three, base five"
25_{nine} is read "two five, base nine"

Each of these numerals represents the number of asterisks in the same set of asterisks. Still another numeral for this number is 35_{six}:

$$35_{\text{six}} = 3 \ sixes + 5 \ \text{ones} = 18 + 5 = 23$$

We call our decimal system of numeration a **base ten** system; when no subscript is used, the numeral is understood to be expressed in base ten. When we group by fives, we have a **base five** system of numeration; that is, we name our system of numeration by the manner in which the grouping is accomplished.

EXAMPLE 1

Draw a diagram for 18 objects and write the corresponding numeral (a) in base five and (b) in base eight notation.

SOLUTION

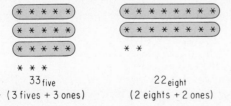

33_{five} 22_{eight}
(3 fives + 3 ones)   (2 eights + 2 ones)

EXAMPLE 2

Write, in base six notation, a numeral for 27.

SOLUTION   We note that $27 = 4 \ sixes + 3 \ \text{ones}$; thus $27 = 43_{\text{six}}$.

exercises

Write numerals for each of the following collections in the bases indicated by the manner of grouping.

1.   2.  3.   4.

Draw a diagram to show the meaning of each of the following.

5. 24_{five} 6. 23_{six}
7. 25_{seven} 8. 32_{nine}
9. 23_{four} 10. 12_{three}

Change to base ten notation.

11. 43_{five} 12. 24_{seven}
13. 32_{eight} 14. 51_{six}
15. 34_{five} 16. 32_{four}
17. Write, in base five notation, a numeral for 17.
18. Write, in base eight notation, a numeral for 43.
19. Write, in base six notation, a numeral for 20.
*20. Change 132_{five} to base ten notation.

EXPLORATIONS

Think of 5 as *1 hand* (5 fingers) and 5^2 as *1 man* (10 fingers, 10 toes, 2 hands, 2 feet, and a head). Then in this hand system

$12 = 2 \text{ (hands)} + 2$
$38 = 1(\text{man}) + 2(\text{hands}) + 3$

which we read as "36 equals one two three in the hand system."

1. How far can you count using the fingers of your right hand to represent ones and the fingers of your left hand to represent hands? Use a closed fist for zero. Do you ever need to use your thumbs? Check your answer by counting from 1 to that answer in this way.

2. Now let's change the system used in Exploration 1. This time use the fingers of your left hand to tell the power of 5 in the expanded notation. Then a closed left fist and two fingers of the right hand represent 2×5^0, that is, two ones. One finger of the left hand and three fingers of the right hand represent 3×5^1, that is, 15. In this system 87 can be represented by displaying in order

> Two fingers of the left hand and three of the right hand
> One finger of the left hand and two of the right hand
> The left fist and two fingers of the right hand

The order is a matter of convenience. Note that

$$87 = 75 + 10 + 2 = (3 \times 25) + (2 \times 5) + 2$$

    (a)   Represent 32 in this system of notation.
    (b)   Represent 94 in this system.
    (c)   Represent 3081 in this system.

4-5

Base Five Notation

In this section we shall explore, in some detail, the manner of writing numerals in another number base. We shall work with a base five system since it is convenient to think of numbers written in base five notation in terms of hands and fingers. Thus 23_{five} may be thought of in terms of two hands and three fingers; that is, as 2 fives and 3 ones.

The numerals 0, 1, 2, 3, 4 have the same meaning in base five notation as in base ten notation. For convenience we shall write all other numbers in base five notation using the numeral 5 as a subscript, such as 23_5. As usual, numbers written without a subscript will be assumed to be in base ten notation. Thus it is correct to write

$$23_5 = 13$$

Recall that these are merely two different names for the same number. (Two fives + three ones represent the same number as one ten + three ones.)

We can draw a diagram to show the meaning of 23_5 by drawing two groups of five and three ones.

In a similar manner we picture 123_5 as one group of 25, two groups of 5, and three ones. Note that the group of 25 represents five groups of 5, that is, one group of 5^2.

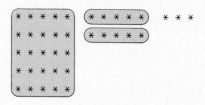

In the decimal system of notation we write numbers in terms of powers of ten using digits 0, 1, 2, 3, 4, 5, 6, 7, 8, and 9. For example,

$$234 = (2 \times 10^2) + (3 \times 10^1) + (4 \times 10^0)$$

In base five notation we write numbers in terms of powers of five using digits 0, 1, 2, 3, and 4. For example,

$$43_5 = (4 \times 5^1) + (3 \times 5^0)$$
$$324_5 = (3 \times 5^2) + (2 \times 5^1) + (4 \times 5^0)$$
$$2143_5 = (2 \times 5^3) + (1 \times 5^2) + (4 \times 5^1) + (3 \times 5^0)$$

Note that $5^0 = 1$.

The numbers 1 to 30 are written in the base five notation in the following table.

Base 10	Base 5	Base 10	Base 5	Base 10	Base 5
1	1	11	21_5	21	41_5
2	2	12	22_5	22	42_5
3	3	13	23_5	23	43_5
4	4	14	24_5	24	44_5
5	10_5	15	30_5	25	100_5
6	11_5	16	31_5	26	101_5
7	12_5	17	32_5	27	102_5
8	13_5	18	33_5	28	103_5
9	14_5	19	34_5	29	104_5
10	20_5	20	40_5	30	110_5

To translate a number that is not listed in the table from base 5 notation to base 10 notation, express the number in terms of powers of 5 and simplify.

EXAMPLE 1

Write 3214_5 in base 10 notation.

SOLUTION
$$3214_5 = (3 \times 5^3) + (2 \times 5^2) + (1 \times 5^1) + (4 \times 5^0)$$
$$= (3 \times 125) + (2 \times 25) + (1 \times 5) + (4 \times 1)$$
$$= 434$$

To translate from base 10 to base 5, any one of several procedures may be adopted. Consider the problem

$$339 = (\quad)_5$$

When a number is expressed to the base 5, it is written in terms of powers of 5:

$$5^0 = 1, \quad 5^1 = 5, \quad 5^2 = 25, \quad 5^3 = 125, \quad 5^4 = 625, \quad \ldots$$

The highest power of 5 that is not greater than the given number is 5^3. This power of 5, namely $5^3 = 125$, can be subtracted from 339 twice. Then the remainder 89 is positive and less than 125.

$$
\begin{array}{r}
339 \\
-125 \\
\hline
214 \\
-125 \\
\hline
89
\end{array}
$$

Thus, we write 2×5^3 in the expansion of 339 to the base 5.

The next power of 5 is 5^2. This number can be subtracted from 89 three times to obtain a nonnegative remainder less than 25.

$$
\begin{array}{r}
89 \\
-25 \\
\hline
64 \\
-25 \\
\hline
39 \\
-25 \\
\hline
14
\end{array}
$$

Thus, we write 3×5^2 in the expansion.

Finally, we subtract 5 from 14 twice, write 2×5 in the expansion, and obtain 4 as a remainder.

$$
\begin{array}{r}
14 \\
-5 \\
\hline
9 \\
-5 \\
\hline
4
\end{array}
$$

$$
\begin{aligned}
339 &= 2(125) + 3(25) + 2(5) + 4 \\
&= (2 \times 5^3) + (3 \times 5^2) + (2 \times 5^1) + (4 \times 5^0) = 2324_5
\end{aligned}
$$

A group of 339 elements can be considered as two groups of 125 elements, three groups of 25 elements, two groups of 5 elements, and 4 elements.

An alternative procedure for changing 339 to the base 5 depends upon successive division by 5:

$$
\begin{aligned}
339 &= 67 \times 5 + 4 \\
67 &= 13 \times 5 + 2 \\
13 &= 2 \times 5 + 3
\end{aligned}
$$

Next, substitute from the third equation into the second. Then substitute from the second equation to the first, and simplify as follows:

$$
\begin{aligned}
13 &= 2 \times 5 + 3 \\
67 &= 13 \times 5 + 2 = (2 \times 5 + 3) \times 5 + 2 \\
339 &= 67 \times 5 + 4 = [(2 \times 5 + 3) \times 5 + 2] \times 5 + 4 \\
&= (2 \times 5^3) + (3 \times 5^2) + (2 \times 5^1) + (4 \times 5^0) = 2324_5
\end{aligned}
$$

The arithmetical steps involved in these computations can be performed as shown in the following array (often called an *algorithm*).

$$
\begin{array}{l}
5 \,\lfloor 339 \\
\quad 5 \,\lfloor 67 \;\text{——} \; 4 \\
\qquad 5 \,\lfloor 13 \;\text{——} \; 2 \qquad \text{Read upward as } 2324_5. \\
\qquad\quad 5 \,\lfloor 2 \;\text{——} \; 3 \\
\qquad\qquad 0 \;\text{——} \; 2
\end{array}
$$

Note that the remainder is written after each division by 5. Then the remainders are read in reverse order to obtain the expression for the number to the base 5. This procedure works for integers only, not for fractional parts of a number.

EXAMPLE 2

Write 423 in base 5 notation.

SOLUTION

$$
\begin{array}{r|l}
5\,|\,423 & \\
5\,|\,84 & 3 \\
5\,|\,16 & 4 \\
5\,|\,3 & 1 \\
0 & 3 \\
\end{array}
$$

Answer: 3143_5.

Check:

$$3143_5 = (3 \times 5^3) + (1 \times 5^2) + (4 \times 5^1) + (3 \times 5^0)$$
$$= 375 + 25 + 20 + 3 = 423$$

After computations in other bases have been considered, the method of successive division by the new base may be used in changing from one base to another. For example, we shall be able to use this procedure to change from base 5 to base 10, successively dividing by 20_5. The computation must be done in base 5 notation. For the present we may change from base 5 to base 10 by expressing the number in terms of powers of 5.

exercises

Write each number in decimal notation.

1.	423_5	**2.**	230_5
3.	444_5	**4.**	4230_5
5.	321_5	**6.**	3403_5
7.	1031_5	**8.**	342_5
9.	1314_5	**10.**	1423_5

Write each number in base 5 notation.

11.	382	**12.**	293
13.	782	**14.**	394
15.	625	**16.**	917
17.	137	**18.**	858
19.	507	**20.**	1000

Extend the concepts of this section by using bases other than ten or five. Explain what is meant by each numeral and write each number in decimal notation.

*21. 437_8 ***22.** 2013_4

*23. 1011_2 ***24.** 321_{12}

*25. 132_{20} ***26.** 312_6

*27. 214_{15} ***28.** 1352_8

EXPLORATIONS

Extend the concepts of this section by using bases other than ten or five.

1. For base 2 notation try to find a rule for divisibility by **(a)** 10_2, that is, 2; **(b)** 100_2, that is, 4; **(c)** 1000_2, that is, 8.

2. For base 3 notation try to find a rule for divisibility by **(a)** 10_3, that is, 3; **(b)** 100_3, that is, 9; **(c)** 2.

3. For base 4 notation try to find a rule for divisibility by **(a)** 2; **(b)** 10_4, that is, 4; **(c)** 3.

4. For base 5 notation try to find a rule for divisibility by **(a)** 10_5, that is, 5; **(b)** 4; **(c)** 2.

4-6
Computation in Base Five Notation

We can form an addition table for the numbers in base 5 notation very easily. Consider, for example, the problem $4_5 + 3_5$. This may be written as $4_5 + 1_5 + 2_5$. Now, $(4_5 + 1_5)$ is one group of 5, or 10_5. We then add 2_5 to obtain 12_5. This is equivalent to

$$(1 \times 5^1) + (2 \times 5^0) = (1 \times 5^1) + (2 \times 1) = 5 + 2 = 7.$$

In a similar manner, $4_5 + 4_5 = 4_5 + 1_5 + 3_5$. Now $(4_5 + 1_5) = 10_5$. We then add 3_5 to obtain the sum 13_5. This result could also have been obtained by grouping: $4_5 + 4_5$ can be represented as (****) + (****), which can be regrouped as (*****) + (***); that is, 13_5.

The next table shows the number facts needed for addition problems in base 5. (You should verify each entry.)

The facts in this table may be used in finding sums of numbers, as illustrated in Example 1.

+	0	1	2	3	4
0	0	1	2	3	4
1	1	2	3	4	10_5
2	2	3	4	10_5	11_5
3	3	4	10_5	11_5	12_5
4	4	10_5	11_5	12_5	13_5

EXAMPLE 1

Find the sum of 432_5 and 243_5. Then check in base 10 notation.

SOLUTION

$$
\begin{array}{r}
432_5 \\
+243_5 \\
\hline
1230_5
\end{array}
$$

Check:

$$432_5 = (4 \times 5^2) + (3 \times 5^1) + (2 \times 5^0) \qquad\qquad = 117$$
$$+243_5 = (2 \times 5^2) + (4 \times 5^1) + (3 \times 5^0) \qquad\qquad = \ \ 73$$
$$1230_5 = (1 \times 5^3) + (2 \times 5^2) + (3 \times 5^1) + (0 \times 5^0) = 190$$

Here are the steps used in Example 1. In each case the familiar symbol 5 has been used in place of 10_{five} to help the reader recognize that powers of 5 are involved. This convention will be followed throughout this chapter. First add the column of 1's.

$$
\begin{array}{c}
(4 \times 5^2) + (3 \times 5^1) + (2 \times 5^0) \\
(2 \times 5^2) + (4 \times 5^1) + (3 \times 5^0) \\
\hline
10_5
\end{array}
$$

Next write the sum 10_5 of the 1's as 1×5^1 in the 5's column, and add the column of 5's.

$$
\begin{array}{c}
1 \times 5^1 \\
(4 \times 5^2) + (3 \times 5^1) + (2 \times 5^0) \\
(2 \times 5^2) + (4 \times 5^1) + (3 \times 5^0) \\
\hline
(13_5 \times 5^1) + (0 \times 5^0)
\end{array}
$$

Then write $10_5 \times 5^1$ from the previous sum in the 5^2 column and add the column of 5^2 entries.

$$1 \times 5^2$$
$$\begin{array}{l} (4 \times 5^2) + (3 \times 5^1) + (2 \times 5^0) \\ \underline{(2 \times 5^2) + (4 \times 5^1) + (3 \times 5^0)} \\ (12_5 \times 5^2) + (3 \times 5^1) + (0 \times 5^0) = 1230_5 \end{array}$$

Note that we "carry" groups of five in base 5 computation, just as we "carry" groups of ten in decimal computation.

Subtraction is not difficult if it is thought of as the inverse of addition. The table of addition facts in base 5 may again be used.

EXAMPLE 2

Subtract in base 5 and check in base 10: $211_5 - 142_5$.

SOLUTION   Think of 211_5 as

$$(2 \times 5^2) + (1 \times 5^1) + (1 \times 5^0)$$

then as

$$(1 \times 5^2) + (11_5 \times 5^1) + (1 \times 5^0)$$

then as

$$(1 \times 5^2) + (10_5 \times 5^1) + (11_5 \times 5^0).$$

Thus:

$$\begin{array}{llll} 211_5 = & (1 \times 5^2) + & (10_5 \times 5^1) + & (11_5 \times 5^0) \\ \underline{-142_5 =} & \underline{(1 \times 5^2) +} & (4 \times 5^1) + & (2 \times 5^0) \\ 14_5 & & (1 \times 5^1) + & (4 \times 5^0) \end{array}$$

Check:

$$\begin{array}{rl} 211_5 = & 56 \\ \underline{-142_5 =} & \underline{47} \\ 14_5 = & 9 \end{array}$$

This problem can be solved by borrowing and thinking in base 5 in the following steps:

$$\textbf{(a)} \quad \begin{array}{r} 211_5 \\ \underline{-142_5} \end{array} \qquad \textbf{(b)} \quad \begin{array}{r} 20^11_5 \\ \underline{-14\ 2_5} \\ 4_5 \end{array} \qquad \textbf{(c)} \quad \begin{array}{r} 1^10^11_5 \\ \underline{-1\ 4\ 2_5} \\ 1\ 4_5 \end{array}$$

Each raised "1" can be interpreted to show that the number 211_5 has not been changed but rather expressed in a different form:

$$211_5 = (2 \times 5^2) + (1 \times 5^1) + (1 \times 5^0)$$
$$20^11_5 = (2 \times 5^2) + (0 \times 5^1) + (1 \times 5^1) + (1 \times 5^0) = 211_5$$
$$1^10^11_5 = (1 + 5^2) + (1 \times 5^2) + (0 \times 5^1) + (1 \times 5^1)$$
$$+ (1 \times 5^0) = 211_5$$

Problems in multiplication can be performed as repeated additions or as in base 10.

EXAMPLE 3

Find the product of 243_5 and 4_5.

SOLUTION

We need the following facts:

$$4 \times 3 = 12_{10} = (2 \times 5^1) + (2 \times 5^0) = 22_5$$
$$4 \times 4 = 16_{10} = (3 \times 5^1) + (1 \times 5^0) = 31_5$$
$$4 \times 2 = 8_{10} = (1 \times 5^1) + (3 \times 5^0) = 13_5$$

The pattern for the computation when multiplying and "carrying" in base 5 notation is then precisely the same as that used in base 10 computation. For example, partial products are indented to represent the powers of 5 involved. In the long form, no attempt is made to "carry" mentally from one place to the next.

243_5	*Condensed form:*	243_5
$\times 4_5$		$\times 4_5$
22_5		2132_5
31_5		
13_5		
2132_5		

Multiplication by more than a one-digit multiplier is possible in other bases. Again, the pattern for computation is similar to that used in base 10 computation. Consider, for example, the product $34_5 \times 243_5$:

$$243_5$$
$$34_5$$
$$2132 = 4_5 \times 243_5$$
$$13340 = 30_5 \times 243_5$$
$$21022_5$$

As in base 10 computation, the product $30_5 \times 243_5$ can be written without the final numeral 0 since the place value is taken care of by indenting.

The pattern used for division in base 10 can also be used for division in base 5. Consider, for example, the problem $121_5 \div 4_5$. As a first step it is helpful to make a table of multiples of 4_5 in base 5:

$$1_5 \times 4_5 = 4_5$$
$$2_5 \times 4_5 = 13_5$$
$$3_5 \times 4_5 = 22_5$$
$$4_5 \times 4_5 = 31_5$$

Since 12_5 is greater than 4_5 and less than 13_5, the first numeral in the quotient is 1.

$$
\begin{array}{r}
1 \\
4_5 \overline{)121_5} \\
4 \\
\hline
31 \\
\end{array}
$$

As in base 10 computation, we multiply $1_5 \times 4_5$ and subtract this from 12_5 in the dividend. The next digit is then brought down. Note that $12_5 = 7$ and thus $12_5 - 4 = 3$.

Next we must divide 31_5 by 4_5. Note that $4_5 \times 4_5 = 31_5$. Thus we can complete the division, and there is no remainder.

$$
\begin{array}{r}
14_5 \\
4_5 \overline{)121_5} \\
4 \\
\hline
31 \\
31 \\
\hline
\end{array}
\qquad
Check: \quad
\begin{array}{r}
14_5 \\
\times 4_5 \\
\hline
121_5 \\
\end{array}
$$

EXAMPLE 4

Divide 232_5 by 3_5 and check by multiplication in base 5.

SOLUTION

$$1_5 \times 3_5 = 3_5$$
$$2_5 \times 3_5 = 11_5$$
$$3_5 \times 3_5 = 14_5$$
$$4_5 \times 3_5 = 22_5$$

$$
\begin{array}{r}
42_5 \\
3_5 \overline{)232_5} \\
22 \\
\hline
12 \\
11 \\
\hline
1 \\
\end{array}
\qquad
Check: \quad
\begin{array}{r}
42_5 \\
\times 3_5 \\
\hline
231_5 \\
+ 1_5 \\
\hline
232_5 \\
\end{array}
$$

$$1_5 \times 2_5 = 2_5$$
$$2_5 \times 2_5 = 4_5$$

The quotient is 42_5 and the remainder is 1_5. Note that in the check we multiply the quotient by the divisor, and add the remainder to obtain the dividend.

exercises

Add in base 5 and check in base 10.

1. 32_5
   $+24_5$

2. 24_5
   $+34_5$

3. 143_5
   $+234_5$

4. 432_5
   $+443_5$

5. 2341_5
   $+3421_5$

6. 4324_5
   $+1442_5$

Subtract in base 5 and check in base 10.

7. 143_5
   -31_5

8. 43_5
   -24_5

9. 312_5
   -123_5

10. 421_5
    -234_5

11. 1321_5
    -403_5

12. 3204_5
    -1342_5

13. 400_5
    -123_5

*14. 2003_5
    -1344_5

Multiply in base 5 and check in base 10.

15. 342_5
    $\times 4_5$

16. 243_5
    $\times 3_5$

17. 1423_5
    $\times 2_5$

18. 2304_5
    $\times 3_5$

19. 24_5
    $\times 32_5$

20. 32_5
    $\times 43_5$

21. 243_5
    $\times 34_5$

*22. 3243_5
    $\times 324_5$

Divide in base 5 and check in base 10.

23. $4_5 \overline{)143_5}$

24. $3_5 \overline{)121_5}$

25. $3_5 \overline{\smash{\big)}\, 1234_5}$ **26.** $4_5 \overline{\smash{\big)}\, 3042_5}$

27. $11_5 \overline{\smash{\big)}\, 143_5}$ **28.** $32_5 \overline{\smash{\big)}\, 3031_5}$

29. $23_5 \overline{\smash{\big)}\, 2043_5}$ **30.** $41_5 \overline{\smash{\big)}\, 43201_5}$

31. Complete a table showing the basic multiplication facts for base 5.

4-7
Other Number Bases

The base 5 system has been explored in detail merely for illustrative purposes; any other positive integer greater than 1 would have served just as well as a base. For each base N, the digits used are $0, 1, 2, 3, \ldots, N - 1$. For each base, powers of that base are used as place values for the digits. Consider, for example, the following numerals for the numbers 0 through 10 in several different bases.

Base 10	Base 3	Base 4	Base 5	Base 6	Base 7	Base 8
1	1	1	1	1	1	1
2	2	2	2	2	2	2
3	10_3	3	3	3	3	3
4	11_3	10_4	4	4	4	4
5	12_3	11_4	10_5	5	5	5
6	20_3	12_4	11_5	10_6	6	6
7	21_3	13_4	12_5	11_6	10_7	7
8	22_3	20_4	13_5	12_6	11_7	10_8
9	100_3	21_4	14_5	13_6	12_7	11_8
10	101_3	22_4	20_5	14_6	13_7	12_8

Note that for any positive integer N the numeral 10_N represents one group of N elements and no groups of one element. For example,

$$10_3 = (1 \times 3) + (0 \times 1) = 3$$
$$10_4 = (1 \times 4) + (0 \times 1) = 4$$
$$10_5 = (1 \times 5) + (0 \times 1) = 5$$
$$10_6 = (1 \times 6) + (0 \times 1) = 6$$
$$10_7 = (1 \times 7) + (0 \times 1) = 7$$
$$10_8 = (1 \times 8) + (0 \times 1) = 8$$

To change from another base to base 10 notation, express the number in terms of powers of that base and simplify as in the following examples.

EXAMPLE 1

Change 324_8 to base 10.

SOLUTION $324_8 = (3 \times 8^2) + (2 \times 8^1) + (4 \times 8^0) = 212$

EXAMPLE 2

Change 1231_4 to base 10.

SOLUTION $1231_4 = (1 \times 4^3) + (2 \times 4^2) + (3 \times 4^1) + (1 \times 4^0) = 109$

Any number may be changed from base 10 to another base by dividing successively by the new base and using the remainders, as in §4-5. This procedure is always performed in the notation of the old base. It has already been used for the base 5 and may be adapted for other bases as well.

EXAMPLE 3

Change 354 to base 8.

SOLUTION $8\,|\,354$

                       $8\,|\,44 \;\underline{\quad}\; 2$

                           $8\,|\,5 \;\underline{\quad}\; 4$ *Answer:* 542_8.

                               $0 \;\underline{\quad}\; 5$

Check:

$542_8 = (5 \times 8^2) + (4 \times 8^1) + (2 \times 8^0)$
$= 320 + 32 + 2$
$= 354$

Number bases greater than 10 are possible, but then new symbols must be introduced. Consider, for example, a base 12 system of numeration. The numeral 10_{12} represents $(1 \times 12) + (0 \times 1)$; that is, 12. Thus new symbols are necessary for 10 and 11. Let us use t for 10 and e for 11. Then we can count in base 12 as follows:

$1, 2, 3, 4, 5, 6, 7, 8, 9, t, e, 10_{12}, 11_{12}, 12_{12}, \ldots$.

We can change a base 12 numeral to base 10 as in the following example.

EXAMPLE 4

Change $2te5_{12}$ to base 10.

SOLUTION

$$2te5_{12} = (2 \times 12^3) + (10 \times 12^2) + (11 \times 12^1) + (5 \times 12^0)$$
$$= 3456 + 1440 + 132 + 5 = 5033$$

The reader should note that we make use of a base 12 system in many of our everyday uses of measurements. For example, linear measurement in terms of feet and inches is based on units of 12. Thus a person who is 5 feet 8 inches tall would be listed as having a height of 58_{12} in base 12 notation. Similarly, the use of dozens and gross is based on units of 12. An order for 3 gross, 5 dozen, 7 pencils could be written as 357_{12}:

$$357_{12} = (3 \times 12^2) + (5 \times 12) + (7 \times 12^0) = 499$$

exercises

Change to base 10.

1. 472_8 **2.** 3213_4

3. 4552_6 **4.** 347_{12}

5. 1212_3 **6.** 523_7

7. 472_{12} **8.** $3t5_{12}$

9. $15et_{12}$ **10.** 1101_2

Write each of the following as a base 12 numeral.

11. $(3 \times 12^3) + (5 \times 12^2) + (9 \times 12^1) + (8 \times 12^0)$

12. $(5 \times 12^3) + (10 \times 12^2) + (6 \times 12^1) + (5 \times 12^0)$

13. $(8 \times 12^3) + (7 \times 12^2) + (11 \times 12^1) + (0 \times 12^0)$

14. $(7 \times 12^3) + (11 \times 12^2) + (10 \times 12^1) + (8 \times 12^0)$

Change to the stated base.

15. $324 = ($ $)_4$ **16.** $576 = ($ $)_8$

17. $427 = ($ $)_6$ **18.** $114 = ($ $)_3$

19. $798 = ($ $)_5$ **20.** $257 = ($ $)_7$

21. $536 = ($ $)_9$ **22.** $182 = ($ $)_{12}$

23. $247 = ($ $)_{12}$ **24.** $2033 = ($ $)_{12}$

*25. $283 = ($ $)_{12}$ ***26.** $27 = ($ $)_2$

*27. $47 = ($ $)_2$ ***28.** $175 = ($ $)_2$

EXPLORATIONS

1. Complete the following tables of addition and multiplication facts for base 4.

+	0	1	2	3
0				
1				
2				
3				

x	0	1	2	3
0				
1				
2				
3				

2. Make up, do, and check at least five three-digit addition problems using base four notation.

3. Repeat Exploration 2 for subtraction.

4. Repeat Exploration 2 for products of a three-digit numeral and a two-digit numeral.

5. Repeat Exploration 2 for quotients of a three-digit numeral divided by a one-digit numeral.

6. Complete the following tables of addition and multiplication facts for base 8.

+	0	1	2	3	4	5	6	7
0								
1								
2								
3								
4								
5								
6								
7								

x	0	1	2	3	4	5	6	7
0								
1								
2								
3								
4								
5								
6								
7								

7-10. Restate and do Explorations 2 through 5 for base 8 numerals.

11. Using t for 10 and e for 11, complete the following tables of addition and multiplication facts for base 12.

+	0	1	2	3	4	5	6	7	8	9	t	e
0												
1												
2												
3												
4												
5												
6												
7												
8												
9												
t												
e												

x	0	1	2	3	4	5	6	7	8	9	t	e
0												
1												
2												
3												
4												
5												
6												
7												
8												
9												
t												
e												

12–15. Restate and do Explorations 2 through 5 for base 12 numerals.

4-8
Binary Notation

Numbers written to the base 2 are of special interest because of their application in modern electronic computers. This system of notation is called **binary notation**, and makes use of only two digits, 0 and 1.

Recent spacecraft that took pictures of the planet Mars transmitted the data back to earth in binary notation. On earth, with the aid of computers, these data were converted into pictures of the surface of the planet that appeared in many local newspapers.

Although there is some evidence that the basic concepts of binary notation were known to the ancient Chinese about 2000 B.C., it is only within recent years that it has been widely applied in computer mathematics, card sorting operations, and other electronic devices. Some of the place values in the binary system are as follows:

$$2^7 \quad 2^6 \quad 2^5 \quad 2^4 \quad 2^3 \quad 2^2 \quad 2^1 \quad 2^0$$
$$128 \quad 64 \quad 32 \quad 16 \quad 8 \quad 4 \quad 2 \quad 1$$

Thus, in binary notation, $11011101_2 = 215$ as shown in the following example.

EXAMPLE 1

Write 11011101_2 in base 10 notation.

SOLUTION

$$11011101_2 = (1 \times 2^7) + (1 \times 2^6) + (0 \times 2^5) + (1 \times 2^4)$$
$$+ (1 \times 2^3) + (1 \times 2^2) + (0 \times 2^1) + (1 \times 2^0)$$
$$= 128 + 64 + 16 + 8 + 4 + 1 = 221$$

Here are the first 16 counting numbers written in binary notation:

Base 10	Base 2	Base 10	Base 2
1	1	9	1001_2
2	10_2	10	1010_2
3	11_2	11	1011_2
4	100_2	12	1100_2
5	101_2	13	1101_2
6	110_2	14	1110_2
7	111_2	15	1111_2
8	1000_2	16	10000_2

Tables for addition and multiplication in binary notation are easy to complete.

+	0	1
0	0	1
1	1	10_2

x	0	1
0	0	0
1	0	1

EXAMPLE 2

Multiply: $101_2 \times 1101_2$.

SOLUTION

$$
\begin{array}{r}
1101_2 \\
101_2 \\
\hline
1101 \\
11010 \\
\hline
1000001_2
\end{array}
$$

Check:

$$1101_2 = 13$$
$$101_2 = \underline{5}$$
$$65$$
$$1000001_2 = (1 \times 2^6) + (1 \times 2^0)$$
$$= 64 + 1 = 65$$

The seven-digit numeral 1000001 is the name for the letter A when that letter is sent over teletype to a computer. The American Standard Code for Information Interchange (ASCII) was adopted in 1967 and is used to convert all letters, numerals, and other symbols into binary notation.

Binary numerals can be shown by means of electric lights flashing on and off. If a light is on, the digit 1 is represented; whereas if it is off, the digit 0 is shown. Similarly, the digits 0 and 1 can be used in a card sorting operation. The reader can gain an appreciation of the process used by construction of a small set of punched cards. First prepare a set of 16 index cards with four holes punched in each and a corner notched, as in the figure.

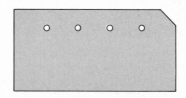

Next represent the numbers 0 through 15 on these cards in binary notation. Cut out the space above each hole to represent 1; leave the hole untouched to represent 0. Several cards are shown in the figure.

$5 = 0101_2$

$6 = 0110_2$

$10 = 1010_2$

$15 = 1111_2$

After all the cards have been completed in this manner, shuffle them thoroughly and align them, making certain that they remain "face up." (The notched corners will help indicate when the cards are right side up.) Then, going from right to left, perform the following operation: Stick a pencil or other similar object through the first hole and lift up. Some of the cards will come up, namely, those in which the holes have not been cut through to the edge of the card (that is, those cards representing numbers whose units digit in binary notation is 0).

Place the cards that lift up in front of the other cards and repeat the same operation for the remaining holes in order from right to left. When you have finished, the cards should be in numerical order, 0 through 15.

Note that only four operations are needed to arrange the sixteen cards. As the number of cards is doubled, only one additional operation will be needed each time to place them in order. That is, 32 cards may be placed in numerical order with five of the described card sorting operations; 64 cards may be arranged with six operations; 128 cards with seven operations; and so forth. Thus a large number of cards may be arranged in order with a relatively small number of operations. For example, over one billion cards may be placed in numerical order with only 30 sortings.

exercises

Write each number in binary notation.

1. 28   **2.** 45

3. 19   **4.** 128

5. 152   **6.** 325

Change each number to decimal notation.

7. 1111_2 **8.** 10101_2

9. 110111_2 **10.** 101010_2

11. 100110_2 **12.** 11011011_2

Perform the indicated operation in binary notation; check in base 10.

13. $\begin{array}{r} 1101_2 \\ +\,1011_2 \end{array}$ **14.** $\begin{array}{r} 10011_2 \\ +\,10101_2 \end{array}$

15. $\begin{array}{r} 11011_2 \\ -\,10110_2 \end{array}$ **16.** $\begin{array}{r} 110101_2 \\ -\,10111_2 \end{array}$

17. $\begin{array}{r} 1101_2 \\ \times\,11_2 \end{array}$ **18.** $\begin{array}{r} 10110_2 \\ \times\,101_2 \end{array}$

19. In the ASCII code the numbers for B, D, and G are 66, 68, and 71, respectively. Find the binary representation of **(a)** B; **(b)** D; **(c)** G.

20. In the ASCII code capital letters A, B, C, . . . in alphabetical order are assigned numbers 65, 66, 67, ... What is the word transmitted by the code 1010010, 1010101, 1001110?

21. Write the number 214 in base 8 and then in base 2 notation. Can you discover a relationship between these two bases?

Base 8 notation is often called **octal notation.** *Write in octal notation.*

22. $101,111,001,010_2$

23. $11,101,011,001_2$

24. $1,001,101,110,010_2$

25. $10,010,101,011,011_2$

26. $111,110,101,101,011,001_2$

27. $1,101,011,001,000,100,110_2$

Write in binary notation.

28. 435_8

29. 6023_8

30. 4257_8

31. 3624_8

32. 12345_8

33. 76543_8

*34. Complete a set of 32 punched cards to represent the numbers 0 through 31. Here five holes are needed on each card and five "lifting" operations are necessary to place the set in numerical order.

EXPLORATIONS

Many recreations are based on the binary system of numeration. Consider, for example, the boxes A, B, C, D within which the numbers 1 to 15 are written according to a certain scheme.

D	C	B	A
8	4	2	1
9	5	3	3
10	6	6	5
11	7	7	7
12	12	10	9
13	13	11	11
14	14	14	13
15	15	15	15

In box A place all numbers that have a 1 in the units place when written in binary notation. In box B place those with a 1 in the second position from the right in binary notation. In C and

D are those numbers with a 1 in the third and fourth positions, respectively.

Next, ask someone to think of a number and tell you in which box or boxes it appears. You then tell that person his number by finding the sum of the first number in each box he mentions. Thus, if his number is 11, he lists boxes A, B, and D. You then find the sum $1 + 2 + 8$ as the number under discussion.

1. Explain why the method given for finding a number after knowing the boxes in which it appears works as it does.

2. Extend the boxes to include all the numbers through 31. (A fifth column, E, will be necessary.) Then explain for the set of five boxes how to find a number if one knows in which boxes it appears.

3. You can use binary notations to identify any one of eight numbers by means of three questions that can be answered as "yes" or "no." Thus consider the numbers from 0 through 7 written in binary notation. The place values are identified by columns A, B, and C.

	C	B	A
0	0	0	0
1	0	0	1
2	0	1	0
3	0	1	1
4	1	0	0
5	1	0	1
6	1	1	0
7	1	1	1

The three questions to be asked are: "Does the number have a 1 in position A?" "Does the number have a 1 in position B?" "Does the number have a 1 in position C?" Suppose the answers are "Yes, No, Yes." Then the number is identified as 101_2; that is, 5. Extend this process to show how you can guess a number that someone is thinking of from 0 through 15 by means of four "yes–no" questions.

4. One can count on one's fingers in base 2 by considering a finger held up as 1 and a finger down as 0. Representations for the first five counting numbers are shown at the top of the next page.

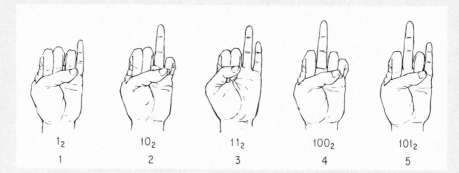

1_2 10_2 11_2 100_2 101_2

1   2   3   4   5

Show how to count through 31 in binary notation using the fingers on one hand.

5. The binary digits 1 and 0 were represented in Exploration 4, respectively, by a finger that is raised and a finger that is not raised. The binary digits can also be represented respectively by a light that is turned on and a light that is not turned on.

Consider five light bulbs in a row on a panel. The bulbs may be labeled so that they can be used in the same manner as fingers in the previous exploration. In the figure, three bulbs are lighted so that 1110_2—that is, 14—is represented.

16   8   4   2   1

Draw a panel of five bulbs and show how they should be lighted to represent 21. Draw another panel of bulbs on which 57 is represented. How many bulbs would you need in a panel on which all whole numbers from 1 to 127 are to be represented?

6. Find a reference to the game of Nim and study its relationship to the binary system of notation.

chapter test

1. Use the Egyptian method of duplation and mediation to find the product 17×31.

2. Multiply, using the "galley" method: 325×437.

3. Write in expanded notation: 23,457.

4. Write a numeral for 58 in: **(a)** base 8 notation; **(b)** base 5 notation.

5. Change to base 10:
   (a) 234_5 (b) 125_6

6. Change to base 10:
   (a) $3te_{12}$ (b) 1011_2

7. Change to the stated base:
   (a) $243 = (\quad)_5$ (b) $47 = (\quad)_4$

8. Compute in base 5:
   (a) $\begin{array}{r} 243_5 \\ +132_5 \\ \hline \end{array}$ (b) $\begin{array}{r} 3021_5 \\ -1314_5 \\ \hline \end{array}$

9. Compute in base 5:
   (a) $\begin{array}{r} 1342_5 \\ \times\ 3_5 \\ \hline \end{array}$ (b) $4_5\overline{\smash{)}3212_5}$

10. Compute in binary notation:
    (a) $\begin{array}{r} 1111_2 \\ +1101_2 \\ \hline \end{array}$ (b) $\begin{array}{r} 1011_2 \\ \times 11_2 \\ \hline \end{array}$

5

Mathematical
SYSTEMS

A **mathematical system** involves a set of elements (such as the set of counting numbers), one or more operations (such as addition, subtraction, multiplication, and division), one or more relations (such as the equality $2 \times 3 = 6$), and some axioms (rules) which the elements, operations, and relations satisfy. For example, we assume that every system involves the relation of equality and the axiom that $a = a$; that is, any quantity is equal to itself. Actually, you have been working with mathematical systems ever since you started school.

We begin this chapter with the study of a specific abstract mathematical system, and relate the properties of this system to those of familiar sets of numbers. Clock arithmetic is considered as a familiar mathematical system with some unusual properties and as an introduction to modular arithmetic. Two by two matrices form another mathematical system and may be used in a wide variety of ways. Still other mathematical systems are considered in the sets of explorations following Sections 5-1 and 5-5.

An Abstract System

Let us define an abstract system composed of the set of elements

$$\{*, \#, \Sigma, [\ ]\}$$

For ease of notation we shall call this set Y; that is,

$$Y = \{*, \#, \Sigma, [\ ]\}$$

Next we define an operation which we call "multition" and denote this with the symbol $\sim$. The operation is defined by means of the following table.

$\sim$	$*$	$\#$	Σ	$[\ ]$
$*$	$*$	$\#$	Σ	$[\ ]$
$\#$	$\#$	Σ	$[\ ]$	$*$
Σ	Σ	$[\ ]$	$*$	$\#$
$[\ ]$	$[\ ]$	$*$	$\#$	Σ

The operation $\sim$ is a **binary operation,** since it associates a single element with any *two* elements of the given set.

To "multify" two elements of Y, we read the first of these elements in the vertical column at the left and the second in the horizontal row across the top. The answer is found where the column and the row intersect within the table. Verify from the table that each of the following statements is correct:

$$\# \sim \Sigma = [\ ]$$
$$[\ ] \sim \# = *$$
$$\Sigma \sim * = \Sigma$$

The task of major importance before us is to examine this system for its basic properties. The following are some of the more important ones.

1. Whenever any two elements of the set Y are combined by the process of multition, the result is a unique element of the original collection Y of elements. In other words, there is one and only one answer whenever two elements of Y are multified, and this answer is always one of the elements of Y. We say that *the set Y is closed under the operation of multition.*

In general a set S is said to be **closed** under a given binary operation if that operation associates a unique element of the set S with any two elements of the set.

Consider, in ordinary arithmetic, all the numbers that we use for counting:

$$C = \{1, 2, 3, 4, 5, \ldots\}$$

This set of counting numbers is closed under the operations of addition and multiplication. That is, the sum of any two counting numbers is again a counting number, and the product of any two counting numbers is a counting number. Is the set of counting numbers closed under subtraction? That is, is the difference of *any* two counting numbers always a counting number?

2. Consider the order in which we combine any two elements of the set Y. Does $\# \sim \Sigma = \Sigma \sim \#$? Does $[\ ] \sim * = * \sim [\ ]$? In general, if a and b represent any two elements of the set Y, does $a \sim b = b \sim a$? We find the answer to be in the affirmative in each case and summarize this property by saying that *the elements of set Y satisfy the commutative property for multition.*

In general, a set of elements is said to satisfy the **commutative property** for a particular operation if the result obtained by combining any two elements of the set under that operation does not depend upon the order in which these elements are combined.

The set of counting numbers is commutative under addition and multiplication. That is, the sum of any two counting numbers is the same, regardless of the order in which the elements are added. Thus $5 + 7 = 7 + 5, 8 + 3 = 3 + 8$, etc. Similarly, the set of counting numbers satisfies the commutative property for multiplication, since the product of any two counting numbers is the same regardless of the order in which the elements are multiplied. Thus $5 \times 7 = 7 \times 5, 8 \times 3 = 3 \times 8$, etc.

EXAMPLE 1

Show that subtraction of counting numbers is not commutative.

SOLUTION

A single *counterexample* is sufficient to show that a property does not hold. Since $7 - 3 = 4$ and $3 - 7 = -4$, $7 - 3 \neq 3 - 7$ and subtraction is not commutative. (The symbol $\neq$ is read "is not equal to.")

3. Next we combine three elements of the set Y. Can you find the answer for $\Sigma \sim [\ ] \sim \#$? Here we see that two possibilities exist. We could combine Σ and $[\ ]$ first and then combine the result with $\#$, or we could combine $[\ ]$ and $\#$ first and then combine Σ with the result obtained. Let us try both ways:

$(\Sigma \sim [\quad]) \sim \# = \# \sim \# = \Sigma$
$\Sigma \sim ([\quad] \sim \#) = \Sigma \sim * = \Sigma$

We obtain the same result in both cases and will do so for any choice of three members of the set Y. We call this the **associative property** for multition, and write, in symbols,

$$a \sim (b \sim c) = (a \sim b) \sim c$$

where a, b, and c may be replaced by any elements of the set Y.

The set of counting numbers is associative with respect to both addition and multiplication. That is, the sum or product of any three counting numbers is the same regardless of the order in which they are associated. For example,

$$3 + (5 + 7) = 3 + 12 = 15 \quad \text{and} \quad (3 + 5) + 7 = 8 + 7 = 15$$
$$3 \times (5 \times 7) = 3 \times 35 = 105 \quad \text{and} \quad (3 \times 5) \times 7 = 15 \times 7 = 105$$

Thus

$$3 + (5 + 7) = (3 + 5) + 7 \quad \text{and} \quad 3 \times (5 \times 7) = (3 \times 5) \times 7$$

EXAMPLE 2

Show that the set of counting numbers is not associative with respect to division.

SOLUTION

Again, one counterexample is sufficient.

$$24 \div (6 \div 2) = 24 \div 3 = 8, \quad \text{whereas} \quad (24 \div 6) \div 2 = 4 \div 2 = 2$$

Thus $24 \div (6 \div 2) \neq (24 \div 6) \div 2$, and division of counting numbers is not associative.

4. The set Y contains a unique element, $*$, which leaves every other element unchanged with respect to multition.

$* \sim * = *$
$\# \sim * = \#$ $* \sim \# = \#$
$\Sigma \sim * = \Sigma$ $* \sim \Sigma = \Sigma$
$[\quad] \sim * = [\quad]$ $* \sim [\quad] = [\quad]$

The element with this property is called the **identity element** for multition. Notice that the identity element is commutative with each element of the set. We call ∗ the *identity element* for multition because it does not change the *identity* of any element under multition.

In ordinary arithmetic, what is the identity element for addition? That is, what can we add to 7 to obtain 7? Does $9 + 0 = 9$? Do you see that 0 is the identity element for addition? However, 0 is not an element of the set of counting numbers. Therefore, we must say that the set of counting numbers does not contain an identity element with respect to addition. Does the set of counting numbers contain an identity element with respect to multiplication?

5. We examine one more property of the set Y. Replace each blank in the following by an element of the set Y to obtain a correct statement.

$$* \sim \underline{} = *$$
$$\# \sim \underline{} = *$$
$$\Sigma \sim \underline{} = *$$
$$[\ \ ] \sim \underline{} = *$$

The correct answers are, respectively, ∗, [ ], Σ, and #. We find that for each element of the set Y there is another element of Y such that the first "multified" by the second produces the identity element ∗. Each of these second elements is called the **inverse with respect to multition** (or simply the **inverse**) of the first element of the statement in which it was used. Thus the inverse of ∗ is ∗, the inverse of # is [ ], the inverse of Σ is Σ, and the inverse of [ ] is #. Note that ∗ and Σ serve as their own inverses.

The set of counting numbers does not contain the inverses for its elements with respect to either addition or multiplication. However, when we extend our system to include *all* integers, then every element has an inverse under addition. Thus the inverse of 5 is −5, in that $5 + (-5) = 0$. Similarly, the inverse of −3 is 3, in that $(-3) + 3 = 0$.

In later extensions of our number system we consider all fractions of the form $\dfrac{a}{b}$, where a and b are integers and $b \neq 0$. This set then contains the inverse for each of its nonzero elements under multiplication. Thus the inverse of 5 is $\frac{1}{5}$, in that $5 \times \frac{1}{5} = 1$. ·Similarly, the inverse of $\frac{2}{3}$ is $\frac{3}{2}$, in that $\frac{2}{3} \times \frac{3}{2} = 1$.

In summary, the set Y

1. Is closed with respect to multition.
2. Is associative with respect to multition.
3. Contains an identity element with respect to multition.
4. Contains an inverse for each of its elements with respect to multition.

We may describe these four properties by saying that the elements of the set Y form a **group** under multition. Since the set Y is also commutative with respect to multition, the elements of the set Y form a **commutative group** under multition.

The set of counting numbers does not form a group under addition since the identity and inverses are missing. The set of counting numbers does not form a group under multiplication since the inverses of numbers different from 1 are missing. However, the set of counting numbers

1. Is closed with respect to addition.
2. Is associative with respect to addition.
3. Is commutative with respect to addition.
4. Is closed with respect to multiplication.
5. Is associative with respect to multiplication.
6. Contains the identity with respect to multiplication.
7. Is commutative with respect to multiplication.

exercises

Find the answer to each of the following from the table given in this section.

1. $\# \sim *$
2. $* \sim \Sigma$
3. $[\ ] \sim [\ ]$
4. $\Sigma \sim \#$

Verify that each of the following is true and state the name of the property that each illustrates.

5. $\Sigma \sim \# = \# \sim \Sigma$
6. $[\ ] \sim (\# \sim \Sigma) = ([\ ] \sim \#) \sim \Sigma$
7. $\# \sim * = \#$
8. $\# \sim [\ ] = *$
9. $(\# \sim \Sigma) \sim * = \# \sim (\Sigma \sim *)$
10. $[\ ] \sim * = [\ ]$
11. $[\ ] \sim \# = *$
12. $[\ ] \sim * = * \sim [\ ]$

Answer Exercises 13 through 22 by using the table on page 164, which defines an operation $\odot$ for the elements of the set $\{\triangle, \square, Q\}$.

13. Find $\triangle \odot \square$.
14. Find $Q \odot \triangle$.
15. Does $Q \odot \square = \square \odot Q$?
16. Does $\triangle \odot Q = Q \odot \triangle$?
17. Does $(Q \odot \square) \odot \triangle = Q \odot (\square \odot \triangle)$?
18. Does $\square \odot (Q \odot \triangle) = (\square \odot Q) \odot \triangle$?
19. Does an affirmative answer to the two preceding exercises prove that the set satisfies the associative property with respect to $\odot$? Explain your answer.

$\odot$	$\triangle$	$\square$	Q
$\triangle$	Q	$\triangle$	$\square$
$\square$	$\triangle$	$\square$	Q
Q	$\square$	Q	$\triangle$

20. Is there an identity element for $\odot$? If so, what is it?

21. Does each element have an inverse with respect to $\odot$? If so, name the inverse of each element.

22. Summarize the properties of the set $\{\triangle, \square, Q\}$ with respect to the operation $\odot$.

23. Show that the set of counting numbers is not commutative with respect to division.

24. Show that the set of counting numbers is not associative with respect to subtraction.

25. What is the identity element with respect to multiplication for the set of counting numbers? Give several specific examples to justify your answer.

26. Show that the set of counting numbers does not form a group under addition.

EXPLORATIONS

Answer Explorations 1 through 4 by using the following table, which defines an operation $\square$ for the set $\{\bullet, [, !, \$\}$.

$\square$	$\bullet$	$[$	$!$	$\$$
$\bullet$	$\bullet$	$[$	$!$	$\$$
$[$	$[$	$!$	$\circ$	$\sim$
$!$	$!$	$\circ$	$\sim$	$[$
$\$$	$\$$	$[$	$!$	$\bullet$

1. Is the set closed with respect to $\square$? Justify your answer.

2. Does the set satisfy the commutative property for $\square$? Explain.

3. Does the set contain an identity element for ☐? If so, what is it?

4. Which elements of the set have inverses with respect to ⬚? Name each of these inverses.

5. Operations may be considered in terms of a **function machine** with three basic parts.

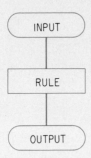

For the rule "add 5" complete this array.

Input	3	7	——	——	21	——
Output	——	——	8	11	——	21

6. The "add 5" rule was used in Exploration 5. Consider the "subtract 5" rule for each element in the output in Exploration 5 and note that in each case the answer is the corresponding element of the input from Exploration 5. This relation between "add 5" and "subtract 5" rules is described by referring to these operations as **inverse operations**; each "undoes" the other.

We may obtain the inverse operation by thinking of the function machine as running backward from output to input. Find the rule for the inverse operation for the operation:
(a) Subtract 2.   **(b)** Add 7.
(c) Multiply by 3.   **(d)** Divide by 2.

Operations in which each input consists of a single number are **unary operations**. The following unary operations are indicated with an arrow to show the correspondence or mapping of the given number (input) onto the new number (output) under the indicated operation. Each operation in the following explorations has been given a name to aid class discussions.

7. This is how we "star" a number.

$6^* \rightarrow 8$ $7^* \rightarrow 9$ $0^* \rightarrow 2$

Can you "star" these numbers?

$5^* \rightarrow ?$ $9^* \rightarrow ?$ $12^* \rightarrow ?$

8. This is how we "quote" a number.

"5" → 26   "3" → 10   "7" → 50

Can you "quote" these numbers?

"6" → ?   "10" → ?   "9" → ?

9. This is how we "bar" a number:

$\overline{2}$ → 7 $\overline{5}$ → 16 $\overline{11}$ → 34

Can you "bar" these numbers?

$\overline{4}$ → ? $\overline{6}$ → ? $\overline{12}$ → ?

Arrows are also often used for binary operations. Try to discover the meaning of the operation # from the examples given. The binary operation # has a different meaning in each exploration.

10. 3 # 4 → 7,   2 # 0 → 2,   5 # 1 → 6,   2 # 2 → 4.

11. 2 # 4 → 3,   5 # 7 → 6,   8 # 10 → 9,   0 # 2 → 1.

12. 5 # 3 → 9,   2 # 4 → 7,   0 # 3 → 4,   1 # 5 → 7.

13. 2 # 3 → 7,   3 # 4 → 13,   1 # 5 → 6,   0 # 3 → 1.

*14. 3 # 4 → 2,   4 # 3 → 5,   5 # 1 → 9,   1 # 1 → 1.

*15. 3 # 5 → 2,   4 # 5 → 1,   2 # 8 → 0,   1 # 6 → 3.

16. Consider the binary operation $\odot$ defined for any two counting numbers a and b in this way:

$a \odot b = 10a + b$

See how many properties you can find for this operation. For example, explore closure, commutativity, and associativity. What relationship, if any, does this operation have to our decimal system of numeration?

17. Let # mean "Select the number if the two given numbers are equal, and the smaller of the two given numbers if they are unequal." Then 5 # 5 = 5 and 5 # 3 = 3. For the set of counting numbers, does $a \# b = b \# a$? Does $a \# (b \# c) = (a \# b) \# c$?

18. Let $ mean "Select the first of two given numbers." Then 5 \$ 2 = 5. For the set of counting numbers, does $a \$ b = b \$ a$? Does $a \$ (b \$ c) = (a \$ b) \$ c$?

*19. Use the definitions given in Explorations 17 and 18 and show that $a \$ (b \# c) = (a \$ b) \# (a \$ c)$ for all counting numbers a, b, and c.

*20. Referring to the definitions given in Explorations 17 and 18, does $a \# (b \$ c) = (a \# b) \$ (a \# c)$?

5-2
The Distributive Property

See if you can discover the meaning of the operations # and $\otimes$ from the following examples:

5 # 3 = 3	8 $\otimes$ 3 = 8
2 # 8 = 2	0 $\otimes$ 3 = 0
3 # 7 = 3	6 $\otimes$ 1 = 6
4 # 4 = 4	3 $\otimes$ 3 = 3

You should have discovered the following meanings of # and $\otimes$:

#: Select the smaller number of the two if the numbers are different; select that number if both are the same.

$\otimes$: Select the first number of the two.

Thus 5 # 9 = 5 because 5 is smaller than 9, whereas 5 $\otimes$ 9 = 5 because 5 is the first of the two numbers in the given expression. Note that where both numbers were the same we wrote 4 # 4 = 4.

We may also combine both operations and determine the meaning of expressions such as

$$a \mathbin{\#} (b \otimes c) \quad \text{and} \quad (a \mathbin{\#} b) \otimes (a \mathbin{\#} c)$$

where a, b, and c represent any of the numbers of ordinary arithmetic. For example, let $a = 3$, $b = 5$, $c = 1$; then the first expression becomes

$$3 \mathbin{\#} (5 \otimes 1) = 3 \mathbin{\#} 5 = 3$$

Using the same values in the second expression, we have

$$(3 \mathbin{\#} 5) \otimes (3 \mathbin{\#} 1) = 3 \otimes 1 = 3$$

Thus we find

$$3 \mathbin{\#} (5 \otimes 1) = (3 \mathbin{\#} 5) \otimes (3 \mathbin{\#} 1)$$

In general, we may write

$$a \mathbin{\#} (b \otimes c) = (a \mathbin{\#} b) \otimes (a \mathbin{\#} c)$$

for all replacements of a, b, and c. We state, formally, that # is **distributive** with respect to $\otimes$. This property, known as the **distributive property**, may be somewhat easier to visualize for the numbers of arithmetic.

Consider, for example, the expression 3(5 + 8). According to the distributive property, we may evaluate this expression in two different ways and obtain the same answer either way. Thus

$$3(5 + 8) = 3(13) = 39$$
$$(3)(5) + (3)(8) = 15 + 24 = 39$$

so that

$$3(5 + 8) = (3)(5) + (3)(8)$$

This example illustrates the fact that because of the distributive property we may either add first and then multiply, or find the two products first and then add. In general, we say that for all replacements of a, b, and c,

$$a(b + c) = ab + ac$$

Formally we call this the **distributive property for multiplication with respect to addition,** or simply the *distributive property.* Note that addition is *not* distributive with respect to multiplication since, for example,

$$3 + (5 \times 8) \neq (3 + 5) \times (3 + 8)$$

that is,

$$3 + 40 \neq 8 \times 11$$

It is the distributive property that allows us, in algebra, to make such statements as

$$2(a + b) = 2a + 2b$$
$$3(x - y) = 3x - 3y$$

It is the distributive property that elementary school youngsters use in multiplication. Consider the problem 7×43. The distributive property is used by thinking of 7×43 as

$$7 \times (40 + 3)$$

as shown in the example:

$$
\begin{array}{r}
43 \\
\underline{\times 7} \\
21 = 7 \times 3 \\
\underline{280} = 7 \times 40 \\
301 = (7 \times 3) + (7 \times 40)
\end{array}
$$

The distributive property can also be used in developing shortcuts in multiplication. Thus the product 8×99 can be found quickly as follows:

$$8 \times 99 = 8(100 - 1) = 800 - 8 = 792$$

It has been said that the distributive property is one of the unifying themes of mathematics. Note that unlike the commutative property or the associative property, the distributive property involves *two* operations of multiplication and addition.

exercises

For Exercises 1 through 6, let $ mean "select the larger of the two given numbers," and let $\otimes$ mean "select the first of two given numbers."

1. Find $3 \, \$ \, 8$, $8 \, \$ \, 3$, $3 \otimes 7$, $7 \otimes 3$.
2. Find $3 \, \$ \, (5 \, \$ \, 2)$, $(3 \, \$ \, 5) \, \$ \, 2$.
3. Find $7 \otimes (2 \otimes 3)$, $(7 \otimes 2) \otimes 3$.
4. Find $5 \, \$ \, (1 \otimes 3)$, $(5 \, \$ \, 1) \otimes (5 \, \$ \, 3)$.
5. Find $3 \otimes (2 \, \$ \, 7)$, $(3 \otimes 2) \, \$ \, (3 \otimes 7)$.
6. Show that $a \otimes (b \, \$ \, c) = (a \otimes b) \, \$ \, (a \otimes c) = a$ for all replacements of a, b, and c.
   Does $a \, \$ \, (b \otimes c) = (a \, \$ \, b) \otimes (a \, \$ \, c)$?
7. In ordinary arithmetic, is addition distributive with respect to addition? That is, does $a + (b + c) = (a + b) + (a + c)$ for all possible replacements of a, b, and c?
8. In ordinary arithmetic, is multiplication distributive with respect to multiplication?
   That is, does $a \times (b \times c) = (a \times b) \times (a \times c)$ for all possible replacements of a, b, and c?

Use the distributive property to find each of these products by means of a shortcut.

9. 7×79 10. 5×58

Find a replacement for n to make each sentence true.

11. $8(2 + 3) = 8 \cdot 2 + 8 \cdot n$ 12. $5(3 + 4) = 5 \cdot 3 + n \cdot 4$
13. $7(4 + n) = 7 \cdot 4 + 7 \cdot 5$ 14. $n(5 + 7) = 3 \cdot 5 + 3 \cdot 7$
15. $(6 + 7)5 = 6 \cdot 5 + n \cdot 5$ 16. $(5 + n)3 = 5 \cdot 3 + 9 \cdot 3$
17. $3 \cdot 5 + 3 \cdot 7 = 3(5 + n)$ 18. $4 \cdot 6 + 4 \cdot n = 4(6 + 3)$
19. $3 \cdot n + 7 \cdot n = (3 + 7)n$ 20. $3(4 + n) = 3 \cdot 4 + 3 \cdot n$

EXPLORATIONS

The distributive property can be used to develop shortcuts in multiplication. For example, the product 8×99 can be found quickly
as follows.

$$8 \times 99 = 8(100 - 1) = 800 - 8 = 792$$

Use the distributive property to find each of these products.

1. 7×98 **2.** 5×399

3. 13×99 **4.** 2×999

5. 5×998 **6.** 25×19

7. 35×21 **8.** 72×11

*Identify the use of the distributive property in each computation
and discuss its usefulness in ordinary arithmetic computations.*

9. $30 + 40 = (3 \times 10) + (4 \times 10) = (3 + 4) \times 10 = 7 \times 10 = 70$

10. $(17 \times 3) + (13 \times 3) = (17 + 13) \times 3 = 30 \times 3 = 90$

5-3
Clock Arithmetic

Let us create another finite mathematical system to establish firmly the ideas
presented thus far. If it is now 9 P.M. as you begin to read this section, what
time will it be in 5 hours? (We hope it won't take you that long to complete
your reading!) Do you see that the statement

$$9 + 5 = 2$$

is a correct statement if we are talking about positions on a 12-hour clock?
Let us consider the numerals 1 through 12 on a 12-hour clock as the
elements of a set T and consider addition on this clock to be based upon
counting in a clockwise direction. Thus to find the sum $9 + 5$ we start at
9 and count 5 units in a clockwise direction to obtain the result 2.
Verify that each of the following is correct.

$8 + 7 = 3$ (on a 12-hour clock)
$5 + 12 = 5$ (on a 12-hour clock)
$3 + 11 = 2$ (on a 12-hour clock)

We may make a table of addition facts on a 12-hour clock as follows:

+	1	2	3	4	5	6	7	8	9	10	11	12
1	2	3	4	5	6	7	8	9	10	11	12	1
2	3	4	5	6	7	8	9	10	11	12	1	2
3	4	5	6	7	8	9	10	11	12	1	2	3
4	5	6	7	8	9	10	11	12	1	2	3	4
5	6	7	8	9	10	11	12	1	2	3	4	5
6	7	8	9	10	11	12	1	2	3	4	5	6
7	8	9	10	11	12	1	2	3	4	5	6	7
8	9	10	11	12	1	2	3	4	5	6	7	8
9	10	11	12	1	2	3	4	5	6	7	8	9
10	11	12	1	2	3	4	5	6	7	8	9	10
11	12	1	2	3	4	5	6	7	8	9	10	11
12	1	2	3	4	5	6	7	8	9	10	11	12

Note that regardless of where we start on the clock, we shall always be at the same place 12 hours later. Thus for any element t of set T we have

$t + 12 = t$ (on a 12-hour clock)

Let us attempt to define several other operations for this arithmetic on the 12-hour clock. What does multiplication mean? Multiplication by an integer may be considered as repeated addition. For example, 3×5 on the 12-hour clock is equivalent to $5 + 5 + 5$. Since $5 + 5 = 10$ and $10 + 5 = 3$, we know that $3 \times 5 = 3$ on the 12-hour clock.

Verify that each of the following is correct.

$4 \times 5 = 8$ (on the 12-hour clock)
$3 \times 9 = 3$ (on the 12-hour clock)
$3 \times 7 = 9$ (on the 12-hour clock)

The following examples provide further illustrations of clock arithmetic.

EXAMPLE 1

Solve the equation $t + 6 = 2$ for t, where t may be replaced by any one of the numerals on a 12-hour clock.

SOLUTION   On the 12-hour clock $8 + 6 = 2$; therefore $t = 8$. Note that we may also solve for t by starting at 6 on the clock and counting, in a clockwise direction, until we reach 2.

EXAMPLE 2

Using the numerals on a 12-hour clock, find a replacement for t such that $\frac{9}{7} = t$.

SOLUTION   We can think of the statement $\frac{9}{7} = t$ as one of the four equivalent statements.

$$9 = 7 \times t \qquad 9 = t \times 7$$

$$\frac{9}{7} = t \qquad \frac{9}{t} = 7$$

Then use the sentence $9 = 7 \times t$ to solve the problem. How many groups of 7 are needed to produce 9 on the 12-hour clock? From a multiplication table, or by trial and error, we find that $7 \times 3 = 9$ (on the 12-hour clock); thus $t = 3$.

EXAMPLE 3

What is $3 - 7$ on the 12-hour clock?

SOLUTION   Let t represent a numeral on the 12-hour clock. The statement "$3 - 7 = t$" is equivalent to "$t + 7 = 3$." From the table of addition facts we find that $8 + 7 = 3$ (on the 12-hour clock); thus $t = 8$.

From a slightly different point of view, we may solve Example 3 by counting in a clockwise direction from 12 to 3, and then counting 7 units in a counterclockwise direction. We complete this process at 8. Thus $3 - 7 = 8$ (on the 12-hour clock).

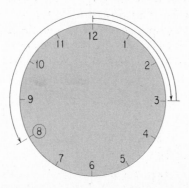

The reader may wish to compare clock arithmetic as developed in this section with the base 12 number system as in §4-7. Note that clock arithmetic is actually a system of remainders. Every multiple of 12 greater than 12 is discarded, and answers are given as a number from 1 through 12. Consider the product $5 \times 8 = 40$. In clock arithmetic and in the base 12 number system

$$5 \times 8 = 40 = (3 \times 12) + 4$$

In clock arithmetic (3×12) represents 3 rotations and is disregarded:

$$5 \times 8 = 4 \quad \text{(on the 12-hour clock)}$$

However, in the base 12 system our aim is to write a numeral for 5×8 in terms of 12:

$$5 \times 8 = 34_{12}$$

EXAMPLE 4

Compare 7×9 in (a) clock arithmetic and in (b) base 12 notation.

SOLUTION

(a) $7 \times 9 = 63; 63 = (5 \times 12) + 3 = 3$ (on the 12-hour clock).
(b) $7 \times 9 = 63; 63 = (5 \times 12) + 3 = 53_{12}$.

exercises

1. Is the set of numbers used in clock arithmetic closed **(a)** with respect to addition? **(b)** with respect to multiplication?
2. Make and complete a table of multiplication facts on the 12-hour clock.
3. Do the set of numbers on the clock contain identity elements with respect to addition and multiplication? If so, what are they?
4. List the numbers of clock arithmetic in a column. Beside each number list the inverse with respect to addition and the inverse with respect to multiplication, if one exists. (Thus the inverse of 3, with respect to addition, is 9, whereas 3 does not have an inverse with respect to multiplication.)

Solve each problem as on a 12-hour clock.

5.	$8 + 7$	6.	$8 + 11$
7.	$9 - 12$	8.	$5 - 9$
9.	3×9	10.	5×5
11.	$1 \div 5$	12.	$2 \div 7$

Find all possible replacements for t for which each sentence is a true sentence for the numerals on a 12-hour clock.

13. $t + 9 = 5$ **14.** $t - 3 = 11$

15. $8 + t = 2$ **16.** $3 - t = 10$

17. $3 \times t = 3$ **18.** $7 \times t = 11$

19. $\dfrac{t}{5} = 8$ **20.** $\dfrac{t}{7} = 9$

*21. $\dfrac{2}{t} = 3$ ***22.** $2 + t = 2 - t$

*23. $t + 12 = t$ ***24.** $3 - t = 5 + t$

Solve each problem (a) as on a 12-hour clock and (b) as in base 12 notation.

25. $9 + 9$ **26.** $8 + 6$

27. 5×9 **28.** 3×8

29. 6×6 **30.** 8×7

EXPLORATIONS

1. Consider arithmetic on a clock with four elements as in the figure. Complete tables for the addition and multiplication facts for a clock with four elements. Explore these systems and list as many properties as you can for each.

2. Use the results of Exploration 1 and tell whether the elements in arithmetic on a clock with four elements form a commutative group with respect to **(a)** addition and **(b)** multiplication. If not, tell which properties are not satisfied under the given operation.

3. Repeat Explorations 1 and 2 for the arithmetic on a clock with seven elements, 0, 1, 2, 3, 4, 5, 6.

5-4
Modular Arithmetic

Let us next consider a mathematical system based on a clock for five hours, numbered 0, 1, 2, 3, and 4, as in the accompanying figure.

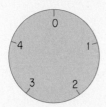

Addition on this clock may be performed by counting as on an ordinary clock. However, it seems easier to think of addition as rotations in a clockwise direction. Thus $3 + 4$ indicates that we are to start at 0 and move 3 units, then 4 more. The result is 2.

We interpret 0 to mean no rotation as well as to designate a position on the clock. Then we have such facts as:

$3 + 0 = 3$ (on a five-hour clock)
$4 + 1 = 0$ (on a five-hour clock)

Verify that the following table of addition facts on a five-hour clock is correct.

+	0	1	2	3	4
0	0	1	2	3	4
1	1	2	3	4	0
2	2	3	4	0	1
3	3	4	0	1	2
4	4	0	1	2	3

Multiplication on the five-hour clock is defined as repeated addition, as in §5-3. For example, 3×4 on the five-hour clock is equivalent to $4 + 4 + 4$. Since $4 + 4 = 3$, and $3 + 4 = 2$, we see that $3 \times 4 = 2$ on the five-hour clock.

We define subtraction and division using equivalent statements as follows.

> *Subtraction:* Since the statements "$a - b = x$" and "$a = b + x$" are equivalent, we define $a - b = x$ *if and only if* $b + x = a$.
> *Division:* Since the statements "$a \div b = x$" and "$a = b \times x$" are equivalent, we define $a \div b = x$ *if and only if* $b \times x = a$.

Regardless of where we start on this five-hour clock, we shall always be at the same place five hours later. Usually the symbols of a mathematical system based on clock arithmetic begin with 0, and the system is called a modular arithmetic. In the illustration of this section we say that we have an arithmetic modulo 5, and we use the symbols 0, 1, 2, 3, 4. Formally, specific facts in this mathematical system are written as follows.

$$3 + 4 \equiv 2 \text{ (mod 5), read "3 + 4 is congruent to 2, modulo 5"}$$
$$4 \times 2 \equiv 3 \text{ (mod 5), read "4 } \times \text{ 2 is congruent to 3, modulo 5"}$$

In general, two numbers are congruent modulo 5 if and only if they differ by a multiple of 5. Thus 3, 8, 13, and 18 are all congruent to each other modulo 5. We may write, for example

$$18 \equiv 13 \text{ (mod 5)}$$
$$8 \equiv 3 \text{ (mod 5)}$$

Notice that every integer in ordinary arithmetic is congruent modulo 5 to exactly one element of the set F, where $F = \{0, 1, 2, 3, 4\}$. Indeed, the set F is the set of all possible remainders when any number is divided by 5. The elements of F are the elements of arithmetic modulo 5.

EXAMPLE 1

Solve for x where x may be replaced by any element in arithmetic modulo 5: $2 - 3 = x$.

SOLUTION   Note that the statement "$2 - 3 = x$" is equivalent to "$3 + x = 2$." Since $3 + 4 = 2$, $x = 4$.

EXAMPLE 2

Solve for x in arithmetic modulo 5: $\frac{2}{3} = x$.

SOLUTION   Note that the statement "$\frac{2}{3} = x$" is equivalent to "$3 \times x = 2$." Since $3 \times 4 = 2$, $x = 4$.

Let us explore the properties of the mathematical system based upon the set $F = \{0, 1, 2, 3, 4\}$ and the operation addition as defined in this section. The following properties are of special interest.

1. *The set F of elements in arithmetic modulo 5 is closed with respect to addition.* That is, for any pair of elements, there is a unique element which represents their sum and which is also a member of the original set. We note, for example, that there is one and only one entry in each place of the table given in this section, and that each entry is an element of the set F.

2. *The set F of elements in arithmetic modulo 5 satisfies the commutative property for addition.* That is,

$$a + b = b + a$$

where a and b are any elements of the set F. Specifically, we see that $3 + 4 = 4 + 3, 2 + 3 = 3 + 2$, etc.

3. *The set F of elements in arithmetic modulo 5 satisfies the associative property for addition.* That is,

$$(a + b) + c = a + (b + c)$$

for all elements a, b, and c of the set F.

As a specific example we evaluate $3 + 4 + 2$ in two ways:

$(3 + 4) + 2 \equiv 2 + 2$ or 4, modulo 5
$3 + (4 + 2) \equiv 3 + 1$ or 4, modulo 5

4. *The set F of elements in arithmetic modulo 5 includes an identity element for addition.* That is, the set contains an element 0 such that the sum of any given element and 0 is the given element. That is,

$$0 + 0 = 0, \qquad 1 + 0 = 1, \qquad 2 + 0 = 2, \qquad 3 + 0 = 3, \qquad 4 + 0 = 4$$

5. *Each element in arithmetic modulo 5 has an inverse with respect to addition.* That is, for each element a of set F there exists an element a' of F such that $a + a' = 0$, the identity element. The element a' is said to be the inverse of a. Specifically, we have the following:

The inverse of 0 is 0, $\qquad 0 + 0 \equiv 0 \pmod{5}$.
The inverse of 1 is 4, $\qquad 1 + 4 \equiv 0 \pmod{5}$.
The inverse of 2 is 3, $\qquad 2 + 3 \equiv 0 \pmod{5}$.
The inverse of 3 is 2, $\qquad 3 + 2 \equiv 0 \pmod{5}$.
The inverse of 4 is 1, $\qquad 4 + 1 \equiv 0 \pmod{5}$.

As in §5-1, we may summarize these five properties by saying that the set F of elements in arithmetic modulo 5 forms a commutative group under addition.

exercises

Each of the following is based upon the set of elements in arithmetic modulo 5.

1. Make and complete a table of multiplication facts.
2. Verify that the commutative property for multiplication holds for at least two specific instances.
3. Verify that the associative property for multiplication holds for at least two specific instances.
4. What is the identity element with respect to multiplication?
5. Find the inverse of each element with respect to multiplication.
6. Verify for at least two specific instances that arithmetic modulo 5 satisfies the distributive property for multiplication with respect to addition.

Solve each of the following for x.

7. $2 + x \equiv 1 \pmod 5$　　　8. $x + 4 \equiv 2 \pmod 5$
9. $x + 3 \equiv 2 \pmod 5$　　　10. $4 + x \equiv 1 \pmod 5$
11. $1 - 3 \equiv x \pmod 5$　　　12. $1 - 4 \equiv x \pmod 5$
13. $2 - 4 \equiv x \pmod 5$　　　14. $3 - x \equiv 4 \pmod 5$
15. $2 \times x \equiv 3 \pmod 5$　　　16. $4 \times x \equiv 2 \pmod 5$
17. $4 \times x \equiv 1 \pmod 5$　　　18. $3 \times x \equiv 1 \pmod 5$
19. $\frac{1}{2} \equiv x \pmod 5$　　　20. $\frac{4}{3} \equiv x \pmod 5$

21. $\dfrac{3}{x} \equiv 2 \pmod 5$　　　22. $\dfrac{2}{x} \equiv 3 \pmod 5$

23. $x + 3 \equiv x \pmod 5$　　　24. $x + 1 \equiv 3 - x \pmod 5$

EXPLORATIONS

In Explorations 1 through 14 find all possible replacements for x for which each sentence is a true statement.

1. $x + 5 \equiv 0 \pmod 7$　　　2. $x - 3 \equiv 2 \pmod 4$
3. $3x \equiv 1 \pmod 7$　　　4. $x \times x \equiv 1 \pmod 8$

5. $\dfrac{x}{4} \equiv 3 \pmod 9$　　　6. $\dfrac{2}{x} \equiv 3 \pmod 7$

7. $1 - x \equiv 4 \pmod 6$　　　8. $4 + x \equiv 1 \pmod 7$
9. $x + 5 \equiv 1 \pmod 8$　　　10. $2 - x \equiv 3 \pmod 6$

11. $2x \equiv 3 \pmod 6$ **12.** $\dfrac{x}{3} \equiv 4 \pmod 8$

13. $x - 5 \equiv 2 \pmod 7$ **14.** $7 + x \equiv 1 \pmod{12}$

15. We describe the property illustrated by the equation $3 \times 4 = 0$ in arithmetic modulo 12, where the product is zero but neither number is zero, by saying that 3 and 4 are zero divisors. Are there other zero divisors in arithmetic modulo 12? If so, list them.

16. Consider a modulo 7 system where the days of the week correspond to numbers as follows: Monday—0; Tuesday—1; Wednesday—2; Thursday—3; Friday—4; Saturday—5; Sunday—6. Memorial Day, May 30, is the 150th day of the year and falls on a Thursday. In that same year, on what day of the week does July 4, the 185th day of the year, fall? On what day does Christmas, the 359th day of the year fall?

5-5
Two by Two Matrices

A matrix is a rectangular array of elements. We consider matrices with two rows and two columns. The matrix

$$\begin{bmatrix} 3 & 5 \\ 1 & 2 \end{bmatrix}$$

has 3 and 5 in its first row, 1 and 2 in its second row, 3 and 1 in its first column, 5 and 2 in its second column. Thus 3 is in the first row and first column, 1 is in the second row and first column, and so forth. The numbers 3, 5, 1, and 2 are the elements of the matrix. Any matrix with two rows and two columns is a two by two matrix.

Matrices are frequently used to record information. For example, we may label the rows and columns to indicate that two newspaper delivery boys Joe and Bill deliver both Sunday and weekday papers. Then the elements of the matrix can be the numbers of papers delivered each day.

$$\begin{array}{c} \\ \text{Joe} \\ \text{Bill} \end{array} \begin{array}{cc} \text{Sunday} & \text{Weekday} \\ \begin{bmatrix} 40 & 50 \\ 30 & 35 \end{bmatrix} \end{array}$$

Two matrices are equal if and only if they have the same (equal) elements. For example, the matrix equation

$$\begin{bmatrix} a & b \\ c & d \end{bmatrix} = \begin{bmatrix} 1 & 0 \\ 2 & 4 \end{bmatrix}$$

is equivalent to the system of four ordinary equations

$$\begin{cases} a = 1 \\ b = 0 \\ c = 2 \\ d = 4 \end{cases}$$

Addition of two matrices is possible only when the matrices have the same number of rows and the same number of columns; the matrices are then said to be **conformable for addition.** When addition is possible, the new matrix is formed with elements that are the sums of the corresponding elements of the matrix addends. For example,

$$\begin{bmatrix} 5 & 7 \\ 0 & 2 \end{bmatrix} + \begin{bmatrix} 1 & 0 \\ 3 & 6 \end{bmatrix} = \begin{bmatrix} 6 & 7 \\ 3 & 8 \end{bmatrix}$$

In general, consider the set S of two by two matrices with integers as elements. Then

$$\begin{bmatrix} a & b \\ c & d \end{bmatrix} + \begin{bmatrix} e & f \\ g & h \end{bmatrix} = \begin{bmatrix} a + e & b + f \\ c + g & d + h \end{bmatrix}$$

and the sum of the two elements of S is an element of S; that is, the set S is closed with respect to addition. The set S is also associative with respect to addition and forms a commutative group under addition. See Exercises 5 through 8.

EXAMPLE 1

Find

$$\begin{bmatrix} 1 & 2 & 5 \\ 7 & x & -2 \end{bmatrix} + \begin{bmatrix} 7 & -11 & 0 \\ x & -x & 3 \end{bmatrix}$$

SOLUTION

$$\begin{bmatrix} 8 & -9 & 5 \\ 7 + x & 0 & 1 \end{bmatrix}$$

Multiplication of matrices is possible only when the number of columns of the first matrix A is the same as the number of rows of the second matrix B; the matrices are then said to be **conformable for multiplication** as AB.

Note that two matrices may be conformable for the product AB and not be conformable for the product BA (see Exploration 2).

We return to our matrix of numbers of newspapers delivered by two newspaper boys to illustrate matrix multiplication.

$$\begin{array}{c} \\ \text{Joe} \\ \text{Bill} \end{array} \begin{array}{cc} \text{Sunday} & \text{Weekday} \\ \begin{bmatrix} 40 & 50 \\ 30 & 35 \end{bmatrix} \end{array}$$

For our second matrix, we use the following matrix of costs in cents and numbers of papers per week.

$$\begin{array}{c} \\ \\ \text{Sunday} \\ \text{Weekday} \end{array} \begin{array}{cc} \text{Cost} & \text{Number} \\ \text{per week} & \text{per week} \\ \begin{bmatrix} 30 & 1 \\ 60 & 6 \end{bmatrix} \end{array}$$

These two matrices enable us to find how much each boy collects (the total cost of the papers that he delivers) each week and the total number of papers that each boy delivers each week. Use arithmetic to find how much each boy collects and how many papers he delivers. Then continue reading and find this information using matrix multiplication.

The product

$$\begin{bmatrix} 40 & 50 \\ 30 & 35 \end{bmatrix} \begin{bmatrix} 30 & 1 \\ 60 & 6 \end{bmatrix}$$

is the matrix

$$\begin{bmatrix} (40 \times 30) + (50 \times 60) & (40 \times 1) + (50 \times 6) \\ (30 \times 30) + (35 \times 60) & (30 \times 1) + (35 \times 6) \end{bmatrix}$$

that is,

$$\begin{bmatrix} 4200 & 340 \\ 3000 & 240 \end{bmatrix}$$

We may interpret this matrix as follows.

$$\begin{array}{c} \\ \\ \text{Joe} \\ \text{Bill} \end{array} \begin{array}{cc} \text{Collection} & \text{Papers delivered} \\ \text{each week} & \text{each week} \\ \begin{bmatrix} \$42.00 & 340 \\ \$30.00 & 240 \end{bmatrix} \end{array}$$

Joe delivers 340 papers and collects $42.00; Bill delivers 240 papers and collects $30.00.

This example illustrates the usefulness of the following unusual appearing definition of multiplication of matrices.

$$\begin{bmatrix} a & b \\ c & d \end{bmatrix}\begin{bmatrix} e & f \\ g & h \end{bmatrix} = \begin{bmatrix} ae + bg & af + bh \\ ce + dg & cf + dh \end{bmatrix}$$

Notice that the element in the first row and the first column of the product matrix is the sum of the products of the corresponding (left-to-right and top-to-bottom) elements in the first row of the first matrix factor and the first column of the second matrix factor. The element in the first row and second column of the product is the sum of the products of the corresponding elements in the first row of the first factor and the second column of the second factor. Similarly, the element in the second row and first column is the sum of products of elements on the second row of the first factor and the first column of the second factor; the element in the second row and second column is the sum of products of elements on the second row of the first factor and the second column of the second factor.

EXAMPLE 2

Find

$$\begin{bmatrix} 1 & 0 & 1 \\ 0 & 1 & 0 \\ 0 & 1 & 1 \end{bmatrix} \times \begin{bmatrix} x & y & z \\ 2 & 4 & 6 \\ 1 & 3 & -1 \end{bmatrix}$$

SOLUTION

$$\begin{bmatrix} x + 1 & y + 3 & z - 1 \\ 2 & 4 & 6 \\ 3 & 7 & 5 \end{bmatrix}$$

Matrices have a very wide range of applications in mathematics and in other subjects. Our interest here is in matrices as a mathematical system—a set of elements with $=$, $+$, and $\times$ defined and with many interesting properties.

exercises

Find **(a)** $A + B$; **(b)** $A \times B$; **(c)** $B \times A$.

1. $A = \begin{bmatrix} 1 & 0 \\ 1 & 1 \end{bmatrix}$, $B = \begin{bmatrix} 1 & 2 \\ 3 & 0 \end{bmatrix}$

2. $A = \begin{bmatrix} 2 & 3 \\ -1 & 0 \end{bmatrix}$, $\quad B = \begin{bmatrix} a & b \\ c & d \end{bmatrix}$

3. $A = \begin{bmatrix} 0 & 0 & 1 \\ 0 & 1 & 0 \\ 1 & 0 & 0 \end{bmatrix}$, $\quad B = \begin{bmatrix} z & y & x \\ w & v & u \\ t & s & r \end{bmatrix}$

4. $A = \begin{bmatrix} 2 & 0 & 0 \\ 0 & 1 & 0 \\ 0 & 0 & 1 \end{bmatrix}$, $\quad B = \begin{bmatrix} a & b & c \\ d & e & f \\ f & h & i \end{bmatrix}$

Consider the set S of two by two matrices with elements from the set of integers. Assume that S is closed under addition and associative with respect to addition. Explain (possibly by examples) each of the following statements.

5. The identity element for S with respect to addition is

$$\begin{bmatrix} 0 & 0 \\ 0 & 0 \end{bmatrix}$$

6. Each element

$$\begin{bmatrix} a & b \\ c & d \end{bmatrix}$$

of S has an inverse element with respect to addition in S

$$\begin{bmatrix} -a & -b \\ -c & -d \end{bmatrix}$$

7. The set S is commutative with respect to addition.

8. The set S is a commutative group under addition.

9. The set S is closed under multiplication.

10. The identity element for S with respect to multiplication is

$$\begin{bmatrix} 1 & 0 \\ 0 & 1 \end{bmatrix}$$

11. $\begin{bmatrix} 1 & 1 \\ 0 & 1 \end{bmatrix}\begin{bmatrix} 0 & 1 \\ 1 & 0 \end{bmatrix} \neq \begin{bmatrix} 0 & 1 \\ 1 & 0 \end{bmatrix}\begin{bmatrix} 1 & 1 \\ 0 & 1 \end{bmatrix}$

12. The set S is not commutative with respect to multiplication.

13. Multiplication of a matrix of S on the right by

$$\begin{bmatrix} 0 & 1 \\ 1 & 0 \end{bmatrix}$$

interchanges the columns; that is, the first column becomes the second column and the second column becomes the first column.

14. Each element

$$\begin{bmatrix} a & b \\ c & d \end{bmatrix}$$

of S such that $ad - bc \neq 0$ has an element

$$\begin{bmatrix} \dfrac{d}{ad-bc} & \dfrac{-b}{ad-bc} \\[2ex] \dfrac{-c}{ad-bc} & \dfrac{a}{ad-bc} \end{bmatrix}$$

as an inverse element under multiplication.

Consider the subset T of S such that each element has an inverse element under multiplication as in Exercise 14. Assume that T is associative with respect to multiplication and explain each statement.

15. The set T is a group under multiplication.

16. The set T is not a commutative group under multiplication.

EXPLORATIONS

1. Beverly is a cashier who recorded the number of coins in her cash drawer at the end of the day as a one by five matrix.

	Half-dollars	Quarters	Dimes	Nickels	Pennies
$C =$	[11	9	21	17	16]

Write an expression for the value of the coins represented by the matrix C.

2. Consider the value matrix V:

$$V = \begin{bmatrix} 50 \\ 25 \\ 10 \\ 5 \\ 1 \end{bmatrix}$$

(a) Are the matrices C and V conformable for multiplication as CV?

(b) If possible, find and interpret the product CV.

(c) Are the matrices C and V conformable for multiplication as VC?

3. Consider the mathematical system represented by the table.

×	a	b	c
a	a	a	a
b	a	b	c
c	a	c	b

(a) Is the system closed under multiplication?
(b) Is multiplication commutative?
(c) Identify the multiplicative identity.
(d) Identify the multiplicative inverse of each element that has a multiplicative inverse.
(e) Does the system form a group under multiplication?

4. Consider the mathematical system represented by the table.

+	a	b	c
a	a	b	c
b	b	c	a
c	c	a	b

(a) Is the system closed under addition?
(b) Is addition commutative?
(c) Identify the additive identity.
(d) Identify the additive inverse of each element that has an additive inverse.
(e) Check 3 of the 27 possible applications of the associative property of addition.
(f) Does the system appear to form a commutative group under addition?

5. Consider the numerals 0, 1, 2, on a three-hour clock. For the arithmetic on a three-hour clock;
(a) Make an addition table.
(b) Does the system appear to form a group under addition?
(c) Make a multiplication table.
(d) Does the system appear to form a group under multiplication?

6. Think of the counterclockwise rotations on the plane of an equilateral triangle PQR about its center. Let R_1 represent a rotation of $0°$, R_2 a rotation of $120°$, and R_3 a rotation of $240°$ as shown in the figures.

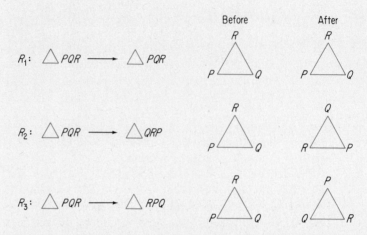

R_1: $\triangle PQR \longrightarrow \triangle PQR$

R_2: $\triangle PQR \longrightarrow \triangle QRP$

R_3: $\triangle PQR \longrightarrow \triangle RPQ$

(a) Let multiplication $R_i \times R_k$ represent the successive application of the rotation R_i and then the rotation R_k and complete the following table.

$\times$	R_1	R_2	R_3
R_1	R_1	—	—
R_2	—	R_3	—
R_3	—	—	—

(b) Is the system closed under this operation?

(c) Identify the identity element for this operation.

(d) Compare the completed table with the table in Exploration 4.

(e) Does the system appear to form a group under the given operation?

7. Discuss the similarities of the tables considered in Explorations 4, 5(a), and 6.

8. Take an index card and label it in each corner as shown. Also write these same letters in the corresponding corners on the reverse side of the card. Call this initial position I. Rotate the card $180°$ to obtain position R. Rotate the card about its horizontal axis, from the initial position, to obtain position H. Finally, rotate the card about its vertical axis, from the initial position, to obtain position V.

A	B
D	C

$$I$$

C	D
B	A

$$R$$

D	C
A	B

$$H$$

B	A
C	D

$$V$$

Use the symbol $\sim$ to mean "followed by." Thus $H \sim R$ means a rotation about the horizontal axis followed by a rotation of $180°$. You should find $H \sim R = V$. Verify also that $V \sim H = R$.

Make and complete a table for this system whose horizontal and vertical headings are I, R, H, and V. List as many properties for this system as you can find.

9. Consider a soldier facing in a given direction. He is then given various commands, such as "right face," "left face," "about face," and "as you were." The last command tells him to retain whatever position he may be in at the time.

A few specific examples of his movements should clarify matters. If he is facing to the north and is given the command "about face," then in his new position he is facing to the south. If we let R represent the command "right face," A represent the command "about face," and L represent the command "left face," then R followed by A is equivalent to the single command L.

Use the symbol $\circ$ to represent the operation "followed by" and verify that each of the following is correct:

$$L \circ L = A$$
$$L \circ A = R$$
$$A \circ R = L$$

Now suppose that our soldier makes a left turn and then we wish him to remain in that position. We make use of the command "as you were," represented by E. Thus $L \circ E = L$; also $E \circ L = L$.

(a) Make a table summarizing all possible movements, using the headings E, R, L, and A.

(b) Find $R \circ R$, $A \circ A$, $L \circ L$.

(c) Does $R \circ L = L \circ R$?

(d) Does $R \circ (L \circ A) = (R \circ L) \circ A$?

(e) Does the set of given commands appear to be commutative and associative with respect to the operation $\circ$?

(f) Find the identity element with respect to $\circ$.

(g) List the inverse of each of the elements of the set.

(h) Is the set of commands closed with respect to $\circ$?

chapter test

Refer to the table to answer Exercises 1 through 4.

$\otimes$	#	*	Σ	Δ
#	Σ	Δ	#	*
*	Δ	#	*	Σ
Σ	#	*	Σ	Δ
Δ	*	Σ	Δ	#

1. Which element is the identity element?
2. What is the inverse of Δ?
3. What is $(\# \otimes \Delta) \otimes *$?
4. For all replacements for a and b for the set $Y = \{ \#, *, \Sigma, \Delta \}$, $a \otimes b$ is an element of Y. What property does this illustrate?

Solve as on a 12-hour clock.

5. **(a)** $9 + 7$ **(b)** 5×11

6. **(a)** $2 - 7$ **(b)** $2 \div 5$

Find all possible replacements for x for which each sentence is a true statement.

7. **(a)** $3 + x \equiv 1 \pmod 5$ **(b)** $2 - x \equiv 4 \pmod 5$

8. **(a)** $x + 2 \equiv 0 \pmod 7$ **(b)** $3 - x \equiv 5 \pmod 8$

9. **(a)** $3x \equiv 2 \pmod 6$ **(b)** $\dfrac{x}{3} \equiv 2 \pmod 5$

10. Use the distributive property to evaluate $3(2 + 4)$ in arithmetic modulo 5 in two different ways.

6

Sets of
NUMBERS

You have been concerned with sets of numbers and operations on numbers since you first started your study of mathematics. In this chapter we shall present a brief development of various collections of numbers and their properties. We are not so much concerned with routine operations on these numbers as we are with the key concepts that form the basis for the structure of our number system. Although many of these ideas may be familiar, they will be presented in a way that should help to develop an appreciation of this structure and the reasons for its development.

The famous mathematician Leopold Kronecker (1823–1891) once said "God created the natural numbers; everything else is man's handiwork." Thus we shall begin our study of the number system with this set of natural numbers, now generally referred to as the set of counting numbers. We consider properties of the counting numbers, pay special attention to prime numbers, and consider the properties of equivalence and order relations. Then we extend the set of counting numbers to obtain the set of whole numbers, integers, rational numbers, real numbers, and complex numbers.

6-1

The Set of Counting Numbers

A number is associated with a set of elements whenever the elements of the set are counted. The set of counting numbers is often represented as the set C:

$$C = \{1, 2, 3, 4, 5, \ldots\}$$

Sums and products of counting numbers may be introduced using cardinal numbers of sets. If A and B are two sets such that $A \cap B = \varnothing$, then

$$n(A) + n(B) = n(A \cup B)$$

Multiplication may be introduced in terms of successive addition. For example,

$$2 \times 3 = 3 + 3, \qquad 3 \times 2 = 2 + 2 + 2$$

Multiplication may also be introduced using equivalent sets that have no common elements, that is, disjoint equivalent sets. For example, let $A = \{a, b\}$, $B = \{p, q\}$, and $C = \{x, y\}$. Then

$$n(A) = n(B) = n(C) = 2$$
$$A \cap B = \varnothing, \qquad A \cap C = \varnothing, \qquad B \cap C = \varnothing$$

and

$$n(A \cup B \cup C) = 3 \times 2$$

The use of equivalent sets to represent multiplication can be extended to obtain a new operation on sets. Let $P = \{1, 2, 3\}$ and $Q = \{a, b\}$. A new set with cardinal number 3 may be formed by taking each element of P with a given element of Q:

$$n\{(1, a), (2, a), (3, a)\} = 3$$
$$n\{(1, b), (2, b), (3, b)\} = 3$$

Notice that each of these sets has pairs of elements such as $(2, b)$ as its elements. Each pair has been purposely written with an element of P first and an element of Q second. Any pair with one element considered as the first element and the other element considered as the second element is an **ordered pair**.

For any two given sets A and B the set of all possible ordered pairs with an element of A as the first element and an element of B as the second element is the **Cartesian product** of A and B, written $A \times B$. For $P = \{1, 2, 3\}$ and $Q = \{a, b\}$,

$$P \times Q = \{(1, a), (2, a), (3, a), (1, b), (2, b), (3, b)\}$$

Notice that for $n(P) = 3$ and $n(Q) = 2$ the set $P \times Q$ may be considered either as two sets of three elements each (one set of pairs including a and one set of pairs including b) or as three sets (one for each element of P) of two elements each. Thus

$$n(P) \times n(Q) = n(P \times Q) = n(Q) \times n(P)$$
$$3 \ \times \ 2 \ = \qquad 6 \qquad = \ 2 \ \times \ 3$$

Commutativity may also be shown from $n(P \times Q) = n(Q \times P)$; that is,

$$n\{(1, a), (2, a), (3, a), (1, b), (2, b), (3, b)\}$$
$$= n\{(a, 1), (a, 2), (a, 3), (b, 1), (b, 2), (b, 3)\}$$

EXAMPLE 1

Use a Cartesian product of sets to illustrate $2 \times 4 = 8$.

SOLUTION

Let $A = \{a, b\}$ and $B = \{p, q, r, s\}$. Then $n(A) = 2$, $n(B) = 4$, $A \times B = \{(a, p), (b, p), (a, q), (b, q), (a, r), (b, r), (a, s), (b, s)\}$, and $2 \times 4 = n(A) \times n(B) = n(A \times B) = 8$.

The elements of a Cartesian product are often represented in an array for the sets A and B of Example 1 as shown.

s	(a, s)	(b, s)
r	(a, r)	(b, r)
q	(a, q)	(b, q)
p	(a, p)	(b, p)
	a	b

The properties of the set of counting numbers under addition and multiplication may be described using the terminology introduced in Chapter 5. For any counting numbers a, b, c we have:

Closure, $+$: There is one and only one counting number $a + b$.
Closure, $\times$: There is one and only one counting number $a \times b$.
Commutative, $+$: $a + b = b + a$; the sum of any two counting

numbers is the same regardless of the order in which the numbers are added.

Commutative, ×: $a \times b = b \times a$; the product of any two counting numbers is the same regardless of the order in which the numbers are multiplied.

Associative, +: $(a + b) + c = a + (b + c)$.

Associative, ×: $(a \times b) \times c = a \times (b \times c)$.

Identity, ×: $a \times 1 = 1 \times a = a$; the counting number 1 is the multiplicative identity and is also called the identity element for multiplication. This property is often referred to as the multiplication property of one.

Distributive Property, × with respect to +:

$$a \times (b + c) = (a \times b) + (a \times c).$$

The set of counting numbers does not include an identity for addition since zero is not a counting number. Also, the set of counting numbers is *not* closed under subtraction or under division. For example, $2 - 5$ and $12 \div 5$ are impossible if you are restricted to using counting numbers. These two counterexamples are sufficient to show that the properties do not hold.

EXAMPLE 2

Apply properties of counting numbers to the left member of the equation $25 \times (11 \times 4) = (25 \times 4) \times 11$ to obtain the right member of the equation. Show each step and name each property.

SOLUTION

$25 \times (11 \times 4) = 25 \times (4 \times 11)$ Commutative, ×

$\qquad\qquad\quad = (25 \times 4) \times 11$ Associative, ×

exercises

1. Find $A \times B$ for $A = \{1, 2\}$ and $B \times \{1, 2, 3, 4\}$.

2. Find $P \times Q$ for $P = \{1, 2, 3\}$ and $Q = \{1, 2, 3\}$.

3. Show that subtraction of counting numbers is not commutative.

4. Show that division of counting numbers is not commutative.

5. Does $8 - (3 - 2) = (8 - 3) - 2$? Is subtraction of counting numbers associative?

6. Does $12 \div (6 \div 2) = (12 \div 6) \div 2$? Is division of counting numbers associative?

7. Show that addition of counting numbers is not distributive with respect to multiplication.

For each arithmetic statement name the property of counting numbers that is illustrated.

 8. $3 \times 2 = 2 \times 3$

 9. $6 + 5 = 5 + 6$

10. $2 + (3 \times 5) = 2 + (5 \times 3)$

11. $2 \times (3 \times 5) = (2 \times 3) \times 5$

12. $1 \times 4580 = 4580$

13. $(25 \times 14) + (25 \times 26) = 25 \times (14 + 26)$

14. $17 \times (15 + 21) = 17 \times (21 + 15)$

15. $48 + (19 + 7) = (19 + 7) + 48$

16. $27 \times (5 + 3) = (5 + 3) \times 27$

17. $17 \times (10 + 1) = (17 \times 10) + (17 \times 1)$

18. $5280 \times (7 + 13) = 5280 \times (13 + 7)$

Identify the property of counting numbers that is used for each step in these applications of our usual addition and multiplication algorithms.

19.
$$
\begin{aligned}
12 + 25 &= (10 + 2) + (20 + 5) && \text{Numeration system.}\\
&= 10 + [2 + (20 + 5)] && \text{(a)}\\
&= 10 + [(20 + 5) + 2] && \text{(b)}\\
&= 10 + [20 + (5 + 2)] && \text{(c)}\\
&= 10 + (20 + 7) && \text{Number fact.}\\
&= (10 + 20) + 7 && \text{(d)}\\
&= [(1 \times 10) + (2 \times 10)] + 7 && \text{Numeration system.}\\
&= [(1 + 2) \times 10] + 7 && \text{(e)}\\
&= (3 \times 10) + 7 && \text{Number fact.}\\
&= 37 && \text{Numeration system.}
\end{aligned}
$$

20.
$$
\begin{aligned}
12 \times 6 &= (10 + 2) \times 6 && \text{Numeration system.}\\
&= (10 \times 6) + (2 \times 6) && \text{(a)}\\
&= 60 + 12 && \text{Number facts.}\\
&= 60 + (10 + 2) && \text{Numeration system.}\\
&= (60 + 10) + 2 && \text{(b)}\\
&= [(6 \times 10) + 10] + 2 && \text{Number fact.}\\
&= [(6 \times 10) + (1 \times 10)] + 2 && \text{(c)}\\
&= (6 + 1) \times 10 + 2 && \text{(d)}\\
&= 7 \times 10 + 2 && \text{Number facts.}\\
&= 72 && \text{Numeration system.}
\end{aligned}
$$

Apply properties of counting numbers to the left member of each equation to obtain the right member. Show each step and name each property.

21. $92 + (50 + 8) = (92 + 8) + 50$

22. $(25 \times 17) \times 4 = 17 \times 100$

23. $37 \times (1 + 100) = 3700 + 37$

24. $(2 + 3) + (8 + 7) = (2 + 8) + (3 + 7)$

25. $(73 + 19) + (7 + 1) = (73 + 7) + (19 + 1)$

26. $(26 \times 1) \times (10 \times 2) = 10 \times (26 \times 2)$

Explain your answer for each question:

27. Note that $6 \div (3 \div 1) = (6 \div 3) \div 1$. Is division of counting numbers associative?

28. Note that $4^2 = 2^4$. Is the operation of raising to a power commutative for counting numbers?

EXPLORATIONS

Computers have come to play a significant role in our mathematics curriculum at the college level. In recent years many secondary schools and some elementary schools have started to make use of computers and calculators in their instructional programs. One of the most widely used languages for the computer is called BASIC. A few of the symbols used in this language are as follows:

Addition: +
Subtraction: —
Multiplication: *
Division: /

A vertical arrow is used to indicate that a number is being raised to a power. For example:

$10 \uparrow 2$ means 10^2 $5 \uparrow 3$ means 5^3

In the language of BASIC we may write:

$(a + b) \uparrow 2$ to mean $(a + b)^2$
$(a \uparrow 2) + (b \uparrow 2)$ to mean $a^2 + b^2$

For further information on BASIC, read "Introduction to an Algorithmic Language," a publication of the National Council of Teachers of Mathematics, 1201 Sixteenth Street, N.W., Washington, D.C.

Compute:

1. $(2 * 3) \uparrow 2$ **2.** $(12/4) \uparrow 3$

3. $(2 \uparrow 3) * 2$ **4.** $(12 \uparrow 3)/4$

5. $(12 \uparrow 2)/(3 * 8)$ **6.** $(12 * 8)/(3 \uparrow 2)$

Before extending the set of counting numbers to other sets, we shall explore an interesting and useful classification of the counting numbers according to several of its subsets. First, however, certain definitions are needed.

The counting number 6 is divisible by 2, since there is a counting number 3 such that $6 = 2 \times 3$; the counting number 7 is not divisible by 2, since there exists no counting number b such that $7 = 2 \times b$. In general, a counting number n is **divisible by** a counting number t if and only if there is a counting number k such that $n = t \times k$. If n is divisible by t, then n is a **multiple of** t and t is a **factor of** n. For example, 6 is a multiple of 2 and 2 is a factor of 6.*

Reproduced by permission of MAD Magazine, page 22, June 1971 issue. © 1971 By E. C. Publications, Inc.

The counting numbers are often considered in terms of the numbers by which they are divisible. The set

$$A = \{2, 4, 6, 8, 10, 12, \ldots\}$$

consists of the numbers that are divisible by 2—that is, the numbers expressible in the form $2k$, where k stands for a counting number. The set

$$B = \{3, 6, 9, 12, 15, 18, \ldots\}$$

consists of the numbers divisible by 3; the set

$$C = \{4, 8, 12, 16, 20, 24, \ldots\}$$

consists of the numbers divisible by 4; the set

$$D = \{5, 10, 15, 20, 25, 30, \ldots\}$$

*The second edition of this book caught the eye of David Berg of MAD Magazine who prepared the above cartoon.

consists of the numbers divisible by 5; the set

$$E = \{6, 12, 18, 24, 30, 36, \ldots\}$$

consists of the numbers divisible by 6; and so forth.

Notice that $C \subset A$; in other words, any number that is divisible by 4 is also divisible by 2. Notice also that $A \cap B = E$; in other words, the set of numbers that are divisible by both 2 and 3 is the set of numbers that are divisible by 6.

The number 1 divides every counting number since $k = 1 \times k$ for every counting number k. Accordingly, the number 1 is called a **unit**. Since the set A does not include all counting numbers, divisibility by 2 is a special property of the elements of the set A.

The number 2 is a member of the set A and is not a member of any other set; that is, 2 is not divisible by any counting number except itself and 1. Any counting number greater than 1 that is divisible only by itself and 1 is called a **prime number**. Thus 2 is prime, 3 is prime, 4 is not prime since 4 is divisible by 2 (that is, 4 is a member of the set A as well as the set C), 5 is prime, and 6 is not prime since 6 is divisible by both 2 and 3. The counting numbers that are greater than 1 and are not prime are called **composite numbers**. Note that every counting number greater than 1 is either prime or composite. The number 1 is neither prime nor composite.

We could extend the list of sets A, B, C, D, E, F to identify other prime numbers, that is, elements that belong to one and only one of these sets. However, the method used to select these sets may be applied to the entire set of counting numbers. We shall illustrate this for the set $\{1, 2, 3, \ldots, 100\}$ and thereby illustrate a method for finding prime numbers that was discovered by a Greek mathematician named Eratosthenes over two thousand years ago. The method is known as the **Sieve of Eratosthenes**. First we prepare a table of the counting numbers through 100. Then selected members of this set are excluded as shown on page 197.

Cross out 1, since we know that it is not classified as a prime number. Draw a circle around 2, the smallest prime number. Then cross out every following multiple of 2, since each one is divisible by 2 and thus is not prime. That is, cross out the numbers in the set $\{4, 6, 8, \ldots, 100\}$.

Draw a circle around 3, the next prime number in our list. Then cross out each succeeding multiple of 3. Some of these numbers, such as 6 and 12, will already have been crossed out because they are multiples of 2. That is, they are members of both sets A and B.

The number 5 is prime and is circled, and we exclude each fifth number after 5. The next prime number is 7, and we exclude each seventh number after 7. Note that 49 is the first multiple of 7 that has not already been excluded as being a member of another set of multiples. The next prime number is 11. Since all multiples of 11 in this set have already been ex-

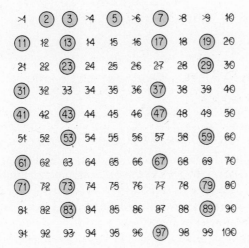

cluded, the remaining numbers that have not been excluded are prime numbers and may be circled.

Notice that 49 is the first number that is divisible by 7 and is not also divisible by a prime number less than 7. In other words, each composite number less than 7^2 has at least one of its factors less than 7. Similarly, we might have observed that each composite number less than 5^2 has at least one factor less than 5. In general, *for any prime number p each composite number less than p^2 has a prime number less than p as a factor.*

We use this property to tell us when we have excluded all composite numbers from a set. In the set of numbers $\{1, 2, \ldots, 100\}$ we have considered the primes 2, 3, 5, and 7. The next prime is 11. Thus by our method we have already excluded all composite numbers up to but not including 11^2; that is, 121. Since this number is greater than any in our table, we have identified the set of prime numbers less than or equal to 100.

EXAMPLE 1

List the set of prime numbers less than 70.

SOLUTION   From the chart we identify this set as

$$\{2, 3, 5, 7, 11, 13, 17, 19, 23, 29, 31, 37, 41, 43, 47, 53, 59, 61, 67\}$$

We have seen that every counting number greater than 1 is either a prime number or a composite number. Now we shall find that every counting number greater than 1 can be expressed in terms of its prime factors in essentially only one way.

Consider the various ways of factoring 24:

$$24 = 1 \times 24$$
$$24 = 2 \times 12$$
$$24 = 3 \times 8$$
$$24 = 4 \times 6$$
$$24 = 2 \times 2 \times 6$$
$$24 = 2 \times 3 \times 4$$
$$24 = 2 \times 2 \times 2 \times 3 = 2^3 \times 3$$

The last factorization in terms of the prime numbers 2 and 3 could be written as $2 \times 3 \times 2^2$ and in other ways. However, these ways are equivalent, since the order of the factors does not affect the product. Thus 24 can be expressed in terms of its prime factors in one and only one way.

One of the easiest ways to find the prime factors of a number is to consider the prime numbers

$$2, 3, 5, 7, 11, 13, 17, 19, 23, 29, 31, \ldots$$

in order and use each one as a factor as many times as possible. Then for 24 we would have

$$24 = 2 \times 12$$
$$= 2 \times 2 \times 6$$
$$= 2 \times 2 \times 2 \times 3$$

Some people prefer to write these steps using division:

```
2 | 24
  2 | 12
    2 | 6
        3
```

Since 3 is a prime number, no further steps are needed and $24 = 2^3 \times 3$.

EXAMPLE 2

Express 3850 in terms of its prime factors.

SOLUTION

```
2 | 3850
  5 | 1925
    5 | 385        3850 = 2 × 5² × 7 × 11
      7 | 77
          11
```

$$3850 = 2 \times 5^2 \times 7 \times 11$$

In general, if a counting number n is greater than 1, then n has a prime number p_1 as a factor. Suppose

$$n = p_1 n_1$$

Then if n is a prime number, $n = p_1$ and $n_1 = 1$. If n is not a prime number, then n_1 is a counting number greater than 1. In this case n_1 is either a prime number or a composite number. Suppose

$$n_1 = p_2 n_2 \quad \text{and thus} \quad n = p_1 p_2 n_2$$

where p_2 is a prime number. As before, if $n_2 \neq 1$, then

$$n_2 = p_3 n_3 \quad \text{and thus} \quad n = p_1 p_2 p_3 n_3$$

where p_3 is a prime number, and so forth. We may continue this process until some $n_k = 1$, since there are only a finite number of counting numbers less than n and

$$n > n_1 > n_2 > n_3 > \cdots > n_k = 1$$

Then we have an expression for n as a product of prime numbers:

$$n = p_1 p_2 p_3 \cdots p_k$$

We call this the **prime factorization** of n, that is, the factorization of n into its prime factors. Except for the order of the factors, the prime factorization of any counting number greater than 1 is unique; that is, *any counting number greater than 1 may be expressed as a product of prime numbers in one and only one way.* As in the examples, we usually write the prime factorization as a product of powers of prime numbers.

EXAMPLE 3

Find the prime factorization of 5280.

SOLUTION

$$
\begin{array}{r|l}
2 & 5280 \\ \hline
2 & 2640 \\ \hline
2 & 1320 \\ \hline
2 & 660 \\ \hline
2 & 330 \\ \hline
3 & 165 \\ \hline
5 & 55 \\ \hline
 & 11
\end{array}
$$

$$5280 = 2^5 \times 3 \times 5 \times 11$$

exercises

Let $A =$ the set of numbers divisible by 2, $B =$ the set of numbers divisible by 3, $D =$ the set of numbers divisible by 5, $F =$ the set of numbers divisible by 12, and $H =$ the set of numbers divisible by 15. Restate each of the statements in Exercises 1 through 6 in terms of divisibility.

1. $H \subset B$ **2.** $H \subset D$

3. $B \cap D = H$ **4.** $F \subset A$

5. $F \subset B$ **6.** $F \subset (A \cap B)$

7. List the composite numbers between 20 and 40.

8. Adapt the Sieve of Eratosthenes to find the prime numbers less than or equal to 200.

9. Is every odd number a prime number? Is every prime number an odd number?

10. Exhibit a pair of prime numbers that differ by 1 and show that there is only one such pair possible.

11. Here is a famous theorem that has not yet been proved: Every even number greater than 2 is expressible as the sum of two prime numbers. (This theorem is often called **Goldbach's conjecture.**) Express each even number from 4 to 40 inclusive as a sum of two prime numbers.

12. Here is another famous theorem. Two prime numbers such as 17 and 19 that differ by 2 are called **twin primes.** It is believed but has not yet been proved that there are infinitely many twin primes. Find a pair of twin primes that are between **(a)** 25 and 35; **(b)** 55 and 65; **(c)** 95 and 105.

13. A set of three prime numbers that differ by 2 is called a **prime triplet.** Exhibit a prime triplet and explain why it is the only possible triplet of primes.

14. It has been conjectured but not proved that every odd number greater than 5 is expressible as the sum of three prime numbers. Verify this for the numbers 7, 9, 11, 13, and 15.

*15. What is the largest prime that you need to consider to be sure that you have excluded all composite numbers less than or equal to **(a)** 200; **(b)** 500; **(c)** 1000?

Express each number as the product of two counting numbers in as many different ways as possible.

16. 12   **17.** 15

18. 18   **19.** 20

20. 13   **21.** 29

Find the prime factorization of each number.

22. 76   **23.** 68

24. 215   **25.** 123

26. 738   **27.** 1425

28. 341   **29.** 818

EXPLORATIONS

1. Some textbooks use the idea of a branching tree to help students think about factors. If the branches terminate with prime numbers, then we have a **prime-factor tree.** Here are two such trees:

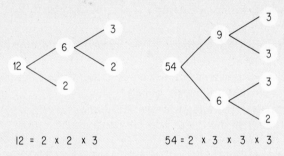

$$12 = 2 \times 2 \times 3 \qquad\qquad 54 = 2 \times 3 \times 3 \times 3$$

Draw prime-factor trees for 30, 60, and 96.

2. No one has ever been able to find a formula that will produce only prime numbers. At one time it was thought that the expression $n^2 - n + 41$ would give only prime numbers for the set of counting numbers as replacements for n. Show that prime numbers are obtained when n is replaced by 1, 2, 3, 4, and 5. Show that the formula fails for $n = 41$.

3. Euclid proved that the set of prime numbers is infinite. Refer to a textbook or a book on the history of mathematics and study this simple, yet elegant, proof.

4. At the time of the writing of this book the largest known prime was $2^{19,937} - 1$, a number with 6002 digits in decimal notation. Look for and read a reference to man's search for large prime numbers through the use of computers.

6-3

Applications of Prime Factorizations

We can make effective use of the concept of prime factorization in many arithmetic situations. Let us first consider the various factors of counting numbers. Recall that a counting number t is said to be a factor of another counting number n if and only if there is a counting number k such that $n = t \times k$. Now consider the set of factors of 12 and the set of factors of 18:

12: $\{1, 2, 3, 4, 6, 12\}$
18: $\{1, 2, 3, 6, 9, 18\}$

The set of **common factors** of 12 and 18 consists of those numbers that are factors of both 12 and 18, that is,

$\{1, 2, 3, 6\}$

The largest member of this set, 6, is called the greatest common factor (G.C.F.) of the two numbers. In general, the **greatest common factor** of two or more counting numbers is the largest counting number that is a factor of each of the given numbers.

We may use the prime factorization of two counting numbers to find their greatest common factor. First express each number by its prime factorization. Then consider the prime numbers that are factors of both the given numbers, and take the product of those prime numbers with each raised to the highest power that is a factor of *both* of the given numbers. For $12 = 2^2 \times 3$ and $18 = 2 \times 3^2$ we have G.C.F. $= 2 \times 3$, that is, 6.

EXAMPLE 1

Find the greatest common factor of 60 and 5280.

SOLUTION

$\begin{array}{r|r} 2 & 60 \\ 2 & 30 \\ 3 & 15 \\ & 5 \end{array}$ $60 = 2^2 \times 3 \times 5$

As in Example 3 of §6-2, we have $5280 = 2^5 \times 3 \times 5 \times 11$. Then the highest power of 2 that is a common factor of 60 and 5280 is 2^2; 3 is a common factor; 5 is a common factor; 11 is not a common factor. The greatest common factor of 60 and 5280 is $2^2 \times 3 \times 5$, that is, 60.

EXAMPLE 2

Find the greatest common factor of 3850 and 5280.

SOLUTION

$3850 = 2 \times 5^2 \times 7 \times 11$
$5280 = 2^5 \times 3 \times 5 \times 11$

The greatest common factor of 3850 and 5280 is $2 \times 5 \times 11$, that is, 110.

EXAMPLE 3

Find the G.C.F. of 12, 36, and 60.

SOLUTION

First write the prime factorization of each number:

$$12 = 2^2 \times 3$$
$$36 = 2^2 \times 3^2$$
$$60 = 2^2 \times 3 \times 5$$

The G.C.F. is $2^2 \times 3$, that is, 12.

We may use the greatest common factor when we reduce (simplify) a fraction. For example,

$$\frac{60}{4880} = \frac{(2^2 \times 5) \times 3}{(2^2 \times 5) \times (2^2 \times 61)} = \frac{3}{2^2 \times 61} = \frac{3}{244}$$

Since 3 is the only prime factor of the numerator and 3 is not a factor of the denominator, the fraction $\frac{3}{244}$ is in **lowest terms.** The numerator and the denominator do not have any common prime factors and are said to be **relatively prime.**

EXAMPLE 4

Reduce the fraction $\frac{60}{168}$ to lowest terms.

SOLUTION

$$60 = 2^2 \times 3 \times 5$$
$$168 = 2^3 \times 3 \times 7$$

The greatest common factor is $2^2 \times 3$,

$$\frac{60}{168} = \frac{(2^2 \times 3) \times 5}{(2^2 \times 3) \times 2 \times 7} = \frac{5}{14}$$

Let us now turn our attention to the concept of a multiple of a number. Recall that if a counting number t is a factor of a counting number n, then n is said to be a multiple of t. Consider the set of multiples of 12 and the set of multiples of 18:

12: $\{12, 24, 36, 48, 60, 72, 84, 96, 108, 120, \ldots\}$
18: $\{18, 36, 54, 72, 90, 108, 126, \ldots\}$

The set of **common multiples** of 12 and 18 consists of those numbers that are multiples of both 12 and 18, that is,

$\{36, 72, 108, \ldots\}$

The smallest member of this set, 36, is called the least common multiple (L.C.M.) of the two numbers. In general, the **least common multiple** of two or more counting numbers is the smallest number that is a multiple of each of the given numbers.

We may use the prime factorization of two numbers to find their lowest common multiple. First express each number by its prime factorization. Then consider the prime factors that are factors of either of the given numbers, and take the product of these prime numbers with each raised to the highest power that occurs in *either* of the prime factorizations. For $12 = 2^2 \times 3$ and $18 = 2 \times 3^2$, we have L.C.M. $= 2^2 \times 3^2$, that is, 36.

EXAMPLE 5

Find the lowest common multiple of 3850 and 5280.

SOLUTION   As in Examples 1 and 2,

$3850 = 2 \times 5^2 \times 7 \times 11$
$5280 = 2^5 \times 3 \times 5 \times 11$

The lowest common multiple of 3850 and 5280 is

$2^5 \times 3 \times 5^2 \times 7 \times 11$, that is, 184,800.

EXAMPLE 6

Find the L.C.M. of 12, 18, and 20.

SOLUTION   First write the prime factorization of each number:

$12 = 2^2 \times 3$
$18 = 2 \times 3^2$
$20 = 2^2 \times 5$

The L.C.M. is $2^2 \times 3^2 \times 5$, that is, 180.

We use the lowest common multiple of the denominators of two fractions when we add or subtract fractions. For example, the lowest common multiple of 12 and 18 is 36:

$$\frac{7}{12} + \frac{5}{18} = \frac{21}{36} + \frac{10}{36} = \frac{31}{36}$$

The answer is in reduced form, since 31 and 36 are relatively prime.

EXAMPLE 7

Simplify: $\frac{37}{5280} - \frac{19}{3850}$

SOLUTION

We use the lowest common multiple as found in Example 5:

$$\frac{37}{5280} - \frac{19}{3850} = \frac{37}{2^5 \times 3 \times 5 \times 11} - \frac{19}{2 \times 5^2 \times 7 \times 11}$$

$$= \frac{37 \times 5 \times 7}{2^5 \times 3 \times 5^2 \times 7 \times 11} - \frac{19 \times 2^4 \times 3}{2^5 \times 3 \times 5^2 \times 7 \times 11}$$

$$= \frac{1295 - 912}{2^5 \times 3 \times 5^2 \times 7 \times 11} = \frac{383}{184,800}$$

The instruction "simplify" is used as in Example 7 to mean "perform the indicated operations and express the answer in simplest form." In the case of fractions, "express in simplest form" means "reduce to lowest terms."

exercises

List the factors of:

1. 20   **2.** 24   **3.** 9   **4.** 16

5. 28   **6.** 48   **7.** 60   **8.** 72

Write the prime factorizations and find the G.C.F. of:

9. 68 and 76   **10.** 123 and 215

11. 76 and 1425   **12.** 123 and 1425

13. 215 and 1425   **14.** 68 and 738

15. 12, 15, and 20   **16.** 18, 24, and 40

17. 12, 18, and 30   **18.** 15, 45, and 60

List the first five elements in the set of multiples of:

19. 7   **20.** 8

21. 15   **22.** 20

Write the prime factorizations and find the L.C.M. of:

23. 68 and 76   **24.** 123 and 215

25. 76 and 1425   **26.** 123 and 1425

27. 215 and 1425   **28.** 68 and 738

29. 12, 15, and 20   **30.** 18, 24, and 40

31. 12, 18, and 30   **32.** 15, 45, and 60

Use the concepts of G.C.F. and L.C.M. to simplify:

33. $\frac{123}{215}$ **34.** $\frac{76}{1425}$

35. $\frac{3}{8} + \frac{5}{12}$ **36.** $\frac{3}{4} - \frac{1}{6}$

37. $\frac{7}{10} - \frac{4}{15}$ **38.** $\frac{11}{12} + \frac{3}{16}$

39. $\frac{11}{12} - \frac{9}{18}$ **40.** $\frac{5}{68} + \frac{11}{76}$

41. $\frac{7}{123} - \frac{2}{215}$ **42.** $\frac{41}{215} + \frac{19}{1425}$

*43. What is the *least* common factor of any two counting numbers?

*44. What is the *greatest* common multiple of any two counting numbers?

EXPLORATIONS

Extend the first three columns of this table for counting numbers from 1 through 17.

Counting numbers	Factors	Number of factors	Sum of factors
1	1	1	1
2	1, 2	2	3
.	.	.	.
.	.	.	.
.	.	.	.

1. Can any counting number greater than 1 have only one factor?

2. Can any prime number be identified in terms of the number of its factors?

3. Can any composite number be identified in terms of the number of its factors?

4. Give a rule for determining whether or not a counting number is a prime number if the number of its factors is known.

5. Find a way to distinguish in terms of the number of factors the counting numbers that are squares of counting numbers.

Extend the fourth column of the table as needed for these explorations.

6. A number is a **deficient number** if the sum of its factors is less than twice the number. List the first five deficient numbers.

7. A number is a **perfect number** if the sum of its factors is equal to twice the number. Find at least one perfect number.

8. A number is an **abundant number** if the sum of its factors is more than twice the number. Find at least one abundant number.

6-4
Equivalence and Order Relations

We have used equalities to state that two symbols or expressions are names for the same number. For example, $2 + 3 = 5$.

The equality relation has the following basic properties, and is therefore called an **equivalence relation:**

> **Reflexive:** $a = a$.
> **Symmetric:** If $a = a$, then $b = a$.
> **Transitive:** If $a = b$ and $b = c$, then $a = c$.

In general, consider a relation R and a set S. Let a, b, and c represent arbitrary elements of S. We read "x R y" as "x is related to y." The relation R is said to be an **equivalence relation** if these three properties are satisfied:

> **Reflexive:** a R a for all elements a in S.
> **Symmetric:** If a R b, then b R a for all a and b in S.
> **Transitive:** If a R b and b R c, then a R c for all a, b, and c in S.

We illustrate these properties for various binary relations in the following examples. A **binary relation** is one that associates a single element with any two given elements of a set.

EXAMPLE 1

Test the relation "is perpendicular to" for lines in a plane to see
if this relation is (a) reflexive; (b) symmetric; (c) transitive.

SOLUTION (a) The relation is not reflexive since a line is not considered
to be perpendicular to itself.

(b) If line m is perpendicular to line n, then line n is perpendicular to line m. The relation is symmetric.

(c) In the figure below, we have $m \perp n$, $n \perp s$, but m is not
perpendicular to s. The relation is not transitive.

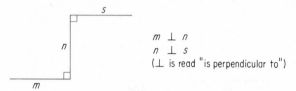

$$m \perp n$$
$$n \perp s$$
($\perp$ is read "is perpendicular to")

EXAMPLE 2

Test the relation "weighs within five pounds of" to determine
whether or not it is an equivalence relation.

SOLUTION The relation is reflexive, since a person weighs within five pounds
of his own weight. Furthermore if A weighs within five pounds
of B, then B weighs within five pounds of A; the relation is symmetric. However, one counterexample is sufficient to show that the
relation is not transitive. Thus assume the weights of A, B, and
C are 145, 149, and 153, respectively. Then A weighs within five
pounds of B, B weighs within five pounds of C, but A does not
weigh within five pounds of C. The relation is therefore not an
equivalence relation.

Relations are widely used for many types of elements. For example,
"being of the same age" is an equivalence relation among people. Consider
three students Don, John, and Bill. Note that even though you do not know
their ages, you do know that Don is the same age that he is. If Don and
John are the same age, then John and Don are the same age. Also, if Don
and John are the same age and John and Bill are the same age, then Don
and Bill are the same age.

The relation of "being younger than" is not an equivalence relation
since, for example, Don is not younger than himself (not reflexive). Also
if Don is younger than Sue, then Sue is not younger than Don (not symmetric). However, if Don is younger than Sue and Sue is younger than

Debbie, then Don is younger than Debbie. The relation "is younger than" is transitive.

Any relation may be considered as a subset of a Cartesian product. Suppose that Don, John, and Bill are each 19 years old, Sue is 20, and Debbie is 21. In the array the dots indicate the elements of $S \times S$, where $S = \{$Don, John, Bill, Sue, Debbie$\}$. For each element (a, b) of $S \times S$ the dot in the first array is circled if a is the same age as b.

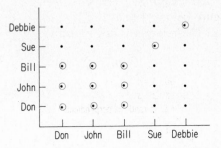

In the next array the dot for (a, b) is circled if a is younger than b.

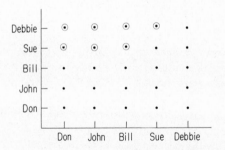

Other arrays could be made for other relations, such as "is at least as old as," "is not the same age as," "is older than," "is taller than," "is a brother or sister of," "has the same color eyes as," and so forth.

For numbers, the **order relations** *is less than* ($<$) and *is greater than* ($>$) have special importance because of their numerous interpretations, such as *is younger than* and *is older than.* You may think of one number such as 2 as less than 5 because the graph of 2 is on the left of the graph of 5 on a number line, such as the following:

$$\mspace{1 \quad 2 \quad 3 \quad 4 \quad 5 \quad 6 \quad 7 \quad 8}$$

Formally, 2 is less than 5 because there is a counting number 3 that can be added to 2 to obtain 5, that is,

$2 < 5$ because $2 + 3 = 5$

We write $5 > 2$, 5 is greater than 2, if and only if $2 < 5$. The symbols $<$ and $>$ are called **symbols of inequality**.

EXAMPLE 3

Show the relation $<$ on the Cartesian product $A \times A$, where $A = \{1, 2, 3, 4, 5\}$.

SOLUTION

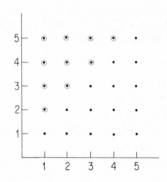

The equivalence and order relations among counting numbers are themselves related by the **trichotomy law**; for any counting numbers a and b, exactly one of these relations must hold:

$$a < b, \qquad a = b, \qquad a > b$$

Observe this property in Exercises 13 and 14 in the following set.

exercises

1. Let $A = \{1\}$, $B = \{1, 2\}$, and $C = \{1, 2, 3\}$. Consider the statements

   (a) $A \, R \, A$ **(b)** $A \, R \, B$ **(c)** $A \, R \, C$
   (d) $B \, R \, A$ **(e)** $B \, R \, B$ **(f)** $B \, R \, C$
   (g) $C \, R \, A$ **(h)** $C \, R \, B$ **(i)** $C \, R \, C$

   Which of these statements is true for the relation $\subseteq$, is a subset of?
2. As in Exercise 1, which of the statements is true for the relation $\subset$, is a proper subset of?
3. As in Exercise 1, which of the statements is true for $\not\subset$, is not a proper subset of?

   For Exercises 4 through 12, tell whether each relation is **(a)** *reflexive,* **(b)** *symmetric,* **(c)** *transitive:*

4. For people, is a daughter of.
5. For students, is a classmate of.

6. For numbers, is less than, $<$.

7. For numbers, is less than or equal to, $\leq$.

8. For books, is heavier than.

9. For meals, is less expensive than.

10. For lines in a plane, is parallel to, $\|$.

11. For numbers, is divisible by.

12. For sets, is a subset of, $\subseteq$.

13. Consider the three relations $<$, $=$, $>$ and tell which relations can be used in place of R in the statement:

    (a) 5 R 4

    (b) 25 R 42

    (c) $(10 + 3)$ R $(3 + 10)$

    (d) $[2 + (3 \times 5)]$ R $[8 + (6 + 2)]$

    (e) $[(5 \times 2) + 1]$ R $[(5 \times 2) + (5 \times 1)]$

14. In Exercise 13 exactly one relation could be used in each case. State which of the three relations hold for:

    (a) $5 + (7 \times 6)$ and $6 + (7 \times 5)$

    (b) b and $b + 1$

    (c) $b + d$ and $a + d$ if $b < a$

    (d) $b \times d$ and $a \times d$ if $b > a$

15. Explain each statement:

    (a) Either $a = b$ or $a \neq b$.

    (b) If $a \neq b$, then either $a < b$ or $a > b$.

EXPLORATIONS

The Secondary School Mathematics Curriculum Improvement Study is a federally financed group that is involved with writing experimental materials for gifted mathematics students. In their materials, they use arrow diagrams to show relations in the following way:

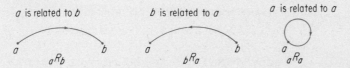

If a relation is reflexive, there is a loop at every element of the given set under discussion. If a relation is symmetric, then all arrows that go from an element a to an element b must also have an arrow that goes from b to a. A relation is transitive if whenever there is an arrow from *a* to *b*, and one from *b* to *c*, then there is also an arrow from *a* to *c*.

1. Here is a graph for the relation "divides" for the set $M = \{2, 3, 6, 12\}$. From the graph, show that the relation is reflexive and transitive but not symmetric.

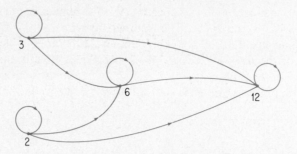

Draw arrow graphs for each of the following relations on the sets given:

2. Is greater than for $A = \{2, 5, 6, 9, 10\}$.

3. Is greater than or equal to for $B = \{3, 8, 9, 12\}$.

4. Is within three pounds of for the set of weights $C = \{130, 132, 134\}$.

6-5
The Set of Whole Numbers

The counting numbers serve man well and are actually all that he needs for many purposes. However, with just this collection of numbers at his disposal he is unable to ask for *half* of a piece of pie, or to say that he has *no* money in his pocket. We need to expand our number system and shall do so by first introducing a numeral 0 to represent the number of elements in the empty set. Zero is then defined as the cardinal number of any empty set.

We also need the number zero to solve sentences such as $5 + n = 5$. If $5 + n = 5$, then $n = 0$ and $5 + 0 = 5$.

Identity, $+$: $a + 0 = a = 0 + a$; the number zero is the **additive identity** and is also called the **identity element for addition**. This property is often referred to as the **addition property of zero**.

When we include 0 with the set of counting numbers, we form a new set W of **whole numbers:**

$W = \{0, 1, 2, 3, 4, 5, \ldots\}$

Note that the only difference between the set of counting numbers and the set of whole numbers is the element 0. Often we say that the set of whole numbers consists of the union of the set of counting numbers and zero.

We can locate this set of numbers on a number line and identify points on the line by numbers. The numbers are the **coordinates** of the points; the points are the **graphs** of the numbers. First draw a line and select any point of the line as the **origin** with coordinate 0, and any other point as the unit point with coordinate 1. Usually the number line is considered in a horizontal position with the unit point on the right of the origin.

The length of the line segment with the origin and the unit point as endpoints is the **unit distance** for marking off a scale on the line. The point representing any given counting number may be obtained by marking off successive units to the right of the origin.

We can now **graph** different sets of numbers on a number line as in the following examples. We shall use the verb "graph" to mean "draw the graph of."

EXAMPLE 1

Graph the set of whole numbers less than 3 on a number line.

SOLUTION   Draw a number line and place a solid dot at the points that correspond to 0, 1, and 2.

EXAMPLE 2

Graph the set of whole numbers between 1 and 5 on a number line.

SOLUTION   The set to be graphed is $\{2, 3, 4\}$.

The set of whole numbers has all of the properties that were described for the set of counting numbers, as well as having an identity element for addition. In summary these properties are

Closure, +   Closure, ×
Commutative, +   Commutative, ×
Associative, +   Associative, ×
Identity, +   Identity, ×
Distributive, × with respect to +

The distinguishing feature between the set of counting numbers and the set of whole numbers is the presence of 0 as a whole number. The number 0 not only serves as the additive identity, but also has the property:

Zero, ×: $a \times 0 = 0 \times a = 0$; the product of any whole number a and 0 is 0. This property is frequently referred to as the **multiplication property of zero**.

exercises

In Exercises 1 through 10 graph each set of numbers on a number line.

1. $\{1, 2, 3\}$ 2. $\{1, 5, 7\}$
3. $\{0, 2, 4, 6\}$ 4. $\{0\}$
5. The set of counting numbers less than 5.
6. The set of whole numbers less than 5.
7. The set of whole numbers between 0 and 5.
8. The set of counting numbers between 1 and 3.
9. The set of counting numbers between 0 and 1.
10. The set of whole numbers greater than or equal to 3 and less than 7.
11. Show that there is a one-to-one correspondence between the elements of the set of whole numbers less than 5 and the elements of the set of counting numbers less than 6.
12. If a and b represent whole numbers and $a \times b = 0$, what can you conclude about a and b?
13. Repeat Exercise 12 if $a \times b = 1$; if $a \times b = 2$; if $a \times b = 3$; if $a \times b = 4$.
14. Does every point on the number line have a whole number as its coordinate? Is every whole number the coordinate of some point on the number line?
15. Let n represent any whole number. How many whole numbers are there between n and $n + 1$?

16. If n represents any whole number, then $n + 1$ is called the **successor** of n. Does every whole number have a successor? Is every whole number the successor of some other whole number?

17. Show that there is a one-to-one correspondence between the elements of the set of whole numbers and the elements of the set of counting numbers.

18. Consider $A \times \emptyset$ and $\emptyset \times A$ for any set A. Identify these sets and the associated property of whole numbers.

EXPLORATIONS

1. Many interesting games and puzzles involve counting numbers only. One interesting puzzle is to represent the set of counting numbers using the digits of a year *in the order in which they appear*. Here are several examples, using the year 1975:

$$1 = 1^{975}, \qquad 3 = 1 + [(\sqrt{9} + 7) \div 5]$$

How many others can you find?

Cross-number puzzles are very popular with many people and serve as a novel and interesting way to provide practice with fundamental skills. The first figure below shows such a puzzle for addition; the object is to find the sum of each row, column, and diagonal. In the second figure the sums are shown for each row and column. These partial sums, $15 + 15$ and $17 + 13$, are then added across and down to give the sum 30 shown in the box in the lower right-hand corner. This serves as a check in that both sums should be the same. The third figure shows a final check made by adding along the diagonals, $7 + 9$ and $8 + 6$. These diagonal sums are placed in the circles and their sum, $14 + 16$, should give the number found in the box, 30.

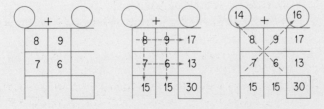

2. Prepare a similar cross-number puzzle for multiplication.

3. Prepare cross-number puzzles for subtraction and division. In these cases the diagonal check does not seem to work. See if you can suggest a modification of the procedures used for addition and multiplication so that the diagonals may be used as checks.

4. Complete the following puzzles that have missing entries.

(a)

	+	
	9	15
8		
		29

(b)

	×	
	3		12
	2		
		20	

6-6

The Set of Integers

Once again we extend our number system. With just the set of whole numbers at our disposal, we are still unable to find replacements for n that make these sentences true:

$$5 + n = 2$$
$$n + 4 = 0$$
$$2 - 7 = n$$

That is, the solution set for each of these sentences is the empty set if only the set of whole numbers may be used as the set of possible replacements for the **variable** *n*.

We begin our extension of the number system by considering points on the number line to the left of the origin 0. For each point on the number line that is the coordinate of a whole number n, locate a point to the left of the origin that is the same distance from 0 as the point with coordinate n. We call the coordinate of this new point negative *n*, written $-n$, and refer to the number that $-n$ represents as the **opposite** of n.

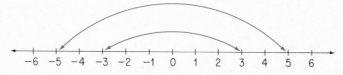

In the figure the point with coordinate negative 3 is located three units to the left of 0. We say that the opposite of 3 is -3; also, the opposite of -3 is 3. Similarly, the point with coordinate -5 is located five units to the left of 0, and we say that 5 and −5 are opposites of each other. Finally we agree that the opposite of 0 is 0; that is, −0 = 0.

We call the set of whole numbers together with the set of the opposites of the whole numbers, the set I of **integers**:

$$I = \{\ldots, -3, -2, -1, 0, 1, 2, 3, \ldots\}$$

Note that the graph of the set of integers continues indefinitely both in the positive direction (to the right of 0) and in the negative direction (to the left of 0).

The set of integers may be considered as the union of these sets of numbers:

Positive integers: $\{1, 2, 3, 4, 5, \ldots\}$
Zero: $\{0\}$
Negative integers: $\{\ldots, -5, -4, -3, -2, -1\}$

EXAMPLE 1

Graph the set of integers between −3 and 2.

SOLUTION

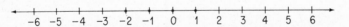

Inverse, +: $n + (-n) = 0$; every integer n has an opposite $-n$ such that $n + (-n) = 0$. This property is sometimes referred to as the **addition property of opposites**.

Each of the integers n and $-n$ is said to be the **additive inverse** or opposite of the other. That is, the additive inverse of 3 is −3, the additive inverse of −3 is 3, and $3 + (-3) = 0$. Indeed, the distinguishing feature between the set of integers and the set of whole numbers is that every integer has an additive inverse *in the set*.

We are now able to solve the sentences such as those given at the beginning of this section.

Sentence	Solution
$5 + n = 2$	$n = -3$
$n + 4 = 0$	$n = -4$
$2 - 7 = n$	$n = -5$

The extension of our number system to include the set of negative integers also makes subtraction of integers always possible. We define the subtraction $a - b$ of two integers, a and b, as follows:

$$a - b = a + (-b)$$

For example, $2 - 7 = 2 + (-7) = -5$. We see that subtraction of integers is always possible and say that the set of integers is closed under subtraction.

Consider, in summary, the following properties of the set of integers under addition:

Closure, +: The sum of any two integers is an integer.

Associative, +: $a + (b + c) = (a + b) + c$ for all integers a, b, and c.

Identity, +: The set of integers contains an element 0 such that $a + 0 = 0 + a = a$ for every integer a. This property is sometimes referred to as the **addition property of zero.**

Inverse, +: For each integer a, the set of integers contains an element $-a$ such that $a + (-a) = 0$.

Commutative, +: $a + b = b + a$ for all integers a and b.

We indicate that the set of integers has the first four properties by saying that the set of integers forms a **group under addition.** Furthermore, since addition of integers is also commutative, we say that the set of integers forms a **commutative group** with respect to addition. (See §5-1.)

EXAMPLE 2

Show that subtraction of integers is not commutative.

SOLUTION We can show this by a single counterexample, such as $5 - 2 = 3$ and $2 - 5 = -3$; $5 - 2 \neq 2 - 5$.

EXAMPLE 3

Show that the set of integers does *not* form a group with respect to multiplication.

SOLUTION Multiplication of integers is closed and associative, and there exists an identity element 1 since $n \times 1 = 1 \times n = n$ for each integer n. However, the set of integers does not contain the multiplicative inverse for each of its elements. For example, there is no integer n such that $3 \times n = 1$.

In general, we have the following properties of the set of integers under multiplication:

Closure, ×: The product of any two integers is an integer.

Associative, ×: $a \times (b \times c) = (a \times b) \times c$ for all integers a, b, and c.

Identity, ×: The set of integers contains an element 1 such that $a \times 1 = 1 \times a = a$ for every integer a. This property is sometimes referred to as the **multiplication property of one.**

Commutative, ×: $a \times b = b \times a$ for all integers a and b.

There are a number of interesting proofs that we are now ready to explore which illustrate the type of reasoning that mathematicians use. First we need several definitions; then we shall consider several proofs and offer others in the exercises.

An integer is *even* if it is a multiple of 2, that is, if it may be expressed as $2k$, where k stands for an integer. Then the set of **even integers** is

$$\{\ldots, -6, -4, -2, 0, 2, 4, 6, \ldots\}$$

An integer that is not even is said to be *odd*. Each odd integer may be expressed in the form $2k + 1$, where k stands for an integer. Then the set of **odd integers** is

$$\{\ldots, -7, -5, -3, -1, 1, 3, 5, 7, \ldots\}$$

EXAMPLE 4

Prove that the sum of any two even integers is an even integer.

SOLUTION Any two even integers m and n may be expressed as $2k$ and $2r$, where k and r stand for integers. Then

$$m + n = 2k + 2r = 2(k + r)$$

where $k + r$ stands for an integer, since the sum of any two integers is an integer. Therefore, $m + n$ is an even integer.

EXAMPLE 5

Prove that the square of any even integer is an even integer.

SOLUTION Any even integer may be expressed as $2k$, where k stands for an integer. Then the square of the integer may be expressed as $(2k)^2$, where

$$(2k)^2 = (2k)(2k) = 2(2k^2).$$

Since k, k^2, and $2k^2$ all stand for integers, $(2k)^2$ stands for an even integer.

exercises

1. Show that there is a one-to-one correspondence between the elements of the set of positive integers and the elements of the set of negative integers.

2. Show that the set of whole numbers does not form a group under addition.

Graph each set of numbers on a number line.

3. The set of integers between -5 and 3.

4. The set of integers -5 through 3 inclusive.

5. The set of integers that are the opposites of the first five counting numbers.

6. The set of integers that are the opposites of the members of the set $M = \{-3, -2, -1\}$.

Classify each of the statements in Exercises 7 through 15 as true or false.

7. Every counting number is an integer.

8. Every whole number is an integer.

9. Every integer is a whole number.

10. Every integer is either positive or negative.

11. Every integer is the opposite of some integer.

12. The set of integers is the same as the set of the opposites of the integers.

13. The set of negative integers is the same as the set of the opposites of the whole numbers.

14. The set of integers is closed under multiplication.

15. The set of integers is closed under division.

16. What is the intersection of the set of positive integers and the set of negative integers?

17. Is the union of the set of positive integers and the set of negative integers equal to the set of integers? Explain your answer.

18. Show that there is a one-to-one correspondence between the elements of the set of whole numbers and the elements of the set of integers.

19. Show that there is a one-to-one correspondence between the set of integers and the set of counting numbers.

Prove each statement.

20. The sum of any two odd integers is an even integer.

21. The product of any two even integers is an even integer.

22. The square of any odd integer is an odd integer.

23. If the square of an integer is odd, the integer is odd; if the square of an integer is even, the integer is even.

EXPLORATIONS

A nomograph is a device for performing computations in a simple manner. The following nomograph can be used to find the sums of integers. Just connect the point representing one addend on the *A* scale, with the corresponding point for the other addend on the *B* scale. The point where the line crosses the *S* scale will give the sum. The figure shows the sum $(+4) + (-6) = -2$.

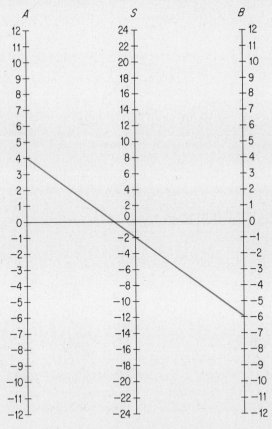

1. Construct your own nomograph, using graph paper, and use it to find various sums of integers.

2. See if you can use your knowledge of elementary geometry to explain why the nomograph shown works as it does.

3. Try to determine how the nomograph shown can be used for subtraction of integers.

4. Prepare a nomograph that will enable you to find $a + 2b$ directly for any two integers a and b.

6-7
The Set of Rational Numbers

With just the set of integers at our disposal, we are still unable to find replacements for n that make these sentences true:

$$2 \times n = 7$$
$$n + \tfrac{1}{4} = \tfrac{7}{8}$$
$$5 \div 3 = n$$

That is, the solution set for each of the preceding sentences is the empty set if only integers may be used as possible replacements for n. To make the solution of such sentences possible, as well as to make division by a counting number possible, we must extend our number system. We start by considering the number line and locating all the points that correspond to "halves"—that is, $\tfrac{1}{2}$, $1\tfrac{1}{2}$, $2\tfrac{1}{2}$, $3\tfrac{1}{2}$, etc., as well as $-\tfrac{1}{2}$, $-1\tfrac{1}{2}$, $-2\tfrac{1}{2}$, $-3\tfrac{1}{2}$, etc.

$$-5 \quad -4 \quad -3 \quad -2 \quad -1 \quad 0 \quad 1 \quad 2 \quad 3 \quad 4 \quad 5$$

Next we locate points on the number line that correspond to the multiplies of $\tfrac{1}{3}$, $\tfrac{1}{4}$, $\tfrac{1}{5}$, etc., both positive and negative. Each of these numbers may be expressed in the form $\dfrac{a}{b}$, that is, a/b, where a is an integer and b is a counting number. We now define any number that can be written in the form a/b, where a is an integer and b is a counting number, to be a **rational number**. The symbol a/b is often called a **common fraction**.

Every integer n is a rational number, since it can be written as $n/1$, the quotient of an integer and the counting number 1. Note that $n/0$ is *not* a rational number in that division by 0 is undefined. (Furthermore, the denominator 0 is not a counting number.)

The set of rational numbers can be classified as being positive (greater than 0), negative (less than 0), or 0. The number 0 is a rational number, but it is neither positive nor negative. As in the case of integers, positive rational numbers are the coordinates of points on the right of the origin and negative rational numbers are coordinates of points on the left of the origin.

Our number line has now become dense with points and "resembles" a complete line. The word "resembles" is used inasmuch as we shall find later that there are still "gaps" in the number line—that is, there are points on the number line that do not have rational numbers as their coordinates. However, the set of rational numbers is said to be **dense**, since between any two elements of the set, there is always another element of the set.

EXAMPLE 1

Locate a rational number between $\frac{17}{19}$ and $\frac{18}{19}$.

SOLUTION

Change each rational number to one with a denominator of 38.

$$\frac{17}{19} \times \frac{2}{2} = \frac{34}{38} \qquad \frac{18}{19} \times \frac{2}{2} = \frac{36}{38}$$

Clearly, $\frac{35}{38}$ lies between the two given rational numbers. Note also that

$$\frac{35}{38} = \frac{1}{2}\left(\frac{17}{19} + \frac{18}{19}\right)$$

EXAMPLE 2

Locate a rational number between $\frac{17}{19}$ and $\frac{35}{38}$.

SOLUTION

Write each rational number as one with a denominator of 76.

$$\frac{17}{19} \times \frac{4}{4} = \frac{68}{76} \qquad \frac{35}{38} \times \frac{2}{2} = \frac{70}{76}$$

Note that $\frac{69}{76}$ lies between the two given rational numbers. Also,

$$\frac{69}{76} = \frac{1}{2}\left(\frac{17}{19} + \frac{35}{38}\right)$$

We can summarize the results of the two preceding examples in this manner:

$$\frac{17}{19} < \frac{35}{38} < \frac{18}{19}$$

$$\frac{17}{19} < \frac{69}{76} < \frac{35}{38}$$

This process can be extended indefinitely. For example, to locate a rational number between $\frac{17}{19}$ and $\frac{69}{76}$, write each with a denominator of 152. Then find one-half the sum of the two numbers. This ability to locate a rational number between any two given rational numbers is what is meant by the density property.

Each of the sets of numbers that we have developed so far in this chapter has the previously considered sets of numbers as subsets. Thus we expect the set of rational numbers to have many of the same properties as the set of integers.

The set of rational numbers forms a commutative group under addition. That is, with respect to addition, the set of rational numbers is closed, associative, commutative, contains an identity element (0), and contains the inverse of each of its elements. (Recall that the inverse of 2 is -2, of -3 is 3, and of $\frac{2}{3}$ is $-\frac{2}{3}$.)

By expanding our number system from the integers to the set of rational numbers, we have included a **multiplicative inverse** or **reciprocal** for each number different from zero. The multiplicative inverse of 2 is $\frac{1}{2}$, of $\frac{2}{3}$ is $\frac{3}{2}$, and of $-\frac{3}{4}$ is $-\frac{4}{3}$. In each case the product of the number and its multiplicative inverse is 1, the identity element for multiplication:

$$2 \times \frac{1}{2} = 1; \qquad \frac{2}{3} \times \frac{3}{2} = 1; \qquad \left(-\frac{3}{4}\right) \times \left(-\frac{4}{3}\right) = 1$$

Inverse, ×: $\dfrac{a}{b} \times \dfrac{b}{a} = 1$; every rational number $\dfrac{a}{b}$, $\dfrac{a}{b} \neq 0$, has a multiplicative inverse $\dfrac{b}{a}$ such that $\dfrac{a}{b} \times \dfrac{b}{a} = 1$. (We need to make the restriction that $\dfrac{a}{b}$ is not equal to 0 because the reciprocal of 0 is undefined.)

We may summarize the properties of the set of rational numbers under the usual operations of addition and multiplication.

Closure, +: The sum of any two rational numbers is a rational number.

Closure, ×: The product of any two rational numbers is a rational number.

Commutative, +: $a + b = b + a$ for all rational numbers a and b.

Commutative, ×: $a \times b = b \times a$ for all rational numbers a and b.

Associative, +: $a + (b + c) = (a + b) + c$ for all rational numbers, a, b, and c.

Associative, ×: $a \times (b \times c) = (a \times b) \times c$ for all rational numbers a, b, and c.

Identity, +: The set of rational numbers contains an element 0 such that $a + 0 = 0 + a = a$ for every rational number a.

Identity, ×: The set of rational numbers contains an element 1 such that $a \times 1 = 1 \times a = a$ for every rational number a.

Inverse, +: For every rational number a, there is another rational number, $-a$, such that $a + (-a) = 0$.

Inverse, ×: For every rational number a, $a \neq 0$, there is another rational number $\frac{1}{a}$ such that $a \times \frac{1}{a} = 1$.

Distributive: For all rational numbers a, b, and c we have $a \times (b + c) = (a \times b) + (a \times c)$.

By studying these properties we note that the set of rational numbers forms a group under addition. However, this set does not form a group with respect to multiplication since zero does not have a multiplicative inverse in the set.

We describe most of the properties of the set of rational numbers by saying that the set of rational numbers forms a field. In general, any set of numbers forms a **field** if:

The set is closed under addition and multiplication.
The associative law holds for addition and multiplication.
The commutative law holds for addition and multiplication.
The identity elements for addition (0) and multiplication (1) are members of the set.
Each element of the set has its inverse under addition in the set, and each element except 0 has its inverse under multiplication in the set.
The elements of the set satisfy the distributive law for multiplication with respect to addition.

In addition to the field properties, we have found that the set of rational numbers has the density property. Between any two rational numbers there is always another rational number. Despite this fact, the rational numbers do not "fill" the number line. Every rational number is the coordinate of some point on the number line, but not every point on the number line can be named by a rational number. In the next section, we shall extend our number system so as to establish a one-to-one correspondence between the points of the number line and a new set of numbers, thereby completing the number line.

exercises

Classify each statement as true or false:

1. Every integer is a rational number.
2. Every rational number is an integer.
3. The reciprocal of every rational number except 0 is a rational number.
4. The number 0 is not a rational number.
5. The multiplicative inverse of a positive rational number is negative.

Give an example illustrating each statement:

6. The sum of two fractions may be an integer.
7. The product of two fractions may be an integer.
8. The quotient of two fractions may be an integer.
9. The difference of two fractions may be an integer.

Copy the following table. Use "√" to show that the number listed at the top of a column is a member of the set listed at the side. Use "×" if the number is not a member of the set.

	10.	**11.**	**12.**	**13.**	**14.**
Set	2	−5	0	$\frac{2}{3}$	$-\frac{3}{4}$
Counting numbers					
Whole numbers					
Integers					
Rational numbers					

15. Name two properties of the set of rational numbers that are not properties of the set of integers.
16. Find three rational numbers between $\frac{7}{9}$ and $\frac{8}{9}$.
17. Find three rational numbers between 0 and $\frac{1}{100}$.
18. Name a rational number between the rational numbers a/b and c/d.
19. Is there a next rational number after 0? Use a property of the set of rational numbers to explain your answer.

Find a replacement for n that will make each of the sentences true where n must be a member of the set named at the top of the column. If there is no such replacement, write "none."

	Sentence	Counting numbers	Whole numbers	Integers	Rational numbers
20.	$n - 3 = 0$				
21.	$n + 3 = 3$				
22.	$n + 3 = 0$				
23.	$3n = 5$				
24.	$n \times n = 2$				

Copy the following table. Use "√" to show that the set of elements named at the top of the columns has the property listed at the side. Use "×" if the set does not have the property.

	Property	Counting numbers	Whole numbers	Integers	Positive rationals	Rational numbers
25.	Closure, +					
26.	Associative, +					
27.	Identity, +					
28.	Inverse, +					
29.	Commutative, +					
30.	Commutative group, +					
31.	Closure, ×					
32.	Associative, ×					
33.	Identity, ×					
34.	Inverse, ×					
35.	Commutative, ×					
36.	Commutative group, ×					
37.	Distributive					

EXPLORATIONS

Here is a flow chart with directions for reducing a fraction.

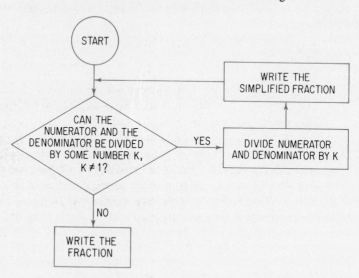

For example, if you start with $\frac{18}{24}$, you could divide numerator and denominator by 2 at first. Then you could divide by 3. Thus,

$$\frac{18}{24} \rightarrow \frac{9}{12} \rightarrow \frac{3}{4}$$

1.   Use the flow chart to reduce each fraction.
(a) $\frac{102}{186}$ **(b)** $\frac{84}{120}$ **(c)** $\frac{96}{144}$ **(d)** $\frac{120}{156}$
2.   The Greek philosopher Zeno proposed several paradoxes in the fifth century B.C. that were based on the density property. Prepare a report on these paradoxes, using a history of mathematics book as your reference.

6-8

The Set of Real Numbers

In mathematics we are concerned with the four fundamental operations $(+, -, \times, \div)$. The rational numbers have been introduced so that these operations are always possible with the exception of division by zero.

However, using just the set of rational numbers, we are still unable to find a replacement for n so as to make this sentence true:

$$n^2 = 2$$

That is, we are unable to find a rational number a/b such that

$$\frac{a}{b} \times \frac{a}{b} = 2$$

We call the positive number whose square is 2 the **square root** of 2, and express this number symbolically as $\sqrt{2}$. We wish to show that it is possible to locate a point on the number line whose coordinate is $\sqrt{2}$, and also that $\sqrt{2}$ does not name a rational number. To do so we will need to make use of the **Pythagorean theorem.** This theorem deals with the relationship among the lengths of the sides of a right triangle. Informally the theorem states that the sum of the squares of the two legs of a right triangle is equal to the square of the hypotenuse. In the figure the hypotenuse has a measure of c units and the two legs have measures of a and b units each.

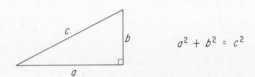

$$a^2 + b^2 = c^2$$

On the number line we have seen that the rational numbers are dense, since there is a rational number between any two given rational numbers. However, the appearance of a solid line of points with rational numbers as coordinates is misleading. For example, if we assume that there is a number that represents the length of a diagonal of a unit square, the square of that number must be 2. We call the number $\sqrt{2}$ and can locate a point with coordinate $\sqrt{2}$ on the number line. However, the number $\sqrt{2}$ cannot be a rational number.

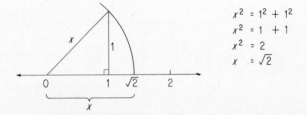

$$x^2 = 1^2 + 1^2$$
$$x^2 = 1 + 1$$
$$x^2 = 2$$
$$x = \sqrt{2}$$

We use an **indirect proof** to prove that $\sqrt{2}$ cannot be a rational number. That is, we assume that $\sqrt{2}$ is a rational number and show that this leads to a contradiction. Thus, suppose that a/b is a rational number, where a and b are not both even numbers and

$$\frac{a}{b} \times \frac{a}{b} = 2, \qquad \text{that is, } \frac{a}{b} = \sqrt{2}$$

Then $a^2 = 2b^2$ and a is an integer whose square is even. Therefore, as in §6-6, Exercise 23, a must be even; that is, $a = 2k$, where k stands for an integer. If we use $2k$ for a, we have

$$(2k)^2 = 2b^2$$
$$4k^2 = 2b^2$$
$$2k^2 = b^2$$

Thus b is an integer whose square is even, and b must be an even integer. This is contrary to the assumption that a and b were not both even. Our assumption that there exists a rational number whose square is 2 has led to a contradiction. In other words, if there is a number whose square is 2, that number cannot be a rational number.

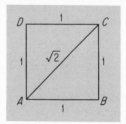

There exist line segments of length $\sqrt{2}$. For example, as we have observed, the diagonal $\overline{AC}$ of a square $ABCD$ with each side 1 unit long has length $\sqrt{2}$. We need numbers to represent lengths of line segments and thus need numbers that are not rational numbers. These numbers that are not rational numbers are called **irrational numbers**. We need irrational numbers such as $\sqrt{2}$ in order to have coordinates for each point on a line. We call the union of the set of rational numbers and the set of irrational numbers the set of **real numbers**.

There is a one-to-one correspondence between the elements of the set of real numbers and the set of points on the number line. Indeed, this is the distinguishing feature between the set of real numbers and the set of rational numbers. Thus every real number is the coordinate of a point on the number line, and every point on the number line is the graph of a real number. Accordingly, we refer to the number line as the real number line and we say that the number line is *complete*.

No further extension of the number system is necessary until one wishes to solve such equations as

$$n \times n = -1$$

We shall explore sentences of this type in §6-9.

Real numbers may be classified as being either rational or irrational by their decimal representations. Each real number may be represented by a decimal; each decimal represents a real number. Real numbers that are rational are represented by terminating decimals or by infinite repeating decimals, such as the following:

$$\frac{1}{4} = 0.25, \qquad \frac{3}{8} = 0.375, \qquad \frac{1}{3} = 0.33\overline{3}, \qquad \frac{3}{11} = 0.2727\overline{27}$$

Note the use of a bar to indicate that a sequence of digits repeats endlessly. The bar is used over the digit or digits that repeat. You may, if you wish, think of any terminating decimal as having repeated zeros and thus as being a repeating decimal. For example,

$$\frac{1}{4} = 0.25\overline{0}, \qquad \frac{3}{8} = 0.375\overline{0}$$

Real numbers that are irrational are represented by nonterminating, nonrepeating decimals, such as

$$\sqrt{2} = 1.414214\ldots, \qquad \pi = 3.1415926\ldots$$

In each of these examples the digits exhibit no repeating pattern no matter how far they are extended. Some irrational numbers can be represented in decimal notation by a sequence of digits that has a pattern but does not repeat any particular sequence of digits. For example, each of the following names an irrational number:

0.20220222022220222220 . . .
0.305300530005300005 . . .
0.404004000400004000004 . . .

EXAMPLE 1

How many zeros are there altogether between the decimal point and the one hundredth 5 in this sequence?

0.05005000500005 . . .

SOLUTION   There is one zero preceding the first 5, two zeros preceding the second 5, etc. The total number of zeros is the sum

$$1 + 2 + 3 + \cdots + 100$$

By the method of §1-1, the sum is 5050.

In summary, every real number can be represented by a decimal. If the real number is a rational number, then it may be represented by a terminating or by a repeating decimal. If the real number is an irrational number, then it may be represented by a nonterminating, nonrepeating decimal.

It is easy to show that every rational number in fractional form can be represented as a repeating (or terminating) decimal. Consider, for example, any rational number such as $\frac{12}{7}$. When we divide 12 by 7, the possible remainders are 0, 1, 2, 3, 4, 5, 6. If the remainder is 0, the division is exact; if any remainder occurs a second time, the terms after it will repeat also. Since there are only seven possible remainders when you divide by 7, the remainders must repeat or be exact by the seventh decimal place. Consider the determination of the decimal value of $\frac{12}{7}$ by long division:

$$
\begin{array}{r}
1.714285 \\
7\,\overline{)12.000000} \\
\underline{7} \\
⑤0 \\
\underline{49} \\
10 \\
\underline{7} \\
30 \\
\underline{28} \\
20 \\
\underline{14} \\
60 \\
\underline{56} \\
40 \\
\underline{35} \\
⑤
\end{array}
$$

The fact that the remainder 5 occurred again implies that the same steps will be used again in the long division process and the digits 714285 will be repeated over and over; that is, $\frac{12}{7} = 1.714285\overline{714285}$. Similarly, any rational number p/q can be expressed as a terminating or repeating decimal, and at most q decimal places will be needed to identify it.

We can also show that any terminating or repeating decimal can be written as a rational number in the form a/b.

If a decimal is terminating, you can write it as a fraction with a power of 10 as the denominator. For example, if $n = 0.75\overline{0}$, then $100n = 75$, and $n = \frac{75}{100}$, which reduces to $\frac{3}{4}$. If a fraction can be expressed as a terminating decimal, its denominator must be a factor of a power of 10.

If a decimal is repeating, it can be written as a rational number. For example, if a decimal n repeats one digit, we can find $10n - n$. Suppose $n = 3.244\overline{4}$; then $10n = 32.444\overline{4}$ and we have:

$$10n = 32.444\overline{4}$$
$$\underline{n = 3.244\overline{4}}$$
$$9n = 29.200\overline{0}$$

$$n = \frac{29.2}{9} = \frac{292}{90} = \frac{146}{45}$$

We can avoid the use of decimals in this way:

$$100n = 324.44\overline{4}$$
$$\underline{10n = 32.44\overline{4}}$$
$$90n = 292$$

$$n = \frac{292}{90} = \frac{146}{45}$$

If a decimal n repeats two digits, we find $10^2 n - n$; if it repeats three digits, we find $10^3 n - n$; and so forth.

EXAMPLE 2

Express $0.\overline{36}$ as a quotient of integers.

SOLUTION   Let $n = 0.\overline{36}$; then $100n = 36.\overline{36}$ and we have:

$$100n = 36.\overline{36}$$
$$\underline{n = 0.\overline{36}}$$
$$99n = 36$$

$$n = \frac{36}{99} = \frac{4}{11}$$

EXAMPLE 3

Express $0.783\overline{46346}$ as a quotient of integers.

SOLUTION

$$1000n = 783.46\overline{346346}$$
$$n = 0.78\overline{346346}$$
$$\overline{999n = 782.68}$$

$$n = \frac{782.68}{999} = \frac{78{,}268}{99{,}900} = \frac{19{,}567}{24{,}975}$$

We may summarize the discussion of this section in this way:

Every rational number can be named by a repeating decimal.
Every repeating decimal is the name of a rational number.

Note that for this summary we classify a terminating decimal, such as 0.25, as a repeating decimal. That is, $0.25 = 0.25\overline{0}$.

exercises

Classify each statement as true or false.

1. Every rational number is a real number.
2. Every real number is a rational number.
3. Every integer is a real number.
4. Every real number is either a rational number or an irrational number.
5. Every point on the number line has a real number as its coordinate.

Copy the following table. Use "$\sqrt{}$" to show that the number at the top is a member of the set listed at the side. Use "$\times$" if it is not a member of the set.

Set	6. 3	7. $-\frac{2}{3}$	8. $\sqrt{9}$	9. 0	10. $\sqrt{5}$
Counting numbers					
Whole numbers					
Integers					
Rational numbers					
Real numbers					

Tell whether each number should be represented by a terminating, a repeating, or a nonterminating, nonrepeating decimal.

11. $\frac{3}{4}$ **12.** $\frac{2}{5}$

13. $\frac{2}{3}$ **14.** $\frac{5}{12}$

15. $\frac{13}{16}$ **16.** $\sqrt{3}$

Write each number as a decimal.

17. $\frac{4}{5}$ **18.** $\frac{9}{20}$ **19.** $\frac{7}{8}$ **20.** $\frac{19}{200}$

21. $\frac{27}{15}$ **22.** $\frac{1}{6}$ **23.** $\frac{3}{7}$ **24.** $\frac{1}{13}$

25. $\frac{9}{13}$ **26.** $\frac{13}{7}$

Express each number as a quotient of integers.

27. $0.\overline{45}$ **28.** $0.\overline{234}$

29. $0.8\overline{1}$ **30.** $0.4\overline{14}$

31. $0.5\overline{22}$ **32.** $0.\overline{9}$

33. $2.11\overline{1}$ **34.** $5.22\overline{2}$

35. $2.14\overline{14}$ **36.** $4.25\overline{25}$

37. $0.123\overline{123}$ **38.** $65.268\overline{268}$

39. Write the first ten digits to the right of the decimal point in the decimal **(a)** $0.\overline{72}$; **(b)** $0.3\overline{571}$; **(c)** $0.8\overline{0}$; **(d)** $0.42\overline{37}$.

40. Write the digit in the fifteenth position to the right of the decimal point in the decimal **(a)** $0.52\overline{0}$; **(b)** $0.\overline{79}$; **(c)** $0.\overline{243}$; **(d)** $0.71\overline{68}$.

41. Arrange in order, from smallest to largest.

    $2.37, \quad $2.98, \quad $2.57, \quad $2.63, \quad $2.08

42. Write each decimal to two decimal places and arrange in order, from smallest to largest.

    4.37,   4.42,   4.4,   4.39,   4.51,   4.3

43. Write each decimal to at least three decimal places and arrange in order, from smallest to largest.

    1.45, $1.\overline{4}$, $1.\overline{5}$, $1.4\overline{5}$, $1.\overline{45}$

44. Write each decimal to at least six decimal places and arrange in order, from smallest to largest.

    0.234, $0.\overline{234}$, $0.2\overline{3}$, $0.2\overline{4}$,   0.24

Name a rational number that lies between each of these pairs of rational numbers.

*45. 0.234 and 0.235   ***46.** $0.\overline{234}$ and $0.\overline{235}$

*47. $0.23\overline{0}$ and $0.24\overline{0}$ ***48.** 0.234 and $0.\overline{234}$

*49. Consider the decimal $0.252252225\ldots$. This is an infinite decimal that has a pattern but that does not repeat a fixed set of digits, and thus does not represent a rational number. **(a)** How many 2's will there be immediately preceding the fifteenth 5? **(b)** How many 2's will there be altogether preceding the fifteenth 5?

*50. Name an infinite decimal for a number between 0.1 and 0.2 that is not a rational number.

EXPLORATIONS

1. Write the decimal representation for each of the rational numbers $\frac{1}{7}$, $\frac{2}{7}$, $\frac{3}{7}$, $\frac{4}{7}$, $\frac{5}{7}$, and $\frac{6}{7}$. See if you can find a pattern that describes the manner in which the digits in each representation are related.

2. Repeat Exploration 1 for the multiples of $\frac{1}{13}$ from $\frac{1}{13}$ to $\frac{12}{13}$.

3. Show that $0.99\overline{9} = 1$. Then try to prepare an explanation of this fact to satisfy a seventh or eighth grader. Consider these possibilities:

$$\frac{1}{3} = 0.33\overline{3}$$
$$+\frac{2}{3} = 0.66\overline{6}$$
$$\overline{1 = 0.99\overline{9}}$$

$$\frac{1}{3} = 0.33\overline{3}$$
$$3 \times \frac{1}{3} = 3 \times 0.33\overline{3}$$
$$\overline{1 = 0.99\overline{9}}$$

Now use the methods of this section to show that $0.99\overline{9} = 1$.

4. If the denominator of a rational number can be expressed as the product of powers of 2 and 5 only, then it can be named by a terminating decimal. **(a)** Show that this is so for several examples, such as $\frac{3}{50}$ and $\frac{19}{40}$. **(b)** Explain why

$$\frac{n}{2^q 5^p}$$

for any whole numbers n, p, and q can be represented by a terminating decimal.

5. Graph $2 + \sqrt{2}$ on the number line.

6. Prove that $2 + \sqrt{2}$ is an irrational number.

6-9
The Set of Complex Numbers

We began our study of the number system with the set of counting numbers, and have now progressed through the set of real numbers. At each step we discussed the various properties of the set of numbers under study, and then made further extensions to be able to solve certain types of problems. As a summary of these developmental stages, use the following chart to note the steps that we have gone through. In each case we wish to find a replacement for *n* so as to make the sentence true, using the set of numbers listed. In each case there is no possible solution, thus justifying as extension of the number system.

Equation	Set of numbers	Solution	Extensions
$n + 2 = 2$	Counting numbers	None	For whole numbers, we have $n = 0$.
$n + 3 = 2$	Whole numbers	None	For integers, we have $n = -1$.
$2n = 3$	Integers	None	For rational numbers, we have $n = 3/2$
$n^2 = 2$	Rational numbers	None	For real numbers, we have $n = \pm \sqrt{2}$
$n^2 = -2$	Real numbers	None	Another extension is needed!

Each extension of the number system has been made so that as many as possible of the properties of the previous set of numbers are retained as properties of the new set of numbers. Furthermore, in each case one or more new properties are acquired. For example, the extension of the set of counting numbers to the set of whole numbers was made to obtain 0, the identity for addition. The extension of the set of whole numbers to the set of integers was made to obtain the additive inverses of the counting numbers, that is, to make subtraction always possible. The extension of the set of integers to the set of rational numbers was made to obtain density and multiplicative inverses for each number different from zero, that is, to be able to divide by any number different from zero. The extension of the set of rational numbers to the set of real numbers was made to obtain completeness, the one-to-one correspondence of the set of real numbers with the set of points on a number line.

In each of these extensions it has been possible to retain all the basic properties of the previous sets of numbers. In particular, the set of rational numbers forms a field (§6-7) and the set of real numbers forms a field.

We now make one more extension of our number system in order to be able to solve equations such as $n^2 + 1 = 0$. For the set of real numbers, there is no replacement for n that will make this sentence true. We therefore invent a number i such that $i^2 + 1 = 0$; that is, $i^2 = -1$ and $i = \sqrt{-1}$. For the sentence $n^2 + 1 = 0$, $n^2 = -1$ and $n = \pm\sqrt{-1}$; that is, $n = \pm i$.

As further illustrations of the notation for the square roots of negative numbers, consider

$$\sqrt{-4} = \sqrt{4} \cdot \sqrt{-1} = 2i, \qquad \sqrt{-5} = \sqrt{5} \cdot \sqrt{-1} = \sqrt{5}\, i$$

A number that may be expressed in the form bi, where b is a real number different from zero and $i = \sqrt{-1}$, is called a **pure imaginary number**. The sum of a real number different from zero and a pure imaginary number, such as $2 + 5i$, is called a **mixed imaginary number**. The union of the set of real numbers and the set of imaginary numbers is called the set of **complex numbers**. In general, a complex number $a + bi$, where a and b are real numbers and $i = \sqrt{-1}$, is defined as follows:

A real number if $b = 0$
An imaginary number if $b \neq 0$
A pure imaginary number if $a = 0$ and $b \neq 0$
A mixed imaginary number if $a \neq 0$ and $b \neq 0$

EXAMPLE 1

Write $2 + \sqrt{-9}$ in the form $a + bi$.

SOLUTION $\sqrt{-9} = \sqrt{9} \cdot \sqrt{-1} = 3i$
$2 + \sqrt{-9} = 2 + 3i$

EXAMPLE 2

Write 5 in the form $a + bi$.

SOLUTION $5 + 0i$

EXAMPLE 3

Write $\sqrt{-3}$ in the form $a + bi$.

SOLUTION $\sqrt{-3} = \sqrt{3} \cdot \sqrt{-1} = \sqrt{3}\,i = 0 + \sqrt{3}\,i$

The importance of recognizing the set of possible replacements being used is shown by the solutions of the equations in the following array. The word "none" implies that for the set of numbers named at the left of the row there is no replacement for n that will make the sentence true.

Replacements	$n + 1 = 1$	$n + 1 = 0$	$2n = 1$	$n^2 - 1 = 0$	$n^2 + 9 = 0$
Counting numbers	none	none	none	1	none
Whole numbers	0	none	none	1	none
Integers	0	-1	none	$1, -1$	none
Rational numbers	0	-1	$\frac{1}{2}$	$1, -1$	none
Real numbers	0	-1	$\frac{1}{2}$	$1, -1$	none
Complex numbers	0	-1	$\frac{1}{2}$	$1, -1$	$3i, -3i$

The relationship of the set of real numbers to some of its subsets that we have studied in this chapter is shown in the following array.

exercises

Express each number in the form a + bi.

1. $2 + \sqrt{-9}$ **2.** $3 + \sqrt{-4}$

3. $\sqrt{9}$ **4.** 0

5. $5 + \sqrt{-3}$

Classify each statement as true or false.

6. Every rational number is a complex number.

7. Every irrational number is a complex number.

8. Every complex number is a rational number.

9. Every real number can be written in the form $a + bi$.

10. Every integer can be written in the form $a + bi$.

Copy the following table. Use "$\checkmark$" to show that the number at the top of the column is a member of the set named at the side. Use "$\times$" if the number is not a member of the set.

	11.	**12.**	**13.**	**14.**	**15.**
Set	-7	$\frac{3}{5}$	0	$\sqrt{5}$	$\sqrt{-9}$
Counting numbers					
Whole numbers					
Integers					
Rational numbers					
Irrational numbers					
Real numbers					
Imaginary numbers					
Complex numbers					

Find a replacement for n that will make each sentence true, if n must be a member of the set named at the top of the column. When there is no such replacement, write "none."

	Sentence	Counting numbers	Integers	Rational numbers	Real numbers	Complex numbers
16.	$n + 2 = 0$					
17.	$n + 5 = 2$					
18.	$3n - 1 = 0$					
19.	$n^2 = 4$					
20.	$n^2 + 4 = 0$					

		Counting numbers	Whole numbers	Integers	Rational numbers	Real numbers
21.	Closure, +					
22.	Closure, x					
23.	Commutative, +					
24.	Commutative, x					
25.	Associative, +					
26.	Associative, x					
27.	Identity, + (0)					
28.	Identity, x (1)					
29.	Inverse, +					
30.	Inverse, x					
31.	Distributive					
32.	Field					
33.	Density					
34.	Completeness					

EXPLORATIONS

Give an example illustrating each statement.

1. The sum of two imaginary numbers may be **(a)** an imaginary number; **(b)** a real number.

2. A difference of two imaginary numbers may be **(a)** an imaginary number; **(b)** a real number.

3. The product of two imaginary numbers may be **(a)** an imaginary number; **(b)** a real number.

4. A quotient of two imaginary numbers may be **(a)** an imaginary number; **(b)** a real number.

chapter test

1. Show a one-to-one correspondence between the set of whole numbers and the set of negative integers.

2. List the set of prime numbers that are less than 15.

3. Write the prime factorization of 300.

4. Find the greatest common factor of 120 and 140.

5. Find the least common multiple of 90 and 1500.

6. Classify each statement as true or false:

(a) Every counting number is a whole number.

(b) The opposite of every whole number is a negative integer.

(c) Every rational number may be expressed as the quotient of two integers.

(d) Every rational number may be expressed as a repeating or a terminating decimal.

(e) Between every two rational numbers there is only one other rational number.

7. Copy the following table. Use "$\checkmark$" to show that the number at the top of the column is a member of the set named at the side. Use "X" if the number is not a member of the set.

	(a) $\frac{3}{4}$	(b) -2	(c) 0	(d) $\sqrt{2}$	(e) $\sqrt{4}$
Set					
Counting numbers					
Whole numbers					
Integers					
Rational numbers					
Real numbers					

8. Express as a quotient of integers: **(a)** $0.\overline{27}$; **(b)** $0.6\overline{12}$.

9. Express as a repeating decimal.

   (a) $\frac{1}{2}$ **(b)** $\frac{1}{9}$

   (c) $\frac{7}{12}$ **(d)** $\frac{1}{13}$

10. Express in the form $a + bi$.

    (a) $3 + \sqrt{-16}$ **(b)** $\sqrt{7}$

    (c) $\sqrt{-4}$ **(d)** $2 + \sqrt{-3}$

7

An Introduction to
ALGEBRA

A symbol that may be replaced by any member of a set of elements is a variable. Variables such as □, n, x, and t are often used in arithmetic. However, the study of sentences involving variables is a part of algebra. In our study of algebra we first consider statements. Statements involving only numbers and operations on numbers can be identified as true or as false. Algebra differs from arithmetic in that statements in algebra also include *variables;* that is, symbols that may be replaced by any member of a set of elements. Statements that involve variables may be true for some replacements for the variables and false for others.

Within this chapter we shall also see the interrelationship of algebra and geometry as we explore graphs of sentences. First we study graphs on a line, and then graphs in a plane. For the latter we are indebted to the discoveries of the French mathematician René Descartes (1596–1650), who provided a geometric interpretation for algebraic sentences in two variables. This interpretation led to the branch of mathematics known as analytic geometry.

7-1

Sentences and Statements

Sentences that can be identified as true or identified as false are called **statements.** Each of the following is a statement.

(a) $7 + 3 = 3 + 7$ (a true statement of equality)
(b) $7 - 3 = 3 - 7$ (a false statement of equality)

Statements of equality, whether true or false, are called **equations.** We may also write **statements of inequality** using any one of these symbols.

$\neq$: is not equal to
$>$: is greater than
$<$: is less than

Each of the following is a statement of inequality.

(a) $7 + 2 > 7 - 2$ (a true statement of inequality)
(b) $5 + 3 \neq 3 + 5$ (a false statement of inequality)
(c) $8 - 2 < 8$ (a true statement of inequality)

We formally define the **order** of any two real numbers a and b, where a is less than b, as follows:

$a < b$ (and $b > a$) if and only if there is a positive number c such that $a + c = b$.

For example, $2 < 5$ (or $5 > 2$) and there is a positive number 3 such that $2 + 3 = 5$. This concept of order is consistent with the ordering of integers in §6-4 and the ordering of rational numbers in §6-7. Indeed, the trichotomy law, cited in §6-4 for counting numbers, holds for all real numbers: If a and b are any two real numbers, then exactly one of the relations.

$a < b, \qquad a = b, \qquad a > b$

must hold.

Many sentences cannot be classified as being true or as being false and are known as **open sentences.** Here are some examples of open sentences:

(a) $x + 2 = 5$ (b) $x - 2 \neq 3$
(c) $x + 1 > 4$ (d) $x - 1 < 3$

In the preceding examples the symbol x represents a number and is called a *variable*. A variable is a placeholder for its replacements. Any open sentence is neither true nor false until a replacement is made for the variable. After such a replacement is made, we have a statement, that is, a sentence that can be classified as either true or false.

EXAMPLE 1

Find a replacement for x that will make the sentence $x + 2 = 5$ a true statement of equality.

SOLUTION

When $x = 3$, we have the true statement $3 + 2 = 5$. When $x \neq 3$, we have a false statement of equality. We call 3 the *solution* of the given open sentence.

Given any open sentence, we need to determine the set of replacements that can be used for the variable. This set of permissible replacements is the **replacement set** of the variable. For example, consider this open sentence:

Christmas is in the month of _____.

The set of possible replacements of the variable (indicated by the blank) in this open sentence is the set of the 12 months. The statement is true when the replacement is December; but the statement is false for any other replacements.

Next let us consider the sentence $x - 1 < 3$ for various replacement sets. Note that for different replacement sets there may be different sets of replacements that make the sentence true. In each case the set of numbers that make the sentence true is called the **solution set** or the **truth set** of the given sentence for that replacement set. Recall that the elements of a set are generally listed within a pair of braces as shown in the array.

Sentence	Replacement Set	Solution Set
$x - 1 < 3$	Counting numbers	$\{1, 2, 3\}$
$x - 1 < 3$	Whole numbers	$\{0, 1, 2, 3\}$
$x - 1 < 3$	Integers	$\{\ldots, -2, -1, 0, 1, 2, 3\}$
$x - 1 < 3$	Real numbers	All real numbers less than 4

Note that the sentence $x - 1 > 3$ has the solution set $\{5, 6, 7, \ldots\}$ whether the replacement set is the set of counting numbers, whole numbers, or integers.

EXAMPLE 2

Find the solution set for $x + 1 > 4$ when x is a whole number.

SOLUTION   When $x = 3$, $3 + 1 = 4$. The solution set S consists of all whole numbers greater than 3; that is,

$$S = \{4, 5, 6, \ldots\}.$$

EXAMPLE 3

Find the solution set for $n + 2 = 2 + n$ when n is a real number.

SOLUTION   This sentence is a true statement of equality for *all* possible replacements of the variable, since it is an application of the commutative property of addition. The solution set is the set of real numbers.

An equation that is true for all possible replacements of the variable (as in Example 3) is called an **identity**. A statement of inequality, such as $n + 2 > n$, may also be classified as an identity.

EXAMPLE 4

For what values of x is the statement $x + 2 = x$ true for an integer x?

SOLUTION   Regardless of the integer selected as the replacement for x, the sentence $x + 2 = x$ is always false. The solution set is the empty set, $\varnothing$.

Statements of inequality may also have the empty set as their solution set. For example, there is no replacement for x that will make the sentence $x + 2 < x$ true; the solution set is the empty set.

exercises

For which of these sentences is the solution set the empty set? Which sentences are identities?

1. $x + 3 > x$ 2. $x + 2 \neq x$
3. $x + 2 \neq 2 + x$ 4. $x > x + 3$

5. $x < x + 1$ **6.** $x + 1 < x + 2$
7. $x + 2 < x + 1$ **8.** $x > x - 1$

Find the solution set for each sentence when the replacement set is the set of whole numbers.

9. $x + 2 = 5$ **10.** $x - 2 = 3$
11. $x + 1 < 4$ **12.** $x - 1 < 4$
13. $x - 2 < 3$ **14.** $x - 2 > 3$

Find the solution set for each sentence when the replacement set is the set of integers.

15. $x + 1 < 4$ **16.** $x - 1 < 4$
17. $x - 2 < 3$ **18.** $x - 2 > 1$

Describe the solution set for each sentence when the replacement set is the set of real numbers.

19. $x + 1 < 4$ **20.** $x + 1 > 4$
21. $x - 2 < 3$ **22.** $x - 2 > 3$
23. $x + 2 \neq 5$ **24.** $x + 1 \neq 1$

Classify each of these statements as **(a)** *a true statement of equality;* **(b)** *a false statement of equality;* **(c)** *a true statement of inequality; or* **(d)** *a false statement of inequality.*

25. $7 - 3 \neq 3 - 7$ **26.** $3 - 7 < 7 - 3$
27. $\frac{23}{25} > \frac{24}{25}$ **28.** $-\frac{23}{25} > -\frac{24}{25}$
29. $19 \times 21 = 20^2 - 1$ **30.** $13^2 = 5^2 + 12^2$
31. $\frac{432}{796} > \frac{432}{795}$ **32.** $-\frac{432}{796} < -\frac{432}{795}$

EXPLORATIONS

The use of two variables in a mathematical expression can be introduced on an intuitive basis through the use of flow charts. Consider, for example, the flow charts on page 248.

Copy and complete each table of values, using the flow charts.

1.

x	3	5	-2	3	-1
y	4	1	7	-2	-4
$2(x+y)$					

2.

a	2	3	-5	-9	-7
b	3	2	4	-1	-5
$2a+3b$					

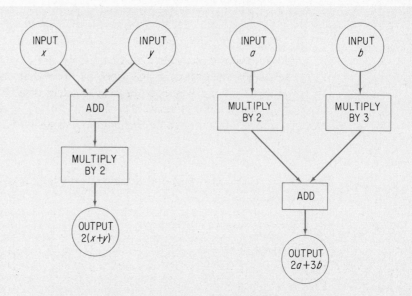

Draw a flow chart to show each of these expressions.

3. $3x + 5y$ **4.** $3(x + y)$

5. $\frac{1}{2}x + 3y$ **6.** $3(2x + 3y)$

7-2

Graphs of Sentences

The solution set of any open sentence in one variable may be graphed on a number line. That is, we may draw a graph to represent the set of points that correspond to the solution set of a sentence. We often refer to this graph simply as the *graph* of the equation or inequality.

The following examples illustrate various types of graphs. In each example we are to find and graph the solution set of the given open sentence for the given replacement set. The graph of the sentence is the graph of its solution set.

EXAMPLE 1

Graph the sentence $x + 3 = 5$, x an integer.

SOLUTION   The solution set consists of a single element, $\{2\}$. We graph this solution set on a number line by drawing a solid dot at 2. The graph consists of this single *point*.

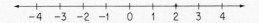

The equation in Example 1 can be thought of as a **set-selector**; it selects from the replacement set of x just those numbers that make the sentence true when used as replacements for x. The selected set is the solution set of the equation. The solution set $\{2\}$ is "the set of all x such that $x + 3 = 5$." We may designate this solution set in **set-builder notation** as

$$\{x \mid x + 3 = 5\}$$

The replacement set of the variable may be indicated in the set-builder notation as follows:

$$\{x \mid x + 3 = 5, x \text{ an integer}\}$$

We may also write

$$\{x \mid x + 3 = 5\} = \{2\}$$

That is, the set of all x such that $x + 3 = 5$ is the set consisting of the element 2.

EXAMPLE 2

Graph the sentence $x + 3 > 5$, x a real number.

SOLUTION
The solution set consists of all real numbers greater than 2. The graph of the solution set is drawn by first placing a hollow dot at 2 on the number line to indicate that this point is not a member of the solution set. We then draw a heavily shaded arrow to show that all numbers greater than 2 satisfy the given inequality.

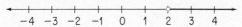

For any real number b the figure formed by the graph of an inequality of the form $x > b$ (see Example 2) is called a **half-line**. (The figure formed by the graph of an inequality of the form $x < b$ is also a half-line.)

EXAMPLE 3

Graph the sentence $x + 3 \geq 5$, x a real number.

SOLUTION
This sentence is read "$x + 3$ is greater than or equal to 5." Therefore, it is true when $x = 2$, and it is true when $x > 2$.

For any real number b the figure formed by the graph of an inequality of the form $x \geq b$ (see Example 3) is called a **ray**. (The figure formed by the graph of an inequality of the form $x \leq b$ is also a ray. The sentence $x \leq b$ is read "x is less than or equal to b."

EXAMPLE 4

Graph the sentence $x + 3 \not< 5$, x a real number.

SOLUTION   This sentence is read "$x + 3$ is not less than 5." This is equivalent to saying that $x + 3$ is greater than or equal to 5; $x + 3 \geq 5$. Thus the solution set is the same as for Example 3.

EXAMPLE 5

Graph the sentence $-1 \leq x \leq 3$, x a real number.

SOLUTION   This sentence is read "-1 is less than or equal to x, which is less than or equal to 3." In other words, "-1 is less than or equal to x and x is less than or equal to 3." That is, "x is greater than or equal to -1 and x is less than or equal to 3." Thus the solution set is the set of real numbers from -1 through 3.

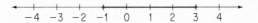

For any real numbers a and b where $a < b$, the graph of the sentence $a \leq x \leq b$ (see Example 5) is called a **line segment.**

EXAMPLE 6

Graph the sentence $-1 < x < 3$, x a real number.

SOLUTION   The solution set consists of all the real numbers *between* -1 and 3.

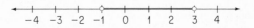

The points with coordinates -1 and 3 are the **endpoints** of the line segment graphed in Example 5. Notice that the graph in Example 6 can be obtained from the graph in Example 5 by removing the endpoints of the line segment. When we wish names for each of these graphs, a line segment with its endpoints is called a **closed line segment;** a line segment without either endpoint is an **open line segment;** a line segment with one endpoint but not both is neither closed nor open. For example, the following graph of $-1 \leq x < 3$ is neither closed nor open.

EXAMPLE 7

Graph the sentence $x + 2 = 2 + x$, x a real number.

SOLUTION   This sentence is true for all replacements of x; it is an identity. The solution set is the set of all real numbers. Thus the graph is the entire number line.

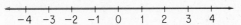

EXAMPLE 8

Graph the sentence $x + 3 \leq 5$, x a whole number.

SOLUTION   The solution set is $\{0, 1, 2\}$ and is graphed as a set of three points.

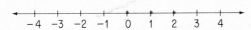

EXAMPLE 9

Graph the sentence $x + 3 > 5$, x a whole number.

SOLUTION   The solution set is $\{3, 4, 5, \ldots\}$, an infinite set of whole numbers. The sentence may be graphed as follows, where the heavily shaded arrowhead is used to indicate that the graph continues in the manner shown. The solution set consists of the set of all whole numbers greater than or equal to 3.

EXAMPLE 10

Graph the sentence $x + 2 = x$, x an integer.

SOLUTION   There is no integer such that the sum of that integer and 2 is equal to the original integer. The solution set is the empty set and has no graph.

We now summarize the concepts considered in this section. An open sentence in one variable divides the replacement set of the variable into two *subsets:* one subset consists of replacements that make the sentence true; the other subset consists of the replacements that make the sentence false. A replacement that makes the sentence true is a *solution* of the sentence. The set of all solutions is the *solution set* of the sentence; the *graph* of an equation or an inequality is the graph of its solution set. Consider again the solution set for the sentence $x - 1 < 3$ (see §7-1) for different replacement sets.

Replacement set	Solution set	Graph of solution set
Counting numbers	$\{1, 2, 3\}$	
Whole numbers	$\{0, 1, 2, 3\}$	
Integers	$\{\ldots, -2, -1, 0, 1, 2, 3\}$	
Real numbers	All real numbers less than 4	

exercises

Identify the graph of the solution set of each sentence as a point, a half-line, a ray, a line segment, or a line.

1. $x + 3 = 7$ **2.** $x - 2 > 7$

3. $x + 2 \geq 5$ **4.** $x < x + 3$

5. $-2 \leq x \leq 5$ **6.** $x + 3 \leq 5$

7. $x + 1 > x$ **8.** $x - 2 = 7$

9. $x - 1 < 5$ **10.** $-3 \leq x \leq 0$

Graph each set for whole numbers x.

11. $\{x \mid x + 2 \leq 5\}$ **12.** $\{x \mid x - 1 < 5\}$

13. $\{x \mid 2 \leq x \leq 7\}$ **14.** $\{x \mid 0 < x \leq 3\}$

15. $\{x \mid x + 2 < 3\}$ **16.** $\{x \mid x + 1 < 5\}$

Graph each sentence for real numbers x.

17. $x + 1 > 2$ **18.** $x - 2 \leq 3$

19. $-2 \leq x \leq 5$ **20.** $-3 < x < 3$

21. $x + 2 \geq 5$ **22.** $x - 1 < 4$

23. $x + 1 > 4$ **24.** $x + 3 \nless 4$

25. $x + 2 > x$ **26.** $x + 2 = x$

27. $x + 1 \not> 3$ 28. $x + 2 \neq 3$
29. $x - 2 < 1$ 30. $x - 1 < 2$

Graph each set of real numbers.

*31. $\{x \mid 3x - 2 \geq -5\}$ *32. $\{x \mid 2x + 3 \leq 1\}$

*33. $\{x \mid x^2 = 25\}$ *34. $\{x \mid x^2 + 3 = 19\}$

*35. $\{x \mid x^2 < 9\}$ *36. $\{x \mid x^2 \geq 16\}$

EXPLORATIONS

Occasionally the symbolism $[a, b]$ is used to represent a set of points on a line in the interval from a to b, with both endpoints a and b included in the set. For integral points, $[2, 6] = \{2, 3, 4, 5, 6\}$. For an interval of real numbers, $[2, 6]$ represents all of the real numbers from 2 through 6 inclusive.

We use the symbolism (a, b) to represent the points on the interval from a to b, exclusive of these two endpoints. A combination of these symbols may be used to indicate that one endpoint is included but not the other. Thus $[a, b)$ is used to represent the points in the interval from a to b, including a but not including b. The symbol $[a, b]'$ is used to represent the complement of the interval $[a, b]$.

List and then show on a number line the integral elements in each set.

1. $[2, 4] \cup [1, 5]$ 2. $[3, 7] \cup [1, 3]$
3. $[2, 5] \cap [3, 7]$ 4. $[3, 5] \cap (5, 8]$
5. $(2, 5] \cup [3, 7)$ 6. $[4, 9) \cap [8, 10)$

For each of the following, let $U = [-10, 10]$, an interval of real numbers. Draw the graph of each set and state the results, using the symbolism for intervals of real numbers.

7. $[-2, 3] \cup [2, 5]$ 8. $[-3, 5] \cap [2, 7]$
9. $(-1, 5] \cap [0, 8)$ 10. $[-5, 2) \cup [2, 4)$
11. $[-10, 3) \cup (5, 10]$ 12. $[-5, 2]'$
13. $\{[-7, 2] \cap [3, 5]\}'$ 14. $[-10, 10]'$
15. $\{[-2, 5] \cup [3, 8]\} \cap \{[-5, 7] \cap [-3, 3]\}$
16. $\{(3, 7] \cap [-1, 2)\} \cup \{[-1, 2] \cap [-2, 1]\}$

7-3

Sentences of the First Degree

So far we have been solving sentences by informal methods. For example, if we are given that

$$x - 3 = 7$$

we recognize that x may be obtained as $(x - 3) + 3$; that is,

$$(x - 3) + 3 = x + (-3 + 3) = x + 0 = x$$

Therefore, we may solve the given equation as follows:

$$x - 3 = 7$$
$$(x - 3) + 3 = 7 + 3$$
$$x = 10$$

We have added 3 to both members of the equation to obtain an **equivalent** sentence, that is, a sentence with the same solution set as the given sentence. Formally, we have used a property of equations. Here are some of the common properties of statements of equality involving real numbers a, b, c, and d:

> **Identity, =:** $a = a$.
> **Addition, =:** If $a = b$ and $c = d$, then $a + c = b + d$.
> **Multiplication, =:** If $a = b$ and $c = d$, then $ac = bd$.

The uses of these properties together with the field properties of Chapter 7 are shown in the examples that follow.

EXAMPLE 1

Solve and give a reason for each step in the solution of $x + 3 = 7$.

SOLUTION

Statements	Reasons
$x + 3 = 7$	Given.
$(x + 3) + (-3) = 7 + (-3)$	Addition, =.
$x + [3 + (-3)] = 7 + (-3)$	Associative, +.
$x + 0 = 4$	Addition.
$x = 4$	Zero, +. (Addition property of zero.)

EXAMPLE 2

Solve $2x - 3 = 7$, and explain each step.

SOLUTION

Statements	Reasons
$2x - 3 = 7$	Given.
$2x + (-3) = 7$	Definition, $-$.
$[2x + (-3)] + 3 = 7 + 3$	Addition, $=$.
$2x + [(-3) + 3] = 7 + 3$	Associative, $+$.
$2x + 0 = 10$	Addition.
$2x = 10$	Zero, $+$.
$\frac{1}{2}(2x) = \frac{1}{2} \times 10$	Multiplication, $=$.
$(\frac{1}{2} \times 2) \cdot x = \frac{1}{2} \times 10$	Associative, $\times$.
$1x = 5$	Multiplication.
$x = 5$	One, $\times$. (Multiplication property of one.)

Notice that subtraction is not needed when solving equations since
$$a - b = a + (-b)$$
Similarly, division is not needed since
$$a \div b = a \times \frac{1}{b}$$

The methods used in Examples 1 and 2 may be used to solve any sentence that is an equation of the **first degree in one variable** x, that is, any sentence that can be expressed in the form

$$ax + b = 0, \qquad a \neq 0$$

The expression $ax + b$ is called an **expression of the first degree** in x.

Sentences involving inequalities may also be solved by making use of certain basic properties. First we list some of the common properties of order relations among real numbers a, b, c, and d:

Addition, $<$:	If $a < b$ and $c = d$, then $a + c < b + d$.
Addition, $>$:	If $a > b$ and $c = d$, then $a + c > b + d$.
Multiplication, $<$:	If $a < b$, then $ac < bc$ if $c > 0$.
	if $a < b$, then $ac > bc$ if $c < 0$.
Multiplication, $>$:	If $a > b$, then $ac > bc$ if $c > 0$.
	If $a > b$, then $ac < bc$ if $c < 0$.

In general, if the same number is added to both members of a statement of inequality, the "sense" of the inequality is preserved. Thus in the following examples the first sentence listed is true, and the second one must also be true.

$$2 < 5, \qquad 2 + 3 < 5 + 3$$
$$8 > 3, \qquad 8 + 5 > 3 + 5$$
$$-3 < 5, \qquad -3 + 1 < 5 + 1$$

If both members of a true statement of inequality are multiplied by the same positive number, the resulting sentence is also true. If both members of an inequality are multiplied by a negative number, it is necessary to reverse the sense of the inequality to obtain an equivalent sentence. Thus consider these sets of equivalent sentences:

$$3 < 7, \qquad 2 \times 3 < 2 \times 7$$
$$5 > -1, \qquad 3 \times 5 > 3 \times -1$$
$$2 < 8, \qquad -3 \times 2 > -3 \times 8$$

The uses of these properties of order relations are shown in the examples that follow.

EXAMPLE 3

Solve $\frac{1}{2}x - 2 > 5$ and explain each step.

SOLUTION

Statements	Reasons
$\frac{1}{2}x - 2 > 5$	Given.
$\frac{1}{2}x + (-2) > 5$	Definition, $-$.
$[\frac{1}{2}x + (-2)] + 2 > 5 + 2$	Addition, $>$.
$\frac{1}{2}x + [(-2) + 2] > 5 + 2$	Associative, $+$.
$\frac{1}{2}x + 0 > 7$	Addition.
$\frac{1}{2}x > 7$	Zero, $+$.
$2(\frac{1}{2}x) > 2 \times 7$	Multiplication, $>$.
$(2 \times \frac{1}{2})x > 2 \times 7$	Associative, $\times$.
$1x > 14$	Multiplication.
$x > 14$	One, $\times$.

EXAMPLE 4

Solve $-2x + 3 > 7$ and explain each step.

SOLUTION

Statements	Reasons
$-2x + 3 < 7$	Given.
$(-2x + 3) + (-3) < 7 + (-3)$	Addition, $<$.

Statements	Reasons
$-2x + [3 + (-3)] < 7 + (-3)$	Associative, $+$.
$-2x + 0 < 4$	Addition.
$-2x < 4$	Zero, $+$.
$-\frac{1}{2}(-2x) > (-\frac{1}{2})4$	Multiplication, $<$.
$(-\frac{1}{2} \times -2)x > (-\frac{1}{2})4$	Associative, $\times$.
$1x > -2$	Multiplication.
$x > -2$	One, $\times$.

Note the change in the sense of the inequality from $<$ to $>$ when both members were multiplied by the negative number, $-\frac{1}{2}$. Also note that both members were multiplied by $-\frac{1}{2}$ inasmuch as the product $(-\frac{1}{2} \times -2)x = 1x$, thus allowing us to solve for x as required.

exercises

State a reason for each step in the following solutions.

1. $x + 2 = 7$
   (a) $(x + 2) + (-2) = 7 + (-2)$
   (b) $x + [2 + (-2)] = 7 + (-2)$
   (c) $x + 0 = 5$
   (d) $x = 5$

2. $x + 2 < 7$
   (a) $(x + 2) + (-2) < 7 + (-2)$
   (b) $x + [2 + (-2)] < 7 + (-2)$
   (c) $x + 0 < 5$
   (d) $x < 5$

3. $2x + 3 = 7$
   (a) $(2x + 3) + (-3) = 7 + (-3)$
   (b) $2x + [3 + (-3)] = 7 + (-3)$
   (c) $2x + 0 = 4$
   (d) $2x = 4$
   (e) $\frac{1}{2}(2x) = \frac{1}{2}(4)$
   (f) $(\frac{1}{2} \times 2)x = \frac{1}{2}(4)$
   (g) $1x = 2$
   (h) $x = 2$

4. $-\frac{2}{3}x + 2 < 8$
   (a) $(-\frac{2}{3}x + 2) + (-2) < 8 + (-2)$
   (b) $-\frac{2}{3}x + [2 + (-2)] < 8 + (-2)$
   (c) $-\frac{2}{3}x + 0 < 6$
   (d) $-\frac{2}{3}x < 6$

(e) $(-\frac{3}{2})(-\frac{2}{3}x) > (-\frac{3}{2}) \times 6$
(f) $[(-\frac{3}{2})(-\frac{2}{3})] \times x > (-\frac{3}{2}) \times 6$
(g) $1x > -9$
(h) $x > -9$

Solve for x.

5. $3x + 2 = 11$ 6. $2x - 3 = 8$
7. $-3x + 1 = 7$ 8. $-2x + 5 = 9$
9. $\frac{1}{2}x - 2 = 5$ 10. $\frac{2}{3}x - 1 = 5$
11. $2x + 3 < 7$ 12. $3x - 2 < 7$
13. $3x - 1 > 5$ 14. $2x + 3 > 8$
15. $-2x + 3 < 7$ 16. $-3x + 1 < 5$
17. $-\frac{1}{2}x + 2 < 5$ 18. $-\frac{1}{3}x + 1 > 2$
19. $-\frac{3}{4}x - 1 > 2$ 20. $-\frac{2}{3}x - 3 < 9$

Determine whether each statement is always true for integers a, b, c, and d, where a < b and c < d. If the statement is not always true, give an example for which it is false.

*21. $a + c < a + d$ *22. $a + c < b + d$
*23. $a - c < b - d$ *24. $a + d > b + c$

EXPLORATIONS

Note the equations described by these flow charts.

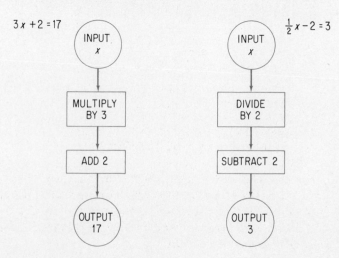

Draw flow charts to describe each of the following equations.

1. $2x + 3 = 7$ **2.** $3x - 5 = 10$

3. $\frac{1}{2}x - 1 = 7$ **4.** $\frac{1}{4}x + 2 = 6$

Reverse flow charts may be used in the solutions of equations. You work backward, using opposite operations, as in these examples for the previously described equations.

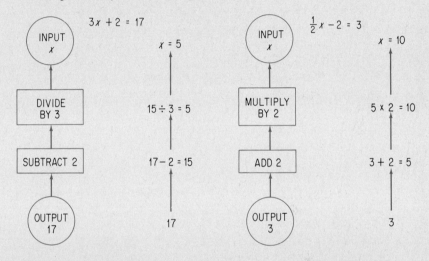

5. Draw reverse flow charts to solve each of the equations in Explorations 1 through 4.

Many mathematical "tricks" can be explained and developed through the use of algebraic techniques. Consider, for example, this set of directions:

Think of a number.
Multiply your number by 2.
Add 9.
Subtract 3.
Divide by 2.
Subtract the number you started with.
Your final answer will always be 3.

6. Use n to represent the original number, and write a mathematical phrase to represent each step in the set of directions just given. For example, the first three steps would give n, $2n$, and $2n + 9$. From this show why the result is always 3, regardless of the number one starts with.

7. Here is a mathematical trick that is frequently used by magicians. The magician asserts that he has memorized a particular telephone book and uses the following procedure for allowing the audience to test his memory.

A "volunteer" selects a number such as 537 that has three different decimal digits. Then a new number is found by reversing the digits of the given number (735) and the smaller of the two numbers is subtracted from the larger (735 − 537 = 198). If this subtraction results in a two-digit number such as 35, the volunteer adds an initial zero to make it a three-digit number (035). In this illustrative example the difference is 198. He then reverses the digits of this number (891) and adds these last two numbers together (198 + 891 = 1089).

The magician's "act" now involves passing the volunteer a local telephone book which he has "memorized," having him turn to the page (89) indicated by the last two digits of the sum just obtained, count down the number (10) of lines in the first column indicated by the first two digits of the sum, and point to the name listed there. The magician then calls out the name.

Try several selections of three-digit numbers to see how the trick works.

7-4

Compound Sentences

Consider this **compound sentence:**

$$x + 1 > 2 \quad \text{and} \quad x - 2 < 1$$

The sentence $x + 1 > 2$ is true for all x greater than 1; the sentence $x - 2 < 1$ is true for all x less than 3. Recall that a **compound sentence** of the form $p \wedge q$ (p and q) is true only when both parts are true. Thus the given compound sentence is true for the set of elements in the *intersection* of the two sets. Graphically, we can show this as follows:

$x + 1 > 2$

$x - 2 < 1$

$x + 1 > 2$ and $x - 2 < 1$

The graph of the compound sentence consists of an open interval and can be described as

$$1 < x < 3$$

The solution set of this compound sentence can be written in set-builder notation as

$$\{x \mid x > 1\} \cap \{x \mid x < 3\}$$

This is read as "the intersection of the set of all x such that x is greater than 1 and the set of all x such that x is less than 3." Note that no replacement set for x was specified here. When such is the case, we shall assume the replacement set to be the set of real numbers.

EXAMPLE 1

Find the solution set

$$x \geq -2 \text{ and } x + 1 \leq 4, \qquad x \text{ an integer}$$

SOLUTION   Here we want the set of integers that are greater than or equal to -2 but at the same time are less than or equal to 3. (If $x + 1 \leq 4$, then $x \leq 3$.) The solution set is $\{-2, -1, 0, 1, 2, 3\}$.

EXAMPLE 2

Find the solution set for the replacement set of real numbers

$$x + 3 < 5 \quad \text{and} \quad 3 < 1$$

SOLUTION   You will note that the second part of this sentence ($3 < 1$) is false. If part of a sentence of the form $p \wedge q$ is false, then, as you may recall, the entire sentence is false. Thus the solution set is the empty set.

Next we consider a compound sentence involving the connective "or":

$$x + 1 < 2 \quad \text{or} \quad x - 2 > 1$$

The sentence $x + 1 > 2$ is true for all x less than 1; the sentence $x - 2 > 1$ is true for all x greater than 3. Recall that a sentence of the form $p \vee q$ (p or q) is true unless both parts are false. Thus the given compound sentence is true for the set of elements in the *union* of the two sets. Graphically we have the following:

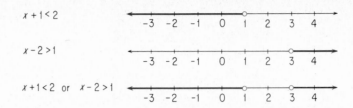

$x + 1 < 2$

$x - 2 > 1$

$x + 1 < 2$ or $x - 2 > 1$

The graph of the compound sentence consists of the union of two half-lines. The solution set can be written in set-builder notation as

$$\{x \mid x < 1\} \cup \{x \mid x > 3\}$$

This is read as "the union of the set of all x such that x is less than 1 and the set of all x such that x is greater than 3."

EXAMPLE 3

Find the solution set for the replacement set of real numbers

$$5 > 1 \quad \text{or} \quad x + 2 < 5$$

SOLUTION   Since the first part of this sentence is true for all real numbers, the whole sentence is true for all real numbers. The solution set is the entire set of real numbers.

The instruction to find the solution set of a sentence may be indicated briefly as "*solve.*" Similarly, the instruction to graph a sentence may be abbreviated as "*graph.*"

EXAMPLE 4

Graph the solution set

$$x \leq -1 \quad \text{or} \quad x \geq 2$$

SOLUTION   The graph is the union of two rays.

EXAMPLE 5

Graph the sentence

$$x = -2 \quad \text{or} \quad x \geq 1$$

SOLUTION   Since no replacement set is specified, we assume that the replace-
                 ment set is the set of real numbers.

EXAMPLE 6

Solve the sentence

$$x + 2 \neq 2 + x \quad \text{or} \quad 7 < 2$$

SOLUTION   Both parts of the sentence are false; the solution set is the empty
                 set.

EXAMPLE 7

Solve the sentence

$$\{x \mid (x - 2)(x + 1) = 0\}$$

SOLUTION   The product of two numbers is zero if and only if at least one
                 of the numbers is zero. That is, for all real numbers a and b, if
                 $a \times b = 0$ then $a = 0$, or $b = 0$, or $a = 0$ and $b = 0$. Therefore,
                 if $(x - 2)(x + 1) = 0$, then $x - 2 = 0$ or $x + 1 = 0$ and $x = 2$ or
                 $x = -1$. The solution set is $\{2, -1\}$.

exercises

List the elements in the solution set for integers x.

1. $x \geq 1$ and $x + 1 \leq 5$ **2.** $x \leq -1$ and $x > -5$
3. $x > 1$ and $x < 3$ **4.** $x \geq 2$ or $x \leq 1$

Graph each sentence for real numbers x.

5. $x \geq 1$ and $x \leq 5$ **6.** $x \leq 1$ or $x \geq 5$
7. $x + 1 < 3$ and $x \geq 0$ **8.** $-5 > 0$ and $x + 2 = 7$
9. $x + 1 > x$ and $5 > 1$ **10.** $x - 1 > 3$ or $x < 0$
11. $x + 2 \geq 5$ and $x - 1 \leq 5$ **12.** $7 > 3$ or $x + 2 \leq 5$
13. $0 < -2$ or $x + 1 < 3$ **14.** $x + 1 \geq 2$ or $x = -1$
15. $x - 2 = 2 - x$ or $3 > 5$ **16.** $x + 1 = x$ or $-2 > 2$

Graph each set for real numbers x.

17. $\{x \mid x > 1\} \cap \{x \mid x < 3\}$ **18.** $\{x \mid x \leq -1\} \cup \{x \mid x \geq 2\}$
19. $\{x \mid x + 2 < 3\} \cup \{x \mid x \geq 5\}$ **20.** $\{x \mid x - 1 < 4\} \cap \{x \mid x \geq -2\}$

21. $\{x \mid x(x + 2) = 0\}$ **22.** $\{x \mid (x - 1)(x + 3) = 0\}$

23. $\{x \mid (x + 1)(x + 5) = 0\}$ **24.** $\{x \mid (x - 3)(x - 2) = 0\}$

*25. $\{x \mid x(x + 2) \geq 0\}$ ***26.** $\{x \mid (x - 1)(x + 3) \leq 0\}$

*27. $\{x \mid (x + 1)(x + 5) < 0\}$ ***28.** $\{x \mid (x - 3)(x - 2) > 0\}$

EXPLORATIONS

The **absolute value** of a number x may be defined on the number line as the distance from the origin of the point with coordinate x. Thus $|3| = 3$, $|-3| = 3$, and in general for any real number x:

$$|x| = x, \qquad \text{if } x \text{ is positive}$$
$$|x| = 0, \qquad \text{if } x = 0$$
$$|x| = -x, \qquad \text{if } x \text{ is negative}$$

Note that if x is negative, then $-x$ is the opposite of x and therefore is a positive number. Thus if $x = -3$, $|-3| = -(-3) = 3$. Accordingly, the absolute value of any real number different from zero is a positive number; $|0| = 0$.

Here is a flow chart for finding the absolute value of a number x.

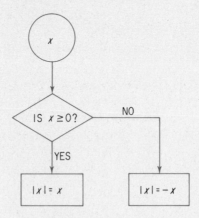

List the set of integers for which each sentence is true.

1. $|x| = 2$ **2.** $|x| \leq 3$

3. $|x| < 5$ **4.** $|x| = -3$

5. $|x - 1| = 3$ **6.** $|x + 1| = 5$

7. $|x + 2| = 1$ **8.** $|x - 2| = 3$

Evaluate.

9. $|-7|$

10. $|-2| + |-3|$

11. $|(-2) + (-3)|$

12. $(|-5|)^2$

13. $|(-5)^2|$

14. $|-5| \times |-3|$

Identify the graph of each sentence on a real number line as two points, a line segment, or the union of two rays.

15. $|x| \leq 7$

16. $|x - 2| = 5$

17. $|x| \geq 1$

18. $|x + 3| \leq 1$

Graph each sentence on a real number line.

19. $|x| \leq 2$

20. $|x| \geq 1$

21. $|x| > 3$

22. $|x| < 5$

23. $|x - 1| \geq 2$

24. $|x + 1| \leq 3$

25. $|x| = -2$

26. $|x| \geq 0$

27. $|x + 2| \geq 3$

28. $|x - 2| \leq 1$

29. $|x - 3| < 2$

30. $|x + 1| > 2$

*31. $2 \leq |x| \leq 3$

*32. $3 \leq |x - 1| \leq 5$

*33. $|x| + 2 = |x + 2|$

*34. $|x| = -x$

*35. $|x| + |x - 3| = 3$

*36. $|x^2 - 10| \leq 6$

7-5

Linear Sentences in Two Variables

Sentences in two variables arise in many ways. For example, the sentence $d = 4h$ may be used to express the distance d in miles that John walks in h hours at 4 miles per hour. The sentence $p = 4s$ may be used to express the perimeter p of a square in terms of the side s. The sentence $I = 0.04P$ may be used to express the simple interest I on P dollars for one year at 4 per cent per year. The sentence $C = \frac{5}{9}(F - 32)$ may be used to express the temperature C in degrees Celsius (or Centigrade) in terms of the temperature F in degrees Fahrenheit.

An open sentence such as $x + y = 8$ remains an open sentence if a replacement is made for only one of the variables. For example, if x is replaced by 5, the sentence becomes $5 + y = 8$, which is an open sentence. Thus, a *pair* of replacements is needed for an open sentence in two variables before we can determine whether it is true or false for these replacements.

If x is replaced by 5 and y is replaced by 3, the sentence $x + y = 8$ is true. Usually we think of the variables as an *ordered pair* (x, y) and speak of the replacements as an ordered pair of numbers $(5, 3)$. By convention, the variable x is assumed to be the first of the two variables, and the first number in the ordered pair of numbers is taken as the replacement for x; the variable y is then taken as the second variable, and the second number in the ordered pair is taken as the replacement for y. For example, the ordered pair $(3, 2)$ implies that $x = 3$ and $y = 2$; the ordered pair $(2, 3)$ implies that $x = 2$ and $y = 3$.

The sentence $x + y = 8$ is true or false for certain ordered pairs of numbers. It is true for $(5, 3)$; it is false for $(2, 7)$. The solution set for the sentence $x + y = 8$ is a set of ordered pairs of numbers for which the sentence is true. This solution set of the sentence may be indicated in the set-builder notation as follows:

$$\{(x, y) | x + y = 8\}$$

This is read as "the set of ordered pairs (x, y) such that $x + y = 8$." The solution set for any sentence in two variables is a set of ordered pairs.

Just as was the case with sentences in one variable, the solution set of a sentence in two variables depends upon the replacement sets of the variables used. For example, if the replacement set for x and y is the set of counting numbers, then the solution set for $x + y = 8$ is

$$\{(1, 7), (2, 6), (3, 5), (4, 4), (5, 3), (6, 2), (7, 1)\}$$

EXAMPLE 1

Find the solution set for the sentence $x + y \leq 2$ if the replacement set for x and for y is the set of whole numbers.

SOLUTION

When $x = 0$, y must be less than or equal to 2. When $x = 1$, y must be less than or equal to 1. When $x = 2$, $y = 0$. Can $x = 3$? The solution set is $\{(0, 0), (0, 1), (0, 2), (1, 0), (1, 1), (2, 0)\}$.

A set of ordered pairs may be obtained from any given set of numbers. For example, consider the set of numbers

$$U = \{1, 2, 3\}$$

We may form a set of all ordered pairs of numbers whose members are both elements of U. A **tree diagram** is helpful in showing the possible pairs of numbers involved. See the figure on page 267.

Each element of U may serve as the first member of an ordered pair and may be matched with each of the elements of U as the second member. This set of ordered pairs is called the **Cartesian product** of U and U, is written

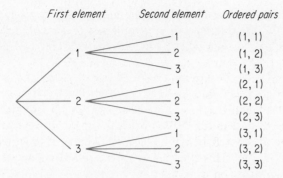

First element   Second element   Ordered pairs

as $U \times U$, and is read "U cross U." The set of all ordered pairs whose coordinates belong to the given set U is $U \times U$:

$$U \times U = \{(1, 1), (1, 2), (1, 3), (2, 1), (2, 2), (2, 3), (3, 1), (3, 2), (3, 3)\}$$

We shall refer to the set U as the **universal set.**

EXAMPLE 2

Let $U = \{1, 2\}$ and list the elements in $U \times U$.

SOLUTION $\{(1, 1), (1, 2), (2, 1), (2, 2)\}$.

EXAMPLE 3

List the elements in the solution set for $x + y = 4$ if the universal set is $U = \{1, 2, 3, 4\}$.

SOLUTION   We are permitted to replace x and y only by elements of U. Thus for $x = 1$, $y = 3$; for $x = 2$, $y = 2$; for $x = 3$, $y = 1$. The solution set is $\{(1, 3), (2, 2), (3, 1)\}$. Why can we not let $x = 4$?

The Cartesian product $A \times B$ is the set of all ordered pairs with first elements from A and second elements from B.

EXAMPLE 4

Let $A = \{1, 2\}$ and $B = \{3, 4. 5\}$. Then list the elements of $A \times B$.

SOLUTION $\{(1, 3), (1, 4), (1, 5), (2, 3), (2, 4), (2, 5)\}$.

The graph of the Cartesian product $U \times U$ for $U = \{1, 2, 3\}$ consists of a set of nine points, as in the next figure.

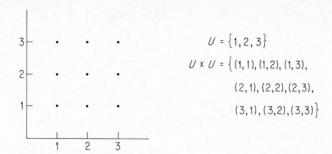

$U = \{1, 2, 3\}$

$U \times U = \{(1, 1), (1, 2), (1, 3),$

$(2, 1), (2, 2), (2, 3),$

$(3, 1), (3, 2), (3, 3)\}$

This figure can now be used to graph the solution sets of sentences for this given universe. For example, the solution set of $x + y = 3$ for $U = \{1, 2, 3\}$ is $\{(1, 2), (2, 1)\}$. The graph of this solution set is drawn by placing heavily shaded dots at the points that correspond to the ordered pairs $(1, 2)$ and $(2, 1)$.

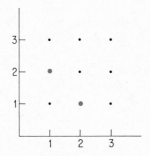

The universal set may include negative numbers. For example, if $U = \{-3, -2, -1, 0, 1, 2, 3\}$, then $U \times U$ consists of a set of 49 ordered pairs of numbers. The graph of $\{(x, y) \mid y = x\}$ for this universal set may be shown as follows:

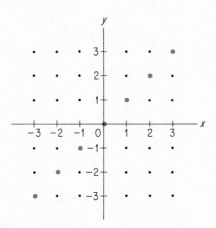

EXAMPLE 5

Graph $\{(x, y) | y \geq x + 2\}$ for the universal set

$U = \{-3, -2, -1, 0, 1, 2, 3\}$

SOLUTION   It is helpful to consider first the corresponding statement of equality, $y = x + 2$.

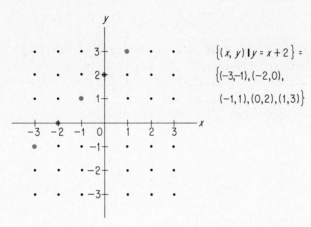

$\{(x, y) | y = x + 2\} =$

$\{(-3,-1), (-2,0),$

$(-1,1), (0,2), (1,3)\}$

The graph of $y \geq x + 2$ consists of the points that are solutions of $y = x + 2$, and also all the points in $U \times U$ that are above the points of the graph of $y = x + 2$.

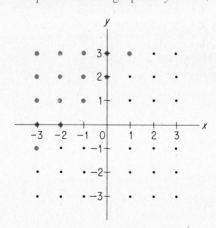

exercises

1. Let $U = \{1\}$ and list the elements in $U \times U$.
2. Let $A = \{1, 2\}$ and $B = \{3, 4\}$. List the elements in $A \times B$.

Graph each set for $U = \{-3, -2, -1, 0, 1, 2, 3\}$.

3. $\{(x, y) \mid y = x + 1\}$ 4. $\{(x, y) \mid y = x - 1\}$
5. $\{(x, y) \mid y \geq x\}$ 6. $\{(x, y) \mid y < x\}$
7. $\{(x, y) \mid y \leq x - 2\}$ 8. $\{(x, y) \mid x + y \geq 4\}$
9. $\{(x, y) \mid y = x - 3\}$ 10. $\{(x, y) \mid y = x + 3\}$
11. $\{(x, y) \mid y \geq x - 3\}$ 12. $\{(x, y) \mid y \leq x + 3\}$

Find the solution set for each sentence for the replacement set $\{1, 2, 3, 4\}$.

13. $x + y = 3$ 14. $2x + y \leq 4$
15. $x + y \leq 4$ 16. $y = x + 1$
17. $x + y < 2$ 18. $y < x$

Find the solution set for each sentence, using the set of counting numbers as the replacement set for x and y.

19. $x + y = 5$ 20. $x + y < 4$
21. $y \leq 3 - x$ 22. $y = 4 - x$
*23. An *ordered triplet* of numbers is a set of three numbers in order, such as $(1, 3, 2)$. List all the possible ordered triplets that can be formed from the set $U = \{1, 2\}$.

EXPLORATIONS

1. The distance d in miles that one walks in h hours at 4 miles per hour is given by the equation $d = 4h$. The cost c in cents of n eight-cent postage stamps is given by the equation $c = 8n$. Give at least three other examples of situations in which one variable is a constant multiple of another.

If $y = kx$ for some constant k and any two variables x and y, then y is said to **vary directly as** x.

Use the correct value of the constant and express each statement as an equation.

2. The perimeter P of a square varies directly as the length s of a side of the square.

3. The simple interest I on a loan of 5 dollars at 1% per month varies directly as the number n of months.

If y varies directly as x, then y is also said to be *proportional to x*. The constant k in the equation $y = kx$ is often called the **constant of proportionality.**

(a) *Express each statement as an equation and* **(b)** *identify the constant of proportionality.*

4. The distance d in miles traveled at 50 miles per hour is proportional to the time t in hours of the trip.

5. The length f of an object in feet is proportional to its length y in yards.

6. The length f of an object in feet is proportional to its length i in inches.

*7. (Suppose that a six-foot man has a four-foot shadow.) The height h of a tree is proportional to the length s of its shadow.

8. Give at least three other examples of statements and constants of proportionality.

7-6
Relations and Functions

A **relation** may be defined as any subset of $U \times U$; thus, a relation is a set of ordered pairs of numbers. It is most often defined by a rule. Consider, for example, $\{(x, y) \mid y > x - 1\}$, where $U = \{1, 2, 3\}$. This set, which *is* the relation, is $\{(1, 1), (1, 2), (1, 3), (2, 2), (2, 3), (3, 3)\}$. It may be graphed as in the accompanying figure.

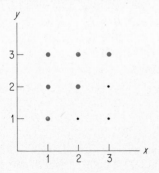

The relation may be defined by the rule, by the graph, or by a table of values for the variable:

x	1	1	1	2	2	3
y	1	2	3	2	3	3

Here is the table and graph for another relation:

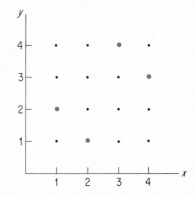

x	1	2	3	4
y	2	1	4	3

This relation is the set of ordered pairs $\{(1, 2), (2, 1), (3, 4), (4, 3)\}$. Note that this second relation differs from the first in that, for any value of x, there is at most one value for y. We call this special type of relation a *function*.

A **function** is a set of ordered pairs (x, y) such that for each value of x there is at most one value of y; that is, no first element appears with more than one second element. You may think of the first element as the **independent variable** and the second element as the **dependent variable**. Each variable has a set of possible values. The set of all first elements of the ordered pairs of numbers is called the **domain** of the function; the set of all second elements is called the **range** of the function. In terms of its graph, any vertical line drawn meets the graph of a function in at most one point.

Here are the graphs of two relations that are also functions:

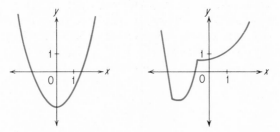

Notice that no vertical line intersects the graph in more than one point.

The graphs of two relations that are not functions are shown on page 273. Notice that in each case there exists at least one vertical line that intersects the graph in two or more points.

Although formulas such as $y = x^2$ may *define* a function, they are not strictly functions. We have defined the function to be a set of ordered pairs

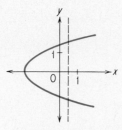 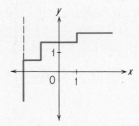

(x, y), such as those obtained from the formula $y = x^2$ for the real variable x. Thus, a formula may provide a rule by which the function may be determined. In other words, a formula may provide a means for associating a unique element in the range with each element in the domain.

When a formula such as $y = x^2 - 2x + 3$ is used to define a function, it is customary to think of y as *a function of* x and to write the formula in **functional notation**:

$y = f(x)$

where

$f(x) = x^2 - 2x + 3$

The symbol $f(x)$ may be read as "f evaluated at x," which is often abbreviated as "f at x." Some books use "f of x" for the "f function evaluated at (of) x." If $y = f(x)$, then the value of y for any value b of x may be expressed as $f(b)$. For example,

$$f(2) = 2^2 - 2(2) + 3 = 3$$
$$f(-1) = (-1)^2 - 2(-1) + 3 = 6$$
$$f(0) = 0^2 - 2(0) + 3 = 3$$
$$f(1) = 1^2 - 2(1) + 3 = 2$$

Notice that for the function $f(x) = (x - 1)^2 + 2$, the domain of the function is the set of all real numbers, and the range of the function is the set of real numbers greater than or equal to 2. The range may be determined from the graph of the function or from the observation that since $(x - 1)^2 \geq 0$ we must have $(x - 1)^2 + 2 \geq 2$. Other letters, as in $g(x)$ and $h(y)$, may be used in designating functions.

exercises

1. Many elementary texts now introduce the concept of function and relation on an intuitive basis through the use of arrow diagrams. For example, each of the following shows a correspondence (relation) between two sets. In each case the arrow indicates which members

of the second set (range) correspond to the given elements of the first set (domain). Which of the relations are functions?

(a)

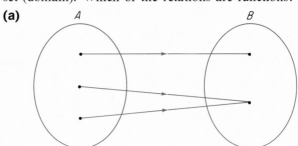

(b)

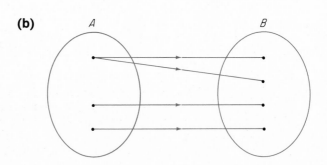

(c)

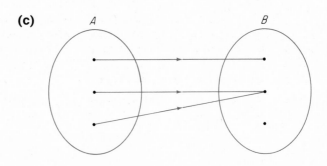

(d)

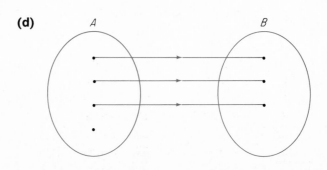

(e)

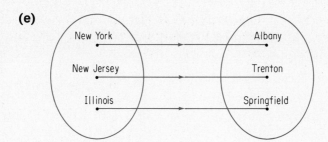

(f)

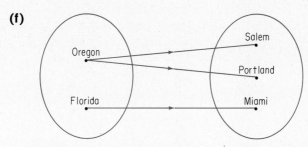

2. A **mapping** of a set A *onto* another set B is a correspondence where each element of A is mapped to a unique element of B, and each element of B is the image of at least one element in A. Draw several different arrow diagrams that illustrate such a mapping, as in Exercise 1. Note that any *mapping* may be considered as a *function*, and vice versa.

3. A *one-to-one* **mapping** of a set A *onto* a set B is one where each element of set A corresponds to a unique element of set B, and each element of set B is the image of a unique element in set A. Draw an arrow diagram that illustrates such a mapping, as in Exercise 1.

4. If $f(x) = x - 2$, find **(a)** $f(1)$; **(b)** $f(2)$; **(c)** $f(17)$.

5. If $f(x) = x^2 - 4x + 5$, find **(a)** $f(0)$; **(b)** $f(2)$; **(c)** $f(-3)$.

6. If $g(x) = x^3 - 7$, find **(a)** $g(2)$; **(b)** $g(-1)$; **(c)** $g(-2)$.

7. If $f(x) = |x| + x$, find **(a)** $f(-2)$; **(b)** $f(0)$; **(c)** $f(3)$.

Graph each relation, identify the relations that are functions, and state the domain and range for each one that is a function. In each case $U = \{1, 2, 3\}$.

8. $y = x$ 9. $y = x + 1$
10. $y < x$ 11. $y < x + 1$

Proceed as in Exercises 8 through 11 for the universe of real numbers.

12. $y = x + 1$ 13. $y = x - 1$
14. $y = |x - 1|$ 15. $y = |x + 1|$
16. $y = |x| + 1$ 17. $y = |x| - 1$
18. $y \geq x$ 19. $y \leq x - 1$
20. $y = -x^2$ 21. $y = (x + 1)^2$

EXPLORATIONS

1. The time t in hours and the rate r in miles per hour for a trip of 100 miles are related by the equation $rt = 100$. Find the rate if t is **(a)** 2 hours; **(b)** 1 hour and 40 minutes; **(c)** 1 hour and 20 minutes.

If y varies directly as the reciprocal of x, then $y = k/x$, $xy = k$, and y is said to vary inversely as x.

Use k as the constant and express each of the following statements as an equation.

2. Other things being equal, a town's tax rate r varies inversely as its total assessed valuation A.

3. At a constant temperature the volume V of an ideal gas varies inversely as its pressure P.

4. The density D of a given mass varies inversely as its volume V.

5. Give at least three other examples of inverse variation.

7-7
Graphs on a Plane

In §7-5, we graphed the solution sets of sentences using finite coordinate systems. Now we shall extend these concepts and consider U as the set of real numbers. Then $U \times U$ becomes an infinite collection of ordered pairs of numbers and its graph consists of an entire plane. In honor of René Descartes, such a plane is frequently referred to as the **Cartesian plane**. The numbers of each ordered pair of real numbers are the **Cartesian coordinates** of a point of the plane. Each point of a Cartesian plane can be represented by (has as its *coordinates*) an ordered pair of real numbers, and each ordered pair of real numbers can be used to identify (locate) a unique point of the plane. We follow the custom of speaking of "the point with coordinate (x, y)" as "the point (x, y)."

Let us now discuss the graphs of sentences for the universe of real numbers. Consider, for example, the sentence $y = x + 2$. We may list in a **table of values** several ordered pairs of numbers that are solutions of the sentence. Thus for $x = -1$, we have $y = -1 + 2 = 1$ and for $x = 2$, we have $y = 2 + 2 = 4$. Confirm the other entries given in this table.

x	-3	-2	-1	0	1	2	3
y	-1	0	1	2	3	4	5

$y = x + 2$:

Each ordered pair of numbers (x, y) from the table can then be graphed. These points may be connected in the order of the x-coordinates and the graph of $y = x + 2$ obtained as in the figure.

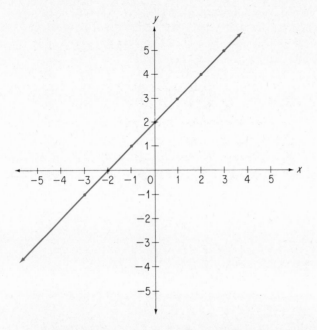

The graph of $y = x + 2$ is a straight line and extends indefinitely as indicated by the arrowheads. Often these arrowheads are omitted. In general, for the universe of real numbers, any equation that can be expressed in the form

$$ax + by + c = 0$$

where a and b are not both zero, is said to be a **linear equation** and has a straight line as its graph. We shall consider U to be the set of real numbers from now on unless otherwise specified.

Inasmuch as a straight line is determined by two points, we can graph a linear equation after locating two of its points. When $a \neq 0$ and $b \neq 0$ the most convenient points to locate are the point $(0, y)$ at which the graph crosses the y-axis and the point $(x, 0)$ at which the graph crosses the x-axis. The graph of $y = x + 2$ crosses the y-axis when $x = 0$, that is, at the point

A: $(0, 2)$, where 2 is the *y-intercept* of the graph. The graph of $y = x + 2$ crosses the x-axis when $y = 0$, that is, at the point B: $(-2, 0)$, where -2 is the *x-intercept* of the graph.

EXAMPLE 1

Graph $\{(x, y) \mid x - 2y = 4\}$.

SOLUTION

The equation may be expressed as $x - 2y - 4 = 0$ and thus has a line as its graph. For $x = 0$, the given equation becomes $-2y = 4$ and $y = -2$. For $y = 0$, the given equation becomes $x = 4$. The x-intercept 4 and the y-intercept -2 determine the line as shown in the graph.

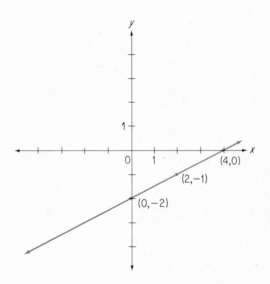

Usually it is desirable to graph a third point as a check of your work. For example, when $x = 2$ we have $2 - 2y = 4$, $-2y = 2$, and $y = -1$. Thus $(2, -1)$ may be graphed and the point should be on the line.

Statements of inequality can also be graphed for the universe of real numbers. Consider the sentence $y \le x + 2$. Here the graph consists of all the points on the line $y = x + 2$, as well as the points in the **half-plane** below the line as indicated by the shaded portion of the graph. Note that for any value b of x the point $(b, b + 2)$ is on the line $y = x + 2$, and for any value of y less than $b + 2$ the point (b, y) is below the line. Thus the graph of $y \le x + 2$ is the union of a line and a half-plane.

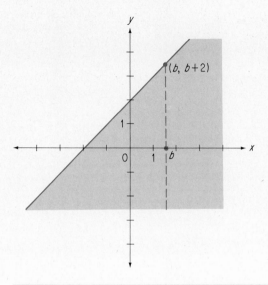

EXAMPLE 2

Graph $y > x - 1$.

SOLUTION

It is helpful to draw first the corresponding statement of equality, $y = x - 1$, as a dotted line. This line divides the plane into two half-planes. To determine which half-plane to shade, we note that

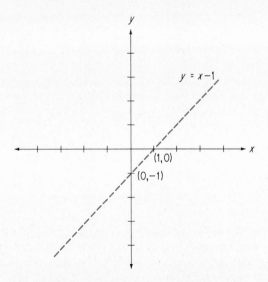

the problem requests all the values of *y greater than x* − 1. Thus we shade the half-plane above the line to represent the solution set. See the graph on page 280.

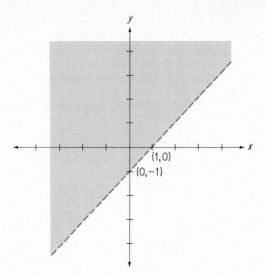

The line in the graph for Example 2 is dotted, since the points of the line *are not* points of the graph. The line in the graph for Example 3 is solid, since the points of the line *are* points of the graph.

EXAMPLE 3

Graph $x - 2y - 4 \geq 0$.

SOLUTION

This sentence is equivalent to $x - 2y \geq 4$. From Example 1 we have the graph of the line $x - 2y = 4$. To determine which half-plane to shade we may proceed in one of two ways. We can place the given sentence in *y*-**form** as follows:

$$x - 2y \geq 4$$
$$-2y \geq -x + 4 \qquad \text{Why?}$$

$$y \leq \frac{1}{2}x - 2 \qquad \text{Why?}$$

This tells us that the half-plane *below* the line is included in the graph of the solution set. An alternate plan for determining which half-plane is included in the solution set is to select a particular point of one half-plane and see whether or not that point belongs in the graph. The origin, $(0, 0)$, is a convenient point to consider. When $x = 0$ and $y = 0$, we have the false statement $0 - 0 \geq 4$. Thus the points of the half-plane that contains $(0, 0)$ are not in the solution set and again we conclude that the graph consists of the line and the half-plane below the line.

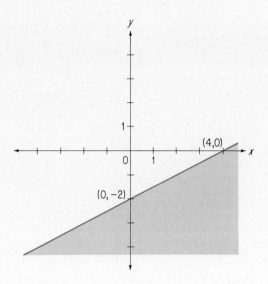

exercises

For the graph of each sentence, find **(a)** *the x-intercept;* **(b)** *the y-intercept.*

1. $x + y = 5$ **2.** $x - y = 2$

3. $2x - 3y = 6$ **4.** $3x - 2y = 6$

5. $y = 2x + 3$ **6.** $y = 3x + 2$

7. $3x - y + 6 = 0$ **8.** $x + 2y - 5 = 0$

9. $5x + 4y = 9$ ***10.** $ax + by + c = 0$

Write each sentence in y-form, that is, solve for y.

11. $2x + y = 7$ **12.** $4x - 2y = 8$ **13.** $3x + y \geq 6$

14. $x + 2y \leq 4$ **15.** $2x - y \leq 4$ **16.** $3x - 2y \geq 6$

Graph each set for the universe of real numbers.

17. $\{(x, y) | y = x + 1\}$ **18.** $\{(x, y) | y = x - 2\}$

19. $\{(x, y) | x + y = 3\}$ **20.** $\{(x, y) | x - y = 3\}$

21. $\{(x, y) | 2x + 3y - 6 = 0\}$ **22.** $\{(x, y) | 3x - 2y - 6 = 0\}$

23. $\{(x, y) | x = 2\}$ **24.** $\{(x, y) | y = -1\}$

25. $\{(x, y) | y \geq x\}$ **26.** $\{(x, y) | y \leq x - 2\}$

27. $\{(x, y) | 2x + y > 4\}$ **28.** $\{(x, y) | x - 2y < 2\}$

29. $\{(x, y) | x - y - 1 \leq 0\}$ **30.** $\{(x, y) | 3x - 4y - 12 \geq 0\}$

EXPLORATIONS

1. The overhead projector is an effective device for showing graphs on a plane, especially through the use of translations and rotations. For example, prepare a set of coordinate axes and grid lines. On a separate sheet of acetate draw the graph of $y = |x|$. Then demonstrate each of the following graphs by appropriate translations of this basic curve:

(a) $y = |x| + 1$
Shift curve one unit up.

(b) $y = |x| - 1$
Shift curve one unit down.

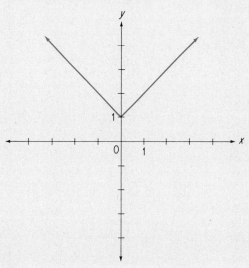

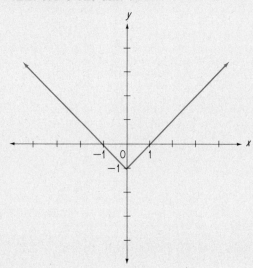

(c) $y = |x + 1|$
Shift curve one unit to the left.

(d) $y = |x - 1|$
Shift curve one unit to the right.

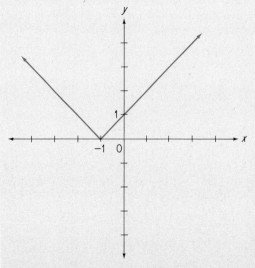

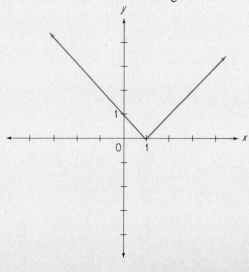

2. A 180° rotation of $y = |x|$ about the x-axis will give the graph of $y = -|x|$. Use this idea to show each of the following by appropriate rotations and translations:

        (a) $\;y = -|x|$ **(b)** $\;y = -|x| + 1$

        (c) $\;y = -|x| - 1$ **(d)** $\;y = -|x + 1|$

Graph the solution set for $U = \{-4, -3, -2, -1, 0, 1, 2, 3, 4\}$.

3. $\;y = |x + 1|$ **4.** $\;y = |x - 1|$

5. $\;y = -|x|$ **6.** $\;y = -|x + 1|$

7. $\;y \geq |x| + 2$ **8.** $\;y \leq |x| - 2$

Graph each set for the universe of real numbers.

9. $\;\{(x, y) \,|\, y = |x| + 1\}$ **10.** $\;\{(x, y) \,|\, y = |x| - 1\}$

11. $\;\{(x, y) \,|\, y = |x + 1|\}$ **12.** $\;\{(x, y) \,|\, y = |x - 1|\}$

13. $\;\{(x, y) \,|\, y > |x + 2|\}$ **14.** $\;\{(x, y) \,|\, y < |x - 2|\}$

15. $\;\{(x, y) \,|\, |x| \leq 2\}$ **16.** $\;\{(x, y) \,|\, |y| \geq 1\}$

*17. $\;\{(x, y) \,|\, |x + y| = 1\}$ ***18.** $\;\{(x, y) \,|\, |x| + |y| = 1\}$

*19. $\;\{(x, y) \,|\, |x| + |y| \leq 1\}$ ***20.** $\;\{(x, y) \,|\, |x| - |y| = 1\}$

7-8

Linear Systems

Consider the compound sentence

$$x + y - 3 = 0 \quad \text{and} \quad x - y - 1 = 0$$

This is frequently written in the following form:

$$\begin{cases} x + y - 3 = 0 \\ x - y - 1 = 0 \end{cases}$$

and is referred to as a **system of linear equations** or as a set of **simultaneous linear equations.** To solve a set of two simultaneous equations we find the set of ordered pairs that are solutions of both equations. Our approach in this section will be a graphical one rather than an algebraic one; that is, we draw the graph of each sentence and identify the point of intersection, if any, of the two lines. For the given system of equations we have the following graph:

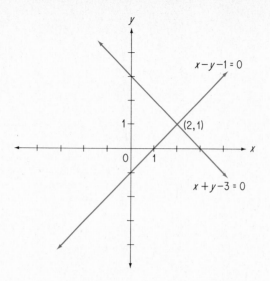

The point located at $(2, 1)$ is on both lines and $(2, 1)$ is the solution of the given system of equations. We may express this fact as

$$\{(x, y) \,|\, x + y - 3 = 0\} \cap \{(x, y) \,|\, x - y - 1 = 0\} = \{(2, 1)\}$$

Consider next the compound sentence

$$x + y = 3 = 0 \quad \text{or} \quad x - y - 1 = 0$$

This word "or" indicates that we are to find the set of ordered pairs of numbers that are solutions of either one or of both of the given equations; that is,

$$\{(x, y) \,|\, x + y - 3 = 0\} \cup \{(x, y) \,|\, x - y - 1 = 0\}$$

The graph of the solution set consists of all the points that are on at least one of the two lines; that is, the graph is the union of the points of the two lines drawn in the preceding figure.

EXAMPLE 1

Graph $x - y - 1 = 0 \quad$ or $\quad x - y + 2 = 0$.

SOLUTION   The graph consists of the union of the points of the two lines.

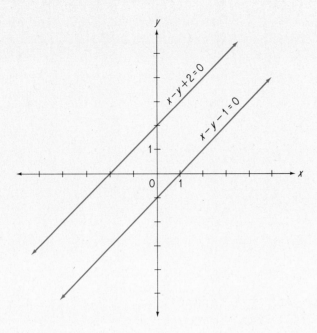

Notice that the two lines in the solution for Example 1 are parallel. The solution set of the sentence "$x - y - 1 = 0$ *and* $x - y + 2 = 0$" would be the empty set.

EXAMPLE 2

Graph $\{(x, y) \mid (x - y - 1)(x - y + 2) = 0\}$.

SOLUTION   First we note that the product $a \cdot b = 0$ implies that $a = 0$ or $b = 0$. (Recall that the use of the word "or" also includes the case that $a = 0$ *and* $b = 0$.) Thus the given sentence can be written in the form $x - y - 1 = 0$ or $x - y + 2 = 0$. But this is precisely the problem in Example 1. Thus the graph of the solution set consists of the two lines drawn in the figure given there.

Systems of inequalities can also be solved graphically as the union or intersection of half-planes. Consider the system:

$$\begin{cases} x + y - 3 > 0 \\ x - y - 1 > 0 \end{cases}$$

The corresponding statements of equality have been graphed earlier in this section. The graphs of the inequalities are shown in the next two figures.

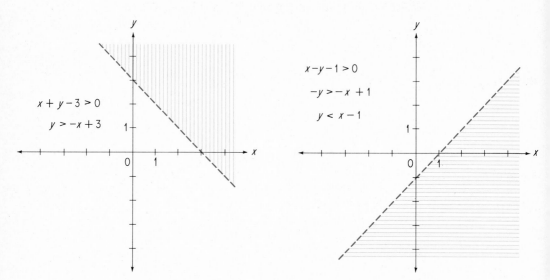

$x + y - 3 > 0$

$y > -x + 3$

$x - y - 1 > 0$

$-y > -x + 1$

$y < x - 1$

Note that the corresponding lines are dotted since the points of the lines are *not* included in the graphs. Also one graph has vertical shading and the other graph has horizontal shading. The graph of the system consists of the intersection of these two graphs—that is, the points of the region that is shaded both vertically and horizontally when the graphs of the two inequalities are drawn on the same coordinate plane.

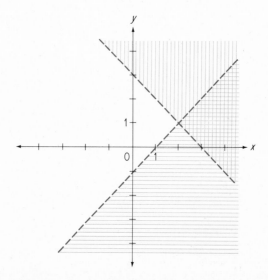

We can also determine from this figure the graph of the sentence $x + y - 3 > 0$ or $x - y - 1 > 0$, namely the union of the points of the two shaded regions. The unshaded region (including the dotted lines) represents the graph of the solution set of the system.

$$\begin{cases} x + y - 3 \leq 0 \\ x - y - 1 \leq 0 \end{cases}$$

EXAMPLE 3

Graph

$$\{(x, y) \mid x - y + 2 \leq 0\} \cap \{(x, y) \mid 2x + y - 4 \geq 0\}$$

SOLUTION

The graph of $x - y + 2 \leq 0$ has horizontal shading in the figure; the graph of $2x + y - 4 \geq 0$ has vertical shading. The graph of the solution set consists of the points in the region with both

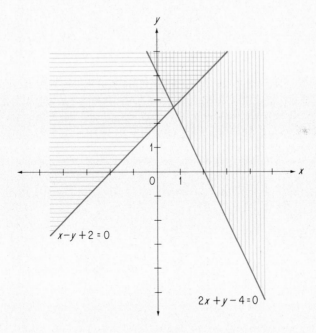

horizontal and vertical shading and includes the points of the two rays that serve as the boundary of this region.

EXAMPLE 4

Graph

$$\{(x, y) \mid x - y + 2 \leq 0\} \cup \{(x, y) \mid 2x + y - 4 \geq 0\}$$

SOLUTION   Consider the graph for Example 3. The solution set in Example
4 consists of all the points in the regions with horizontal shading
or vertical shading, or both, together with the points of the boundary of this region. That is, all points are included *except* those in
the unshaded region.

exercises

Graph the solution set for each system.

1. $\begin{cases} x + y - 2 = 0 \\ x - y - 4 = 0 \end{cases}$ **2.** $\begin{cases} x + y - 5 = 0 \\ x - y - 1 = 0 \end{cases}$

3. $\begin{cases} x + y = 1 \\ x + 2y = 4 \end{cases}$ **4.** $\begin{cases} 2x - y = 6 \\ x + 2y = -2 \end{cases}$

5. $\{(x, y) | x + y - 5 = 0\} \cup \{(x, y) | x + y + 2 = 0\}$

6. $\{(x, y) | x + 2y - 1 = 0\} \cup \{(x, y) | 2x + 4y - 2 = 0\}$

7. $\{(x, y) | 2x - y - 4 = 0\} \cap \{(x, y) | 4x - 2y + 4 = 0\}$

8. $\{(x, y) | (x - y + 3)(2x - 3y + 6) = 0\}$

9. $\begin{cases} x - y + 2 > 0 \\ x + y - 2 > 0 \end{cases}$ **10.** $\begin{cases} 2x + y - 4 < 0 \\ x - 2y + 4 < 0 \end{cases}$

11. $\begin{cases} 3x - 2y - 6 \geq 0 \\ 2x + 3y + 6 \leq 0 \end{cases}$ **12.** $\begin{cases} x + 2y - 4 \leq 0 \\ 2x - y + 2 \geq 0 \end{cases}$

13. $\{(x, y) | x - y - 3 \leq 0\} \cap \{(x, y) | x + y - 3 \leq 0\}$

14. $\{(x, y) | x + 2y - 6 \geq 0\} \cap \{(x, y) | 2x - y + 2 \leq 0\}$

15. $\{(x, y) | 3x - y + 3 \leq 0\} \cup \{(x, y) | 3x + 4y - 12 \geq 0\}$

16. $\{(x, y) | x + 4y - 4 \geq 0\} \cup \{(x, y) | 4x - 2y + 8 \geq 0\}$

*17. $\{(x, y) | (2x - 3y + 6)(x + 2y - 4) \leq 0\}$

*18. $\{(x, y) | y \geq |x + 1|\} \cap \{(x, y) | y \leq 3\}$

*19. $\{(x, y) | y \geq |x| - 1\} \cap \{(x, y) | y \leq -|x| + 1\}$

*20. $\{(x, y) | |x + y| \leq 2\} \cap \{(x, y) | |x| \leq 3\}$

7-9
Linear Programming

Graphs of linear statements in two variables provide a very important tool
for solving modern problems. Although mathematicians are developing
theories for statements that are not necessarily linear, we shall consider only

the linear case. Thus we assume that the conditions of a problem have been represented by or approximated by linear statements. Then the solution of the problem depends upon the solution of a system (that is, a set) of linear statements. The usual method of solution is by graphing; that is, the method is geometrical. Accordingly, we first consider two examples of graphs of systems of linear statements in two real variables.

EXAMPLE 1

Graph the system $x \geq 0, y \geq 0, y \geq x - 1$.

SOLUTION

The solution set of the given system consists of the ordered pairs of real numbers (x, y) that satisfy all three of the statements. To graph the system, we graph each one of the three statements and then take the intersection of their graphs.

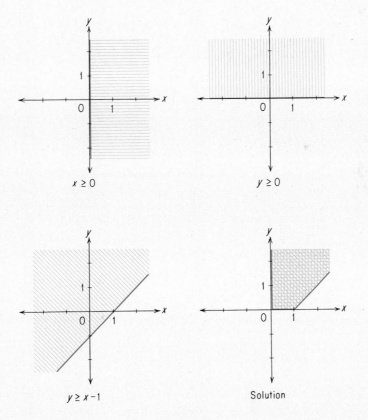

As in §7-7, we graph an inequality by first graphing the equality (using a solid line if its graph is part of the solution set, a dashed line otherwise).

EXAMPLE 2

Graph the system $x \geq 0$; $y \geq 0$; $x \leq 6$; $y \leq 7$; $x + y \leq 10$

SOLUTION

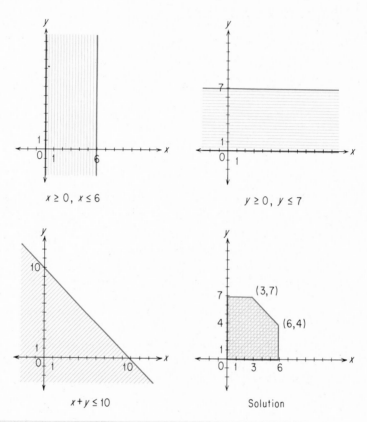

$x \geq 0, \, x \leq 6$

$y \geq 0, \, y \leq 7$

$x + y \leq 10$

Solution

In Examples 1 and 2, the solutions of the systems of statements are known as **polygonal convex sets.**

In a linear programming problem we not only must set up our conditions, but we also must maximize or minimize an expression for profit, cost, or other quantity. Suppose the conditions for Example 2 represent the manufacture of x metal boxes and y glass jars in a given time, and $x + 2y$ represents the manufacturer's profit. If the manufacturer wishes to have a profit of $14, we then graph the equation $x + 2y = 14$ over the conditions of Example 2 as on the top of page 291.

Note that there are now many ways in which the manufacturer can earn a profit of $14. In particular, consider the points $(0, 7)$ and $(6, 4)$. Thus he can earn $14 by manufacturing 0 metal boxes and 7 glass jars, or 6 metal boxes and 4 glass jars. Indeed, any point on the line $x + 2y = 14$ that is within the polygonal region represents an ordered pair (x, y) under the stated restrictions, and such that the profit is $14.

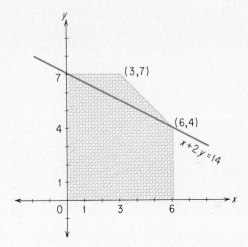

Next consider the same example, but this time we wish the profit to be k. For each value k the graph of $x + 2y = k$ is a straight line. As k takes on different values, we have a set of parallel lines. When several of these lines are graphed with the solution of the conditions in Example 2, we see that under these conditions k may have any value from 0 to 17 inclusive. The maximum (largest) value that is possible for k under these conditions is 17 and occurs at $(3, 7)$. The minimum (smallest) possible value for k is 0 and occurs at $(0, 0)$. Recall that the conditions are for the manufacture of x metal boxes and y glass jars in a given time and that $x + 2y$ represents the manufacturer's profit. Then if we assume that the manufacturer can sell all that he can make, he would make the most profit by manufacturing 3 boxes and 7 jars per unit of time.

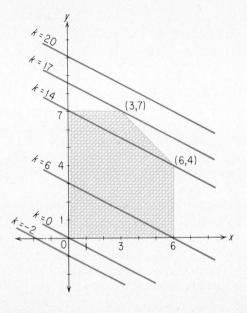

In any linear programming problem the maximum and minimum always occur at a vertex (possibly at two vertices, that is, along a side) of the polygonal region. Intuitively, the reason is that the region is convex and thus the lines of a set of parallel lines first intersect the region either by passing through a vertex, as in our example, or by passing along a side of the region. Accordingly, in our example, we could have found the maximum value of $x + 2y$ for points of the region by testing the values corresponding to the vertices $(0, 0)$, $(6, 0)$, $(6, 4)$, $(3, 7)$, and $(0, 7)$ of the region. The corresponding values of $x + 2y$ are 0, 6, 14, 17, and 14, respectively. Thus, as we observed before, the minimum value of $x + 2y$ is 0 and occurs as $(0, 0)$; the maximum value of $x + 2y$ is 17 and occurs at $(3, 7)$.

EXAMPLE 3

Graph the system

$$x \geq 0, \qquad x \leq 3, \qquad y \leq 0, \qquad x - y \leq 5$$

SOLUTION

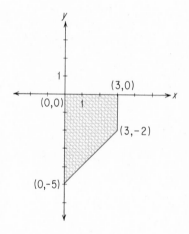

EXAMPLE 4

Find the minimum and maximum values of the expression $2x + y$ defined over the system of Example 3.

SOLUTION

Let f represent the expression $2x + y$. Then at $(0, 0)$, $f = 0$; at $(0, -5)$, $f = -5$; at $(3, 0)$, $f = 6$; at $(3, -2)$, $f = 4$. The minimum value, -5, occurs at $(0, -5)$. The maximum value, 6, occurs at $(3, 0)$.

EXAMPLE 5

A manufacturer produces gidgets and gadgets, and has his machines in operation 24 hours a day. To produce a gidget requires
2 hours of work on machine A and 6 hours of work on machine
B. It takes 6 hours of work on machine A and 2 hours on machine
B to produce a gadget. The manufacturer earns a profit of $5 on
each gidget and $2 on each gadget. How many of each should
he produce each day in order to earn the maximum profit possible?

SOLUTION   If we let x represent the number of gidgets to be produced, and
y the number of gadgets, then the conditions of the problem may
be stated and graphed as follows.

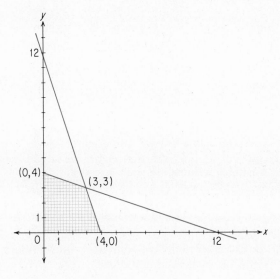

$$x \geq 0$$
$$y \geq 0$$
$$2x + 6y \leq 24$$
$$6x + 2y \leq 24$$
$$\text{Profit} = P = 5x + 2y$$

We test the profit expression, $5x + 2y$, at the vertices of the polygon. At $(0, 0)$, $P = 0$; at $(0, 4)$, $P = 8$; at $(3, 3)$, $P = 21$; and at
$(4, 0)$, $P = 20$. Thus to insure a maximum profit, the manufacturer
should produce 3 gidgets and 3 gadgets daily.

exercises

Graph each system.

1. $x \geq 2, x \leq 4, y \geq 0, x + y \leq 5$

2. $x \geq 0, y \geq 0, x + 3y \leq 9, 2x + y \leq 8$

Find the values of x and y such that under the set of conditions in the specified exercise the given expression has **(a)** *a maximum value;* **(b)** *a minimum value.*

3. $x + 2y$; Exercise 1.   **4.** $3x + y$; Exercise 1.

5. $x + y$; Exercise 2.   **6.** $x + 5y$; Exercise 2.

Use linear programming to solve these hypothetical problems.

7. A college is experimenting with a combination of teaching methods, using both teachers in the classroom and closed-circuit television. The college has facilities for handling five sections of a class at once using closed-circuit television. For the five sections of a class that meets three clock hours per week, the conditions appear to be as follows: The cost per minute of regular teaching is $5; the cost per minute of closed-circuit television is $3. For a certain week at most $750 can be spent on the instruction of these classes. Assume that the class hour can be spent in part with a teacher and in part with television. When neither the teacher nor the television is on, the students are free to discuss anything they wish. If the value to the students of x minutes of regular teaching and y minutes of closed-circuit television may be expressed as $3x + 2y$, how many minutes of regular teaching and how many minutes of closed-circuit television would be best for the students during the week?

8. Repeat Exercise 7 with the additional condition that the instructor must be present at least 30 minutes each week.

9. Use the conditions as for Exercise 8 and find the number of minutes of regular teaching and of closed-circuit television when the value to the students may be expressed as $x + 2y$.

10. Repeat Exercise 9 when the value to the students may be expressed as $2x + y$.

chapter test

1. Which sentences have the empty set as their solution set? Which sentences are identities?

   (a) $x + 1 = 1 + x$ (b) $x + 3 \neq x$

   (c) $x > x + 1$ (d) $x - 1 < x$

2. Describe the graph of the solution set for each sentence in terms of a specific geometric figure.

   (a) $x - 2 < 5$ (b) $x + 2 = 2 + x$

   (c) $2 \leq x \leq 5$ (d) $x + 1 \geq 3$

3. List the elements in the solution set of each sentence for whole numbers x.

   (a) $x + 1 \leq 5$ (b) $1 < x \leq 5$

   (c) $x - 1 > 2$ (d) $|x| \leq 3$

4. Classify as true or false:

   (a) $3 > 1$ and $-2 < -3$ (b) $3 < 7$ and $|-2| = 2$

   (c) $-1 > 0$ or $0 > -2$ (d) $2 < 0$ or $|5| = -5$

5. Solve:

   (a) $3x + 1 = 7$ (b) $-\frac{3}{4}x + 1 > 7$

6. Graph each solution set for $U = \{1, 2, 3\}$.

   (a) $\{(x, y) | y = x - 1\}$ (b) $\{(x, y) | y > x\}$

   Graph each solution set for the universe of real numbers.

7. (a) $y = x - 3$ (b) $x + y = 1$

8. (a) $y \geq x - 2$ (b) $y < |x|$

9. $\{(x, y) | x + y - 4 \geq 0\} \cap \{(x, y) | x - y + 1 \geq 0\}$

10. $\{(x, y) | x + y - 4 \geq 0\} \cup \{(x, y) | x - y + 1 \geq 0\}$

An Introduction to
GEOMETRY

Geometry has evolved from a concern for earth measure (geo-metry), through the use of line segments and other figures to represent physical magnitudes, to a study of properties of sets of geometric elements. The figures serve as the elements of geometry. Relations among these elements and proofs of their properties from given sets of postulates are considered in more advanced courses.

All geometric figures are usually considered to be sets of points. Although other basic elements are considered in abstract geometry, we shall restrict our consideration to points. This is not a serious restriction, since points may be *interpreted* in many ways. For example, we usually think of a point as a position on a line, on a plane, or in space. We may also think of a point on a number scale in terms of its coordinate. We may even think of cities as points on a map, with the air routes joining them considered as lines. This freedom to interpret points in a variety of ways provides a basis for more abstract geometries.

Geometric figures may be studied in terms of unions and intersections of points, lines, rays, and planes (*nonmetric geometry*), in terms of measures of geometric figures (*metric geometry*), or in terms of coordinates (*coordinate geometry*). We consider each of these three approaches.

8-1
Points, Lines, and Planes

We have an intuitive idea of what is meant by a point, but we have not and shall not *define* a point. To define a term, we need to distinguish it from other terms. In doing this, we must describe our term in words that are already known to us (that is, in simpler terms) in such a way that whenever the description is applicable, the term being defined is also appropriate.

The early Greeks described a point as "that which has no part." Today we recognize such a description as an aid to our interpretations, but we do not consider it a definition. We object to it as a formal definition because the terms used are not simpler terms and, indeed, simpler terms are not available.

Lines also are left undefined. In any logical system there must be some undefined terms. In other words, it is not possible to define everything; we must start with something. In geometry we start with points and lines. Then we define other figures in terms of these.

Even though lines are undefined, they do possess certain properties. Here are four of these properties:

1. *A line is a set of points.* Each point of the line is said to be *on* the line.

2. *Any two distinct points determine a unique line.* In other words, there is one and only one line *AB* through any two given points *A* and *B*. Thus a line may be named by any two of its points. If *C* and *D* are points on the line *AB*, then the line *CD* is the same as the line *AB*.

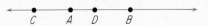

3. *Every line is a real number line.* Then each line is a dense set of points and has infinitely many points.

4. *Any point A on a line separates the line into three parts: the point A and two half-lines, one on each side of A.* Each half-line is a set of points (§7-2). The point *A* is not a point (member) of either half-line (page 298).

Three points *A*, *B*, *C* either lie on a line or do not lie on a line. Note that we think of (interpret) a "line" as a "straight line." If the points do not lie on the same straight line, then, since any two points determine a unique line, the three points determine three lines *AB*, *BC*, and *AC*, as in the next figure.

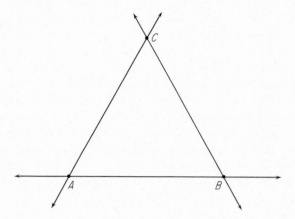

Each of the lines *AB*, *BC*, and *AC* is a set of points. Any two points on a line determine that line; for example, the points *D* and *E* on the line *BC* determine the line *BC*, and we say that *BC* and *DE* are two names for the same line. Any two points of the figure consisting of the lines *AB*, *BC*, and *AC* determine a line. If the points are not on the same one of these three lines, then the points determine a new line; for example, *D* and *F* determine a new line in the figure.

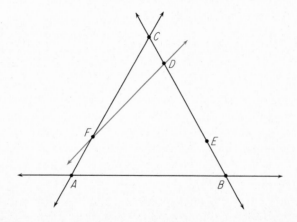

Many new lines can be determined in this way. A few such lines are shown in the next figure. The set of all points on such lines is the set of points of the plane determined by the three points *A*, *B*, and *C*.

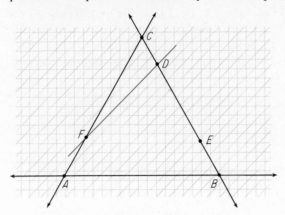

Planes are often accepted as undefined. However, all planes have certain basic properties. Here are five of these properties:

1. *A plane is a set of points.* Each point of the plane is said to be *on* the plane.

2. *Any three points that are not on the same line determine a unique plane.* In other words, there is one and only one plane through any three noncollinear points. For example, a piece of cardboard will balance firmly on the points of three thumbtacks; a three-legged stool is more stable than many four-legged chairs. We use this property when we name planes such as *ABC* by any three points that are on the plane but not on a straight line.

3. *If two distinct points of a line are on a plane, then every point of the line is on the plane.*

4. *Any line m on a plane separates the plane into three parts: the line m and two half-planes.* The points of the line *m* do not belong to either half-plane. However, the line *m* is often called the **edge** of both half-planes even though it is not a part of either.

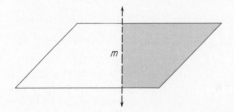

Note that we may think of the points of the plane that are on the right of *m* as forming one half-plane and the points of the plane that are on the left of *m* as forming the other half-plane.

5. *Two distinct planes have either a line in common or no points in common.* In other words, any two distinct planes either intersect in a line or are parallel.

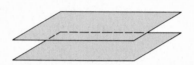

In ordinary geometry (often called **Euclidean geometry**) it is possible for two planes to fail to have a point in common. Two planes that do not have any point in common are said to be **parallel**. A line is parallel to a plane if it does not have any point in common with the plane.

Think of an ordinary classroom. On any given wall the line along the ceiling and the line along the floor do not appear to intersect, that is, do not have any point in common no matter how far they are extended. The line along the ceiling and one wall also does not appear to intersect the line of intersection of two other walls. Any distinction between these two situations must take into consideration the fact that in the first case the two lines were on the same wall (plane), whereas in the second case there could not be a single plane containing both lines. In general, two lines that are on the same plane and do not have any point in common are also said to be **parallel lines**; two lines that are not on the same plane are called **skew lines**; two distinct lines that have a point in common are called **intersecting lines**; lines that have all their points in common may be visualized as two names for the same line and are called **coincident lines**.

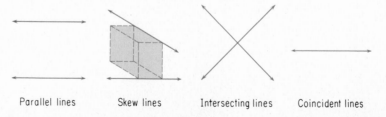

Parallel lines   Skew lines   Intersecting lines   Coincident lines

Note that the arrows indicate that the lines may be extended indefinitely, that is, that the figure involves representations for lines rather than actual lines. These representations are on the printed page and thus on a plane. Thus the figure for skew lines must be visualized as a "picture" on a plane of skew lines in space.

Points, lines, and planes are the building blocks of geometry. We use these elements and relations such as intersections, parallelism, coincidence, skewness, and separation as we define other geometric figures, that is, as we name other sets of points.

exercises

1. Label three points A, B, and C that are not on the same line. Draw and name each of the lines that are determined by these points.

2. Repeat Exercise 1 for four points A, B, C, and D such that no three of the points are on the same line.

3. Assume that the four points in Exercise 2 are not on the same plane. Name each of the planes that is determined by these points.

4. Assume that all of the given points are on the same plane and no three of the given points are on the same line. State how many lines are determined by the given number of points: **(a)** 2; **(b)** 3; **(c)** 4; **(d)** 5; **(e)** 6; *(f) 10; *(g) n.

5. Assume that all of the given lines are on the same plane, that each line intersects each of the other lines, and that no three of the given lines are on the same point. State how many points are determined by the given number of lines: **(a)** 2; **(b)** 3; **(c)** 4; **(d)** 5; **(e)** 6; *(f) 10; *(g) n.

6. Consider the lines along the edges of the rectangular solid shown in the given figure and state whether the specified lines are parallel lines, skew lines, or intersecting lines:

   (a) $\overleftrightarrow{AB}$ and $\overleftrightarrow{CD}$ 　　　　**(b)** $\overleftrightarrow{AB}$ and $\overleftrightarrow{EF}$

   (c) $\overleftrightarrow{AB}$ and $\overleftrightarrow{EH}$ 　　　　**(d)** $\overleftrightarrow{CD}$ and $\overleftrightarrow{FG}$

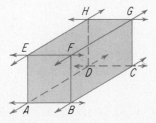

Use pencils, table tops, etc. to represent lines and planes and determine whether each statement appears to be true or false in Euclidean geometry.

7.   Given any line *m* and any point *P* that is not a point of *m*, there is exactly one line *t* that is parallel to *m* and contains *P*.

8.   Given any line *m* and any point *P* that is not a point of *m*, there is exactly one line *q* such that *q* contains *P* and *m* and *q* are intersecting lines.

9.   Given any line *m* and any point *P* that is not a point of *m*, there is exactly one line *s* such that *s* contains *P* and *s* and *m* are skew lines.

10.   Given any line *m* and any point *P* that is not a point of *m*, there is exactly one plane that is parallel to *m* and contains *P*.

11.   Given any plane *ABC* and any point *P* that is not a point of *ABC*, there is exactly one plane that contains *P* and intersects *ABC*.

12.   Given any plane *ABC* and any point *P* that is not a point of *ABC*, there is exactly one plane that contains *P* and is parallel to *ABC*.

13.   Given any plane *ABC* and any point *P* that is not a point of *ABC*, there is exactly one line that contains *P* and intersects *ABC*.

14.   Given any plane *ABC* and any point *P* that is not a point of *ABC*, there is exactly one line that contains *P* and is parallel to *ABC*.

15.   Given any plane *ABC* and any line *m* that is parallel to *ABC*, there is exactly one plane that contains *m* and intersects *ABC*.

16.   Given any plane *ABC* and any line *m* that is parallel to *ABC*, there is exactly one plane that contains *m* and is parallel to *ABC*.

17.   Given any plane *ABC* and any line *m* that is parallel to *ABC*, there is for any given point *P* that is not on *m* and not on *ABC* exactly one line that is parallel to *m* and also parallel to *ABC*.

EXPLORATIONS

Many interesting figures may be obtained by paper folding. Wax paper is especially well suited for these exercises.

1.   Draw a line *m* and select a point *P* that is not on *m*. Fold the point *P* onto points of *m* in at least 20 different positions spaced as evenly as possible. Describe the pattern formed by the folds.

2.   Draw a circle *C* and select a point *P* outside the circle. Fold the point *P* onto points of *C* in at least 20 different positions spaced as evenly as possible. Describe the pattern formed by the folds.

3.   Repeat Exploration 2 for a point *P* that is inside the circle but not at the center of the circle.

4.   Repeat Exploration 2 for the point *P* at the center of the circle.

5.   Repeat Exploration 2 for a point *P* on the circle.

Suppose that we are given a line m and two points A and B on m. There is one and only one line m on both A and B. We call this the line AB and write it as $\overleftrightarrow{AB}$. As in §8-1, the point A separates the line m into two half-lines. Note that one of these half-lines contains the point B and one does not.

The set of points consisting of the half-line AB and the point A is called a **ray** (§7-2). The point A is the **endpoint** of the ray. There is one and only one ray with endpoint A and containing the point B. We call this the ray AB and write it as $\overrightarrow{AB}$. There is also a ray BA with endpoint B and containing A.

The **line segment** AB, written as $\overline{AB}$, consists of the points which $\overrightarrow{AB}$ and $\overrightarrow{BA}$ have in common. (See §7-2.) We may use the symbol $\cap$ for the intersection of two sets of elements (in this case, points) and write

$$\overline{AB} = \overrightarrow{AB} \cap \overrightarrow{BA}$$

A line AB has no endpoints and may be named by any two of its points. A ray AB has one endpoint and may be named by that point A and any other point on the ray. A line segment AB has the two endpoints A and B and is named by those two points. However, for a line segment those two points may be stated in either order; that is, $\overline{AB} = \overline{BA}$.

Any line AB is a set of points. If C is a point of the line such that A is between C and B, then the line is the union of the rays AB and AC. These two rays have a common endpoint A and are both on the line AB. Any figure formed by two rays that have a common endpoint is a **plane angle**. These rays may be on the same line but need not be on the same line. Each angle in the next figure may be designated as $\angle BAC$;

$$\angle BAC = \overrightarrow{AB} \cup \overrightarrow{AC}.$$

The rays AB and AC are called **sides** of $\angle BAC$. For the angle in the following figure there exist points P and Q of the angle such that the open line segment $\overset{\circ\!\!-\!\!\circ}{PQ}$ does not intersect the angle. All points of such open line segments are **interior points** of the angle.

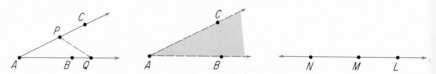

Notice that an angle such as $\angle LMN$ does not have any interior points, since $\overline{PQ} \subset \angle LMN$ for all points P and Q of the angle. Persons who have previously studied geometry may observe that any angle that has an interior has a measurement between $0°$ and $180°$. For this reason some elementary books consider only angles whose measurements are greater than $0°$ and less than $180°$.

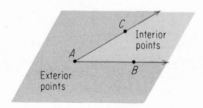

If an angle has an interior, then the angle separates the plane into three disjoint sets: the points of the angle, the interior points of the angle, and the **exterior points** of the angle.

Consider the four figures in the accompanying diagram, where each is a union of line segments. Notice that in each case there seem to be three separate parts. A part may consist of one line segment or be the union of

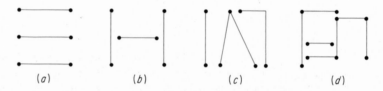

(a) (b) (c) (d)

two or more line segments. However, you could pick out the three parts of each figure. You recognized that any two distinct parts were not "connected," that you could not draw them both without removing the point of the pencil from the paper to move from one to the other. We now assume that the word "connected" is understood even though it has not been defined.

Each of the three figures in the next diagram is a union of line segments, and each figure is connected. Each may be drawn by starting at one of the points A, B, and ending at the other.

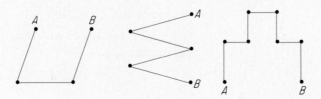

Each of the next five figures is a union of line segments, each is connected, each may be drawn as a continuous line by starting at one of the points R, S and ending at the other, and each differs from the figures A, B in the preceding set. See if you can identify that difference.

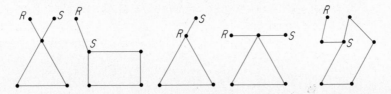

Any connected union of line segments is a broken line. Compare the three broken lines AB with the five broken lines RS. Each broken line AB may be drawn starting at A and ending at B without retracing any line segments or arriving at any point a second time. Each broken line RS may be drawn starting at R and ending at S without retracing any line segments, but for each broken line RS some point, possibly the starting point, must be used twice in the drawing. Briefly, we say that each broken line RS intersects itself. That is, each broken line RS contains at least one point that can be approached along at least three "paths." Such a point may appear, for example, as in any of these figures:

The word "simple" is used to distinguish between figures that intersect themselves and figures that do not intersect themselves. A figure such as one of the broken lines AB which does not intersect itself is a **simple figure**. Figures such as triangles, squares, circles, and rectangles are also simple figures. A figure such as one of the broken lines RS that does intersect itself is *not a simple figure*.

The broken lines *AB* and *RS* differ from each of the next two figures. Each of these next figures is simple; each is connected; each is a union of line segments. Each differs from the previous figures in that when it is drawn one starts at a point and returns to that point. We say that such a figure is **closed**.

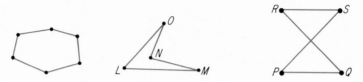

Figures may be either simple or not simple; they may also be either closed or not closed. For example, the figure *LMNO* is simple and closed; the figure *PQRS* is closed and not simple; the broken lines *AB* are simple and not closed; the broken lines *RS* are neither simple nor closed.

A simple closed broken line on a plane is called a **polygon**. The line segments are called **sides** of the polygon; the endpoints of the line segments are called **vertices** of the polygon. A polygon is a **convex polygon** if every line segment *PQ* with points of the polygon as endpoints is either a subset of a side of the polygon or has only the points *P* and *Q* in common with the polygon. A polygon that is not a convex polygon is a **concave polygon**.

Convex polygon   Concave polygon

If *P* and *Q* are points of a convex polygon and $\overline{PQ}$ is not a subset of a side of the polygon, then the points of the open line segment *PQ* are **interior points** of the convex polygon. The union of the points of a convex polygon and its interior points is a **polygonal region**. The points of the plane of a convex polygon that are not points of the polygonal region are **exterior points** of the polygon.

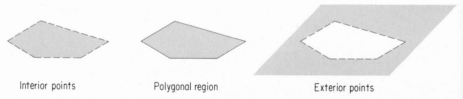

Interior points   Polygonal region   Exterior points

A polygon with three sides is a **triangle**. For example, triangle *ABC* consists of the points of $\overline{AB} \cup \overline{BC} \cup \overline{CA}$. The **interior** of the triangle *ABC* consists of the points in the intersection of the interiors of the three angles, ∠ABC, ∠BCA, and ∠CAB. The **exterior** of triangle ABC consists of the points of the plane ABC that are neither points of the triangle nor points of the interior of the triangle.

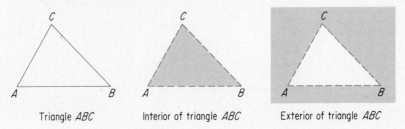

| Triangle *ABC* | Interior of triangle *ABC* | Exterior of triangle *ABC* |

In general, plane polygons are classified according to the number of sides. The common names are shown in this array.

Number of Sides	Name of Polygon
3	Triangle
4	Quadrilateral
5	Pentagon
6	Hexagon
7	Heptagon
8	Octagon
9	Nonagon
10	Decagon
12	Dodecagon

EXAMPLE

Consider the given figure and identify each set of points: (a) $\overrightarrow{AB} \cap \overrightarrow{ED}$; (b) ∠ BAE ∩ ∠ BCD; (c) (interior △ ACE) ∩ (exterior ∠EFD); (d) three angles such that the intersection of their interior points is the interior of △ABF; (e) (exterior △ABF) ∩ $\overrightarrow{ED}$; (f) $\overline{AC} \cap \overrightarrow{BC}$; (g) $\overleftrightarrow{AC} \cap \overleftrightarrow{EF}$.

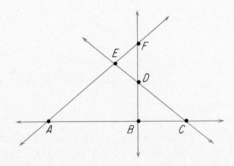

SOLUTION　(a) $\{C\}$; (b) $\overleftrightarrow{AC} \cup E$; (c) interior $\triangle BCD$; (d) $\angle ABF$, $\angle BFA$, and $\angle FAB$; (e) $\overrightarrow{DC}$; (f) $\overline{BC}$; (g) $\varnothing$.

exercises

Sketch a union of line segments that has this property.

1. Connected.
2. Not connected.

Sketch a broken line that has these properties.

3. Simple but not closed.
4. Closed but not simple.
5. Simple and closed.
6. Neither simple nor closed.
7. A convex polygon of five sides.
8. A concave polygon of six sides.

Consider the given figure and identify each set of points.

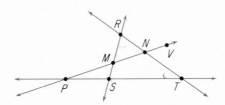

9.　$\overrightarrow{MN} \cap \overrightarrow{TS}$ 　　　　　　　　**10.**　$\overrightarrow{SR} \cap \overrightarrow{ST}$

11.　$\triangle RST \cap \overline{PM}$ 　　　　　　　**12.**　(Exterior $\triangle RST$) $\cap \overline{MN}$

13.　$\overrightarrow{TS} \cap \overline{PM}$ 　　　　　　　　**14.**　$\overrightarrow{PV} \cap \triangle RST$

15.　(Interior $\triangle RST$) $\cap \overline{MN}$ 　　**16.**　$\overline{RS} \cap \overline{ST}$

17.　$\overline{RM} \cup \overline{MN} \cup \overline{RN}$ 　　　　　**18.**　(Exterior $\triangle RST$) $\cap \overrightarrow{NM}$

19.　(Exterior $\triangle RST$) $\cap \overleftrightarrow{MN}$ 　　**20.**　$\overrightarrow{PM} \cup \overrightarrow{PS}$

21.　$\overrightarrow{PV} \cap \triangle MRN$ 　　　　　　　**22.**　$\angle SRT \cap \angle RTS$
23.　(Interior $\triangle RST$) $\cap$ (interior $\triangle MRN$)
24.　(Interior $\triangle RST$) $\cup$ (interior $\triangle MRN$)
25.　(Interior $\angle VPT$) $\cap$ (interior $\triangle RST$)
26.　(Interior $\angle SRT$) $\cap$ (interior $\angle RTS$)
27.　(Interior $\angle RST$) $\cap$ (interior $\angle STR$) $\cap$ (interior $\angle TRS$)
28.　(Interior $\triangle RMN$) $\cup$ (exterior $\triangle RMN$) $\cup \triangle RMN$

Consider the given figure and identify each set of points.

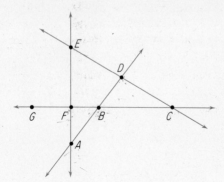

29. $\overrightarrow{GF} \cap \overrightarrow{DE}$ **30.** $\overrightarrow{CG} \cup \overrightarrow{CE}$

31. $\triangle CFE \cap \overleftrightarrow{AB}$ **32.** $\triangle ABF \cap \triangle BCD$

33. $\triangle ABF \cap \overrightarrow{CG}$ **34.** $\triangle CBD \cap \triangle CFE$

35. (Interior $\triangle CBD$) $\cap$ (interior $\triangle CFE$)

36. (Interior $\triangle ABF$) $\cap$ (exterior $\triangle CBD$)

37. (Exterior $\triangle ABF$) $\cap$ $\overrightarrow{BF}$

38. (Exterior $\triangle ABF$) $\cap$ $\overleftrightarrow{CD}$

39. (Interior $\triangle CFE$) $\cap$ $\overleftrightarrow{BD}$

EXPLORATIONS

Name as many different types of convex polygons (triangles, rectangles, trapezoids, octagons, etc.) as possible with vertices at the indicated number of points on a plane.

1. 2   **2.** 3   **3.** 4

4. 5   **5.** 6   **6.** 10

8-3

Space Figures

Any two points A and B that are on a line but not on the same point determine a unique line segment AB. Any three points A, B, and C that are on a plane but not on the same line determine a unique triangle ABC.

Any four points A, B, C, and D that are in space but not on the same plane determine a unique **tetrahedron**. The four points A, B, C, D are the **vertices** of the tetrahedron; the six line segments $\overline{AB}$, $\overline{AC}$, $\overline{AD}$, $\overline{BC}$, $\overline{BD}$, $\overline{CD}$ are the **edges** of the tetrahedron.

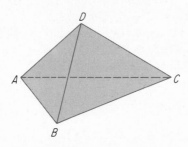

The triangular regions of the four triangles ABC, BCD, CDA, and ABD are the **faces** of the tetrahedron. The **interior points** of the tetrahedron are the points of the open line segments PQ, where P and Q are points of the tetrahedron (points of the union of the faces of the tetrahedron) and $\overset{\circ\!-\!\circ}{PQ}$ does not intersect the tetrahedron. The points of the faces of the tetrahedron form a space figure that is also called a **triangular pyramid.** Any one of the triangular regions may be taken as the base of the pyramid with the remaining vertex as the **vertex** of the pyramid.

On a plane a polygon is a simple, closed union of line segments, that is, a union of line segments (sides of the polygon) such that exactly two sides have each endpoint of a line segment (vertex of the polygon) in common and no side contains an interior point of another side. In space a **polyhedron** is a simple, closed union of polygonal regions (**faces** of the polyhedron) such that exactly two faces have each side of a polygonal region (**edge** of the polyhedron) in common and no face contains an interior point of another face. The vertices of the polygonal regions are also called **vertices** of the polyhedron.

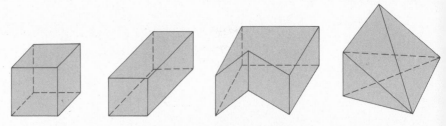

Each of these figures is a polyhedron. The **cube** has square faces and is a convex polyhedron. The rectangular box **(parallelepiped)** has rectangular faces and is a convex polyhedron. The third figure has two faces (top and bottom) that are not convex polygons and is not a convex polyhedron. The last figure appears to be formed by two triangular pyramids with their common base removed. This figure has six triangular faces and is a convex polyhedron. This last figure differs from the others in that each of the others has all of its vertices in two parallel planes, whereas this figure does not.

A polyhedron with all of its vertices in two parallel planes is a **prisma-toid.** If the edges that are not in the parallel planes are parallel edges, then the polyhedron is a **prism.** The faces of the prism in the two parallel planes are the **bases** of the prism. The parallel edges joining the two bases are **lateral edges.** In more advanced courses it can be shown that the bases of a prism are congruent polygons. Notice in the sketches that the "hidden" edges are indicated by dotted line segments.

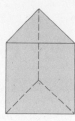

Prism

A polyhedron with all but one of its vertices on a plane is a **pyramid.** The vertices on a plane are the vertices of the **base** of the pyramid. Prisms and pyramids are often classified as triangular, rectangular, square, and so forth according to their bases.

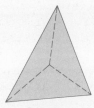

Pyramid

exercises

1. The tetrahedron *MNOP* is a triangular pyramid. Name its **(a)** vertices; **(b)** edges; **(c)** faces.

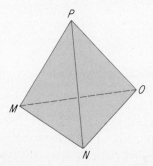

2. Repeat Exercise 1 for the **square pyramid** *PQRST*.

3. Repeat Exercise 1 for the cube *ABCDEFGH*.

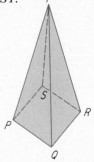

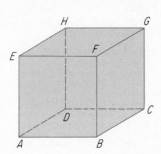

4. Think of the lines along the edges of the cube in Exercise 3 and identify the lines that appear to be **(a)** parallel to $\overleftrightarrow{AB}$; **(b)** skew to $\overleftrightarrow{AB}$; **(c)** parallel to the plane *ABFE*.

5. A cube has all of its vertices on two parallel planes and has square faces in those planes. Accordingly, a cube is an example of a **square prism.** Repeat Exercise 1 for the **triangular prism** *JKLMNO*.

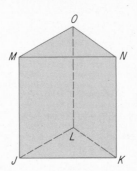

6. Let *V* be the number of vertices, *E* the number of edges, and *F* the number of faces. Then copy and complete the following table.

Figure	V	E	F
Triangular pyramid			
Square pyramid			
Cube			
Triangular prism			

In Exercises 7 through 13 extend the table in Exercise 6 for each figure.

7. A pyramid with a pentagon as a base.

8. A prism with a pentagon as a base.

  9. A prism with a hexagon as a base.

  10. A pyramid with a hexagon as a base.

  11. The given figure which represents two triangular pyramids with their common base removed.

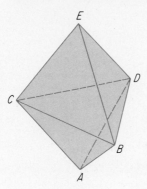

  12. The given figure *FGHIJK* which represents two square pyramids with their common base removed.

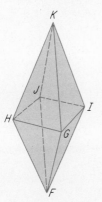

  13. The given figure *ABCDEFGHI* which represents a cube with one face replaced by the four triangular faces of a square pyramid.

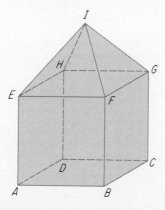

14. In the tables for Exercises 6 through 13 study the values of V, E, and F for each polyhedron. Compare the value of E with the value of $V + F$. Then conjecture a formula for E in terms of V and F. You should obtain the famous *Euler formula for polyhedrons*.

Consider the given figures and identify the:

15. Prisms.   **16.** Pyramids.

17. Triangular pyramids.   **18.** Square prisms.

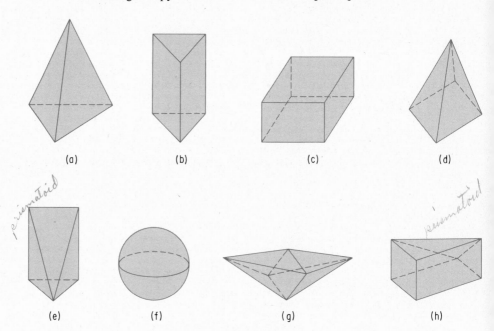

(a)   (b)   (c)   (d)

(e)   (f)   (g)   (h)

Sketch each figure.

19. A triangular pyramid.   **20.** A triangular prism.

21. A square pyramid.   **22.** A square prism.

23. A prismatoid with one square base and one triangular base.

24. A convex polyhedron that is not a prismatoid.

25. A polyhedron that is not convex.

26. A plane that intersects a cube in **(a)** a square; **(b)** a triangle; **(c)** a rectangle that is not a square.

27. A plane that intersects a square pyramid in **(a)** a square; **(b)** a triangle; **(c)** a trapezoid that is not a square.

28. A polyhedron with eight triangular regions as its faces.

29. A polyhedron with six triangular regions as its faces.

EXPLORATIONS

1. The "pictures" (sketches, drawings) of space figures on a sheet of paper never completely represent the figure and are often hard to visualize. Consider the accompanying pattern and make a model of a tetrahedron.

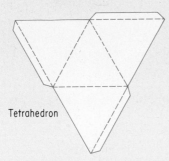

Tetrahedron

2. As in Exploration 1, consider the pattern and make a model of a cube.

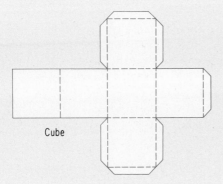

Cube

Name as many different types of polyhedrons as possible with vertices at the indicated number of points.

3. **(a)** 3 **(b)** 4

4. **(a)** 5 **(b)** 6

Consider cubes and their interior points.

5. A cube has two-inch edges and is painted red. Think of this two-inch cube as cut into eight one-inch cubes using three planes, one parallel to each pair of parallel faces of the cube. How many of the one-inch cubes have six painted faces? Five painted faces?

Four painted faces? Three painted faces? Two painted faces? One painted face? No painted faces?

6. Repeat Exploration 5 for a painted three-inch cube divided into 27 one-inch cubes.

7. Repeat Exploration 5 for a painted four-inch cube divided into 64 one-inch cubes.

8. Repeat Exploration 5 for a painted five-inch cube divided into 125 one-inch cubes.

9. Repeat Exploration 5 for a painted n-inch cube, $n \geq 2$, divided into n^3 one-inch cubes.

8-4
Plane Curves

Throughout this chapter we have considered points, lines, planes, rays, line segments, figures composed of line segments and rays, and regions of planes determined by line segments, rays, and half-planes. We now consider plane figures that may be approximated by broken lines and polygons; that is, we consider **plane curves.** Each of the following figures is a **simple closed curve.** Notice that a polygon is a special case of a simple closed curve.

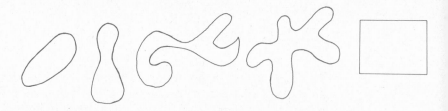

Any two simple closed curves have a common property. Can you find that property? Size, shape, angles, and straight lines are certainly not common properties here. What is left to be considered? Some readers may consider the common property that we are seeking trivial because of its simplicity. We prefer to believe that the really basic properties of mathematics are inherently simple. The property that we have in mind may be stated very easily: Any simple closed curve has an inside and an outside.

In advanced mathematics we say that *any simple closed curve in a plane divides the plane into two regions.* This is the **Jordan curve theorem.** The curve is the common boundary of the two regions, and one cannot cross from one region to the other without crossing the curve. The Jordan curve theorem is a very powerful theorem and yet a very simple one. Notice that it is independent of the size or shape of the curve.

How can such a simple theorem have any significance? It provides a basis for Euclid's assumption that any line segment joining the center of a circle to a point outside the circle must contain a point of the circle. In more advanced courses it provides a basis for the existence of a zero of a polynomial on any interval on which the polynomial changes sign. It provides a basis for Euler diagrams and Venn diagrams in any two-valued logic. In §2-6 we assumed that all points of the universal set were either points of a circular region A or points of A', the complement of A.

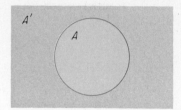

In §3-1 we considered statements p that are either true or false (in which case $\sim p$ is true) but not both true and false. Because of the Jordan curve theorem we were able to represent the situations under which statements were true (or false) by Euler diagrams and by Venn diagrams.

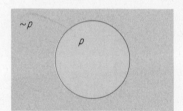

This simple theorem regarding the existence of an inside and an outside of any simple closed curve may also be used to answer questions raised by the next problem. This problem is a popular one, and many people have spent hours working on it. Consider three houses ($\times$) in a row and three utilities ($\circ$) in a second row on a plane surface. The problem is to join each house to each utility by an arc on the plane in such a way that no two arcs cross or pass through houses or utilities except at their endpoints. As in the figure on the left it is easy to designate paths from one house to each of the three utilities. One can also designate paths to each utility from the second house as in the second figure. Then one can designate two of the paths from the third house, but it is not possible on an ordinary plane to draw the path from the remaining house to the remaining utility. This assertion is based upon the fact that the simple closed curve indicated in the third figure divides the plane into two regions; the third house is inside the curve (shaded region), the remaining utility is outside the curve, and the two cannot be joined without crossing the curve.

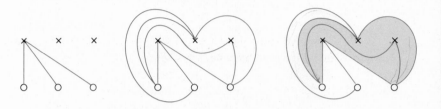

As in the case of broken lines, plane curves may be either simple or not simple and either closed or not closed. Each of the figures that follow is a plane curve that may be approximated by a closed broken line that is not simple. Each curve is a closed curve that is not simple.

Closed curves that are not simple divide the plane into three or more regions. The points at which a curve intersects itself are **vertices;** also, any other points may be designated as vertices. The simple curves with vertices as endpoints and not containing any other vertices are **arcs.** Each of the following curves has three points designated as vertices. The numbers V of vertices, A of arcs, and R of regions into which the plane is divided are shown below each figure. The last figure leaves the plane as a single region, since, for any two points P and Q that are points of the plane but not points of the figure, there is a simple plane curve (arc) with endpoints P and Q that does not intersect the given figure.

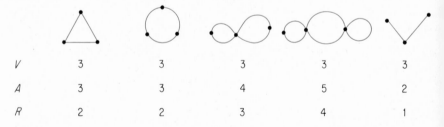

V	3	3	3	3	3
A	3	3	4	5	2
R	2	2	3	4	1

The next figures may be approximated by simple broken lines that are not closed. They are called simple curves that are not closed. Notice that a line and a ray are special cases of simple curves that are not closed; thus in geometry curves do not have to be "crooked."

The next figures may be approximated by a broken line that is neither simple nor closed; each is called a curve that is neither simple nor closed.

exercises

Sketch three curves that are different in appearance and are each:

1. Simple and closed. **2.** Simple and not closed.
3. Closed and not simple. **4.** Neither simple nor closed.

Sketch a broken line that closely approximates a given:

5. Simple closed curve.
6. Simple curve that is not closed.
7. Closed curve that is not simple.
8. Polygon.

In Exercises 9 through 14, find the numbers V of vertices, A of arcs, and R of regions for each figure.

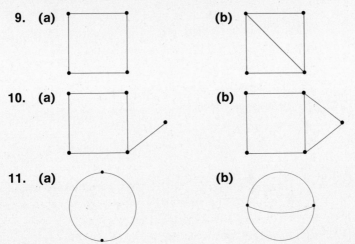

9. (a)   **(b)**

10. (a)   **(b)**

11. (a)   **(b)**

12. (a)   **(b)**

13. (a)   **(b)**

14. (a)   **(b)**

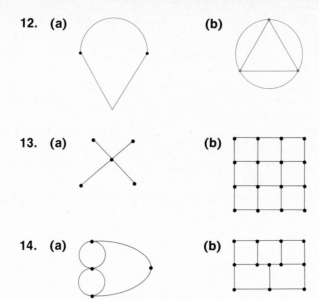

15. Study the values of V, A, and R obtained in Exercises 9 through 14 and find an expression for A in terms of V and R. You should obtain a form of the *Euler formula for networks*.

EXPLORATIONS

1. What is the smallest possible number of sides for a convex polygon?

2. What is the smallest possible number of sides for a polygon that is not a convex?

3. What is the smallest possible number of segments for a closed broken line that is not simple?

Describe a procedure that may be used to sketch a figure with the following properties.

4. A convex polygon with six sides.

5. A regular convex polygon with six sides.

6. A convex polygon with twenty-five sides.

*7. A convex polygon with n sides for any given positive integer $n \geq 3$.

Any set of line segments (or arcs) forms a **network.** If the network can be drawn by tracing each line segment exactly once without removing the point of the pencil from the paper, the network is **traversable.** For example, any connected simple network is traversable. If such a network is closed, any point may be selected as a starting point for the drawing and this point will also be the terminating point of the drawing. If the connected simple network is not closed, the drawing must start at one of the endpoints and terminate at the other.

The study of the traversability of networks probably stemmed from a problem concerning the bridges in the city of Königsberg. There was a river flowing through the city, two islands in the river, and seven bridges as in the figure. The people of Königsberg loved a Sunday stroll and thought

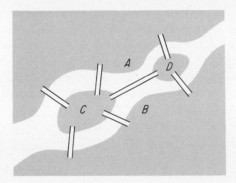

it would be nice to take a walk that would take them across each bridge exactly once. They found that no matter where they started or what route they tried, they could not cross each bridge exactly once. Gradually it was observed that the basic problem was concerned with paths between the two sides of the river *A*, *B*, and the two islands *C*, *D* as in the figure. With this

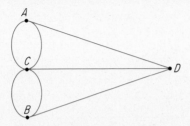

representation of the problem by a network, it was no longer necessary to discuss the problem in terms of walking across the bridges. Instead one could discuss whether or not the network associated with the problem was traversable. The problem was solvable if and only if its network was traversable.

When is a network traversable? One can walk completely around any city block, and it is not necessary to start at any particular point to do so. In general, one may traverse any simple closed broken line in a single trip. We next consider walking around two blocks and down the street $\overline{BE}$ separating them. This problem is a bit more interesting in that it is necessary to start at B or E. Furthermore, if one starts at B, one ends at E, and conversely.

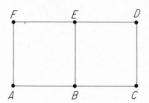

Note that it is permissible to pass through a vertex several times, but one may traverse a line segment only once. The peculiar property of the vertices B and E is based upon the fact that each of these endpoints is an endpoint of three line segments, whereas each of the other vertices is an endpoint of two line segments. A similar observation led a famous mathematician named Euler to devise a complete theory for traversable networks.

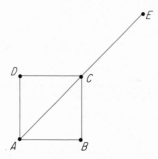

Euler classified the vertices of a network as odd or even. For example, in the given figure the vertex A is an endpoint of three arcs $\overline{AB}$, $\overline{AC}$, and $\overline{AD}$ and thus is an odd vertex; B is an endpoint of two arcs $\overline{BA}$ and $\overline{BC}$ and is an even vertex; C is an endpoint of four arcs $\overline{CB}$, $\overline{CA}$, $\overline{CD}$, and $\overline{CE}$ and is an even vertex; D is an endpoint of two arcs $\overline{DA}$ and $\overline{DC}$ and is an even vertex; E is an endpoint of one arc $\overline{EC}$ and is an odd vertex. Thus the figure has two odd vertices A and E and three even vertices B, C, and D. For any network a vertex that is an endpoint of an odd number of arcs is an **odd vertex**; a vertex that is an endpoint of an even number of arcs is an **even vertex**. Since each arc has two endpoints, there must be an even number of odd vertices in any network.

Any network that has only even vertices is traversable, and the trip may be started at any vertex. Furthermore, the trip will terminate at its starting point. If a network has exactly two odd vertices, it is traversable, but the trip must start at one of the odd vertices and will terminate at the other. If a network has more than two odd vertices, it is not traversable. In general, a network with $2k$ odd vertices may be traversed in k distinct trips. The network for the Königsberg bridge problem has four odd vertices and thus is not traversable in a single trip. Notice that the Königsberg bridge problem is independent of the size and shape of the river, bridges, or islands.

The study of simple closed curves, networks, and other plane curves that we have considered is a part of a particular geometry called **topology**.

exercises

In Exercises 1 through 8 identify **(a)** *the number of even vertices;* **(b)** *the number of odd vertices;* **(c)** *whether or not the network is traversable and, if it is traversable, the vertices that are possible starting points.*

1.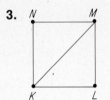

2.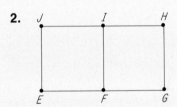

3.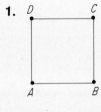

4.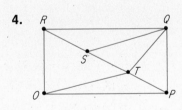

5.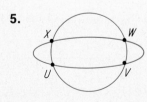

6.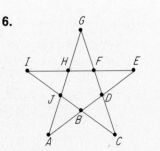

7. The network formed by the edges of a triangular pyramid.

8. The network formed by the edges of a cube.

9. Explain how a highway inspector can use a network to determine whether or not he can inspect each section of a highway without traversing the same section of the highway twice.

10. Use a related network to obtain and explain your answer to the following question: Is it possible to take a trip through a house with the floor plan indicated in the figure and pass through each doorway exactly once?

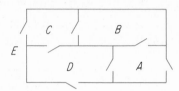

11. Explain whether or not it is possible to draw a simple connected broken line cutting each line segment of the given figure exactly once.

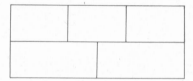

12. Consider the use of highway networks by a salesman who wants to visit each town exactly once. **(a)** Does the salesman always need to travel each highway at least once? **(b)** Consider each of the networks in Exercises 1 through 4 and indicate whether or not the salesman can visit each town exactly once without retracing any highway. **(c)** On the basis of your answer for part (b), does traversability of the network appear to be of interest to the salesman?

*13. Describe a modification that would make the Königsberg bridge problem possible.

*14. Describe a modification of the network for Exercise 11 such that the possibility of the desired construction is changed.

EXPLORATIONS

Many figures in topology have very unusual properties. One of these is a surface that is one sided. A fly can walk from any point on this surface to any other point without crossing an edge. Unlike a table top or a wall, it does not have a top and a bottom or a front and a back. This surface is called a **Möbius strip,** and it may be very easily constructed from a rectangular piece of paper such as a strip of gummed tape. Size is theoretically unimportant, but a strip an inch or two wide and about a foot long is easy to handle. We may construct a Möbius strip by twisting the strip of gummed tape just enough (one half-twist) to stick the gummed edge of one end to the gummed edge of the other end. If we cut across this strip, we again get a single strip similar to the one we started with. But if we start with a rectangular strip and cut around the center of the Möbius strip (see the dotted line in the second figure), we do not get two strips. Rather, we get one strip with two half-twists in it.

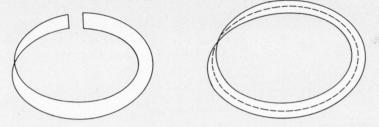

On one occasion one-sided surfaces of this sort were used as place cards at a seven-year-old's birthday party. While waiting for dessert, the youngsters were encouraged to cut the strip down the middle while guessing what the result would be. They were suitably impressed when they found only one piece, and were anxious to cut it again. Once more they were impressed when they found two pieces linked together. Almost a year after the party, one of the boys asked about the piece of paper that was in only one piece after it was cut in two.

When confronted with unusual properties such as those of the Möbius strip, both children and adults may ask questions that the nearest teacher cannot answer and that many college mathematics professors cannot answer. This is good for all concerned, since it impresses upon them that there is more to mathematics than formal algebraic manipulations and classical geometric constructions.

1. Construct a Möbius strip and cut around the center to obtain one strip with two half-twists in it.

2. Repeat Exploration 1 and then cut around the center again.

3. Construct a Möbius strip and cut along a path that is about one-third of the distance from one edge to the other.

4. Construct a Möbius strip, mark a point A on it, and draw an arc from A around the strip until you return to the point A.

5. Explain why a Möbius strip is called a one-sided surface.

6. Does a Möbius strip have one or two simple closed curves as its edge? Explain the reason for your answer.

8-6
Linear Measures

Distances and lengths of objects are often called **linear measurements** since the shortest distance from one point to another is the length of the line segment joining them. Whenever distances or lengths are stated, some unit of measure needs to be given or assumed. For example, the length of a certain walking stick might be described by various people as

> A little more than a yard
> About 4 feet
> About 1 meter
> 1.15 meters
> 45 inches
> $1\frac{1}{4}$ yards
> 5 spans

In each case, the number used is a **measure;** the number of units is a **measurement.**

A variety of units of length are considered in schools along with the relations among these units. Since very few people have difficulties with the study of units of linear measure, we simply assume that some unit has been selected and use a scale as on a number line.

Any line may be considered as a number line (§8-1). Then each point P_x on the line has a unique real number x as its *coordinate;* each real number x has a unique point P_x as its *graph.* Since each of the points of a number line may be identified by *one* real number, a line is often called a **one-space.**

Sets of points on a line form **line figures.** For example, any two points P_r and P_s on a number line are endpoints of a unique line segment $\overline{P_r P_s}$ with the absolute value of $s - r$ as its **linear measure.** We denote the linear measure of $\overline{P_r P_s}$ by $m(\overline{P_r P_s})$ (read as "measure of $\overline{P_r P_s}$") and write

$$m(\overline{P_r P_s}) = |s - r|$$

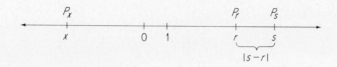

EXAMPLE 1

Find the linear measure of (a) $\overline{P_4P_1}$; (b) $\overline{P_{-2}P_3}$.

SOLUTION

(a) $m(\overline{P_4P_1}) = |1 - 4| = |-3| = 3$

(b) $m(\overline{P_{-2}P_3}) = |3 - (-2)| = |5| = 5$

Any two line segments with equal measures are called **congruent line segments**. Since any line segment has a unique linear measure, any line segment is congruent to itself; for example, $\overline{AB} \cong \overline{BA}$ (read as "line segment AB is congruent to line segment BA").

Measures of line segments on a number line are often used in the discussion of addition of whole numbers. In general, any line segment $\overline{AB}$ has a nonnegative real number $m(\overline{AB})$ as its linear measure. The **length** of the line segment is $m(\overline{AB})$ units for some unit of linear measure such as foot, inch, meter, mile, or kilometer. For example, a pencil with a linear measure of 3 in inches has length 3 inches.

A point has linear measure 0. In the figure $m(\overline{AB}) = 3$, $m(\overline{CD}) = 4$, $\overline{AB} \cap \overline{CD} = \overline{CB}$, $m(\overline{CB}) = 1$, $\overline{AB} \cup \overline{CD} = \overline{AD}$, $m(\overline{AD}) = 6$. Notice that $m(\overline{AB}) + m(\overline{CD}) \neq m(\overline{AD})$, since $m(\overline{AB} \cap \overline{CD}) \neq 0$.

$$m(\overline{AD}) = m(\overline{AB} \cup \overline{CD}) = m(\overline{AB}) + m(\overline{CD}) - m(\overline{AB} \cap \overline{CD})$$

In general, if the measure of the intersection of two line segments is 0, then the linear measure of the union of the two line segments is equal to the sum of their linear measures.

EXAMPLE 2

Show the relationship among the linear measures of $\overline{P_1P_2}$, $\overline{P_2P_5}$, and $\overline{P_1P_5}$.

SOLUTION

$$m(\overline{P_1P_2}) = 1, \; m(\overline{P_2P_5}) = 3$$

$$m(\overline{P_1P_2} \cap \overline{P_2P_5}) = m(\overline{P_2P_2}) = 0$$

$$m(\overline{P_1P_5}) = 4 = m(\overline{P_1P_2}) + m(\overline{P_2P_5})$$

EXAMPLE 3

Show the relationship among the linear measures of $\overline{P_{-2}P_3}$, $\overline{P_2P_5}$, and $\overline{P_{-2}P_5}$.

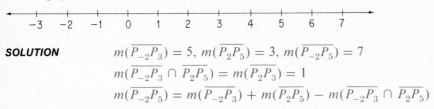

SOLUTION

$$m(\overline{P_{-2}P_3}) = 5, \; m(\overline{P_2P_5}) = 3, \; m(\overline{P_{-2}P_5}) = 7$$

$$m(\overline{P_{-2}P_3} \cap \overline{P_2P_5}) = m(\overline{P_2P_3}) = 1$$

$$m(\overline{P_{-2}P_5}) = m(\overline{P_{-2}P_3}) + m(\overline{P_2P_5}) - m(\overline{P_{-2}P_3} \cap \overline{P_2P_5})$$

The *linear measure* of a figure is a number n; the *length* of the figure is n units. The length of a simple closed curve, such as a polygon, is its **perimeter.** Since any two sides of a polygon have at most one point in common, the perimeter of the polygon is the sum of the lengths of its sides.

EXAMPLE 4

Find the perimeter of the given polygon.

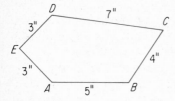

SOLUTION

The perimeter in inches of polygon $ABCDE$ is

$$m(\overline{AB}) + m(\overline{BC}) + m(\overline{CD}) + m(\overline{DE}) + m(\overline{AE}) =$$
$$5 + 4 + 7 + 7 + 3 = 22$$

Thus the perimeter is 22 inches.

The measurement of a line segment to determine its length is done with some standard scale such as a ruler marked off in inches. The reading of the linear measure from the ruler requires an estimation to the nearest unit used by the person making the measurement. For example, consider the measurement of $\overline{PQ}$ using the scale below it.

To the nearest unit $m(\overline{PQ}) = 6$. We write $m(\overline{PQ}) \approx 6$ (read as "the measure of the line segment PQ is approximately 6") to indicate that the length of $\overline{PQ}$ is about 6 units. Some people mentally subdivide the unit of length on a scale and *estimate* measures to smaller units. For example, different people might say:

$m(\overline{PQ}) \approx 6$ to the nearest half-unit

$m(\overline{PQ}) \approx 6\frac{1}{4}$, that is, 6 and $\frac{1}{4}$ to the nearest quarter unit

$m(\overline{PQ}) \approx 6.2$, that is, 6 and $\frac{2}{10}$ to the nearest tenth of a unit

Whatever subdivisions are used, the measure is based upon someone's "reading the scale" and thus is approximate. All linear measurements are approximate since they are based upon estimation from reading a scale. Indeed, all measurements are approximate. Even though a measurement is made with a very precise instrument, there is always a final estimation and thus an approximation.

If the length is 6 inches to the nearest inch, the greatest possible error is $\frac{1}{2}$ inch. If the length is $6\frac{1}{4}$ inches to the nearest quarter of an inch, the greatest possible error is one-half of the unit $\frac{1}{4}$ inch, and thus $\frac{1}{8}$ inch. If the length is 6.2 inches to the nearest tenth of an inch, the greatest possible error is one-half of the unit 0.1 inch, and thus 0.05 inch. For any measurement the **greatest possible error** is one-half of the smallest unit used.

There exist special techniques for computations with approximate numbers, that is, numbers that represent approximations. We shall leave such special techniques for more advanced courses. Thus we treat numbers obtained from measurements the same as other numbers. We simply use the **approximation symbol** $\approx$ to indicate that the numbers are approximate rather than cardinal numbers of sets of individual elements such as 5 books or 72 people.

exercises

On a number line find the linear measure of:

1. $\overline{P_1 P_5}$ **2.** $\overline{P_{-1} P_4}$ **3.** $\overline{P_{-1} P_{-4}}$

4. $\overline{P_2 P_{-3}}$ **5.** $\overline{P_{-7} P_{-1}}$ **6.** $\overline{P_3 P_{-2}}$

Show the relationship among the linear measures of:

7. $\overline{P_2P_5}$, $\overline{P_5P_7}$, and $\overline{P_2P_7}$

8. $\overline{P_{-3}P_2}$, $\overline{P_2P_5}$, and $\overline{P_{-3}P_5}$

9. $\overline{P_1P_4}$, $\overline{P_2P_7}$, and $\overline{P_1P_7}$

10. $\overline{P_{-4}P_0}$, $\overline{P_{-1}P_3}$, and $\overline{P_{-4}P_3}$

11. $\overline{P_1P_{-3}}$, $\overline{P_{-2}P_4}$, and $\overline{P_{-3}P_4}$

12. $\overline{P_5P_2}$, $\overline{P_5P_6}$, and $\overline{P_2P_6}$

13. $\overline{P_4P_{-1}}$, $\overline{P_{-1}P_2}$, and $\overline{P_2P_4}$

14. $\overline{P_{-1}P_{-5}}$, $\overline{P_{-5}P_{-2}}$, and $\overline{P_{-3}P_{-1}}$

15. $\overline{P_2P_{-3}}$, $\overline{P_{-3}P_5}$, and $\overline{P_0P_5}$

Find the perimeter of each given polygon.

16.
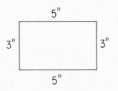

17.

18.
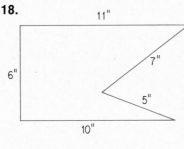

19.
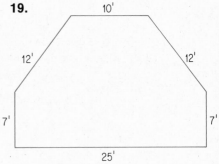

Express the measure of $\overline{PQ}$.

20. To the nearest unit.

21. To the nearest half-unit.

22. To the nearest quarter unit.

23. Find the greatest possible error of the measurements in Exercises 20 through 22.

Which of these measurements is probably approximate?

24. I have 27 students in my class.

25. My breakfast has 627 calories.

26. Portland has a population of 67,500.

27. There are 527 pages in my history book.

28. Bobby is 18 months old.

EXPLORATIONS

Arcs and vertices provide the basis for the new game of **sprouts.** There is even a more complicated version called "Brussels sprouts." Both games are presented in Martin Gardner's "Mathematical Games" in the July 1967 issue of the *Scientific American*, pages 112 to 115.

To play the game of sprouts, the players start with one or more given points (vertices) called *spots.* Each player in turn draws an arc from a spot to another or the same spot and places a new spot at some point on this arc. There are two basic rules:

I.   Each arc may have any shape but must not cross itself, cross a previously drawn arc, or pass through a previously made spot.

II.   No spot may be an endpoint of more than three arcs.

The last person able to draw an arc subject to these rules is the winner.

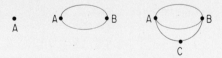

Suppose that two players start with a single spot *A*. Then the first player draws an arc from *A* to *A* and adds a spot *B* on this arc. The second player draws an arc from *A* to *B* (or *B* to *A*) and adds a spot *C* on this arc. Now *A* is an endpoint of three arcs, cannot be used again, and is called a *dead spot.* Similarly, *B* is a dead spot. The spot *C* is already an endpoint of two arcs and thus cannot be used for two ends of another arc. Thus it is not possible to draw another arc and the second player is the winner.

Consider the game of sprouts with two given spots A, B, that is, two-spot sprouts, and two players.

1. Suppose that the first three moves are as shown in the figures.

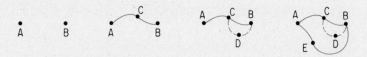

(a) Copy the last figure and draw an arc that makes the second player the winner immediately, that is, makes it impossible for the first player to continue.

(b) Copy the last figure and show how the first player can be the winner (assuming certain plays from the second player).

2. Suppose that the first player draws arc *ACB* and the second player draws *BDC* as in the figure.

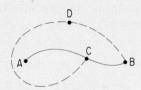

(a) Copy the figure and show how the second player can win.

(b) Copy the figure and show how the first player can win.

3. Since each spot becomes a dead spot as soon as it is used three times as an endpoint of arcs, each of the original spots may be regarded as having three lives. Then in a game starting with two spots there are six lives in all originally.

(a) How many lives are there after the first player has completed his move?

(b) How many lives are there after the second move?

(c) What is the largest possible number of moves in two-spot sprouts?

Consider the game of sprouts with more than two given spots.

4. What is the largest possible number of moves in three-spot sprouts?

5. What is the largest possible number of moves in four-spot sprouts?

Area measure is a surface measure and is used for rectangular regions and other regions. The unit of area measure is the measure of a square region having an edge with length 1 unit of linear measure. For example, 1 square inch, 1 square foot, and 1 square mile may each be used as units of area measure.

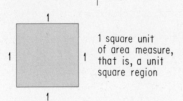

1 square unit of area measure, that is, a unit square region

Consider a rectangular region $ABCD$ with $m(\overline{AB}) = b$ and $m(\overline{BC}) = h$. When b and h are whole numbers, the rectangular region has the same area measure as bh unit square regions. In the second figure $b = 3$, $h = 2$, and the rectangular region $PQRS$ has area measure 6; that is, the area of rectangular region $PQRS$ is 6 square units.

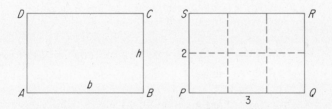

We use the symbol $\square ABCD$ for rectangle $ABCD$ and follow the custom of speaking of the area of the rectangle when we mean the rectangular region. Thus we write $\mathcal{C}(\square ABCD)$ for the **area measure** of rectangle $ABCD$. Then

$$\mathcal{C}(\square PQRS) = 6$$

and, in general,

$$\mathcal{C}(\square ABCD) = bh$$

where $m(\overline{AB}) = b$ and $m(\overline{BC}) = h$. We have illustrated the area formula for whole numbers b and h. We *define* the area measure of the rectangle to be bh in square units for all linear measures b and h.

A square is a special case of a rectangle, $b = h$. Thus the area of a square region with an edge e units long is e^2 square units. (See page 334.) As in the case of rectangles we often speak of the areas of squares or other polygons and mean thereby the areas of the polygonal regions.

A point has area measure 0 and a line segment has area measure 0. If two plane regions have area measures and the area measure of their

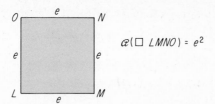

$\mathscr{C}(\square\ LMNO) = e^2$

intersection is 0, then the area measure of the union of the two regions is equal to the sum of their area measures.

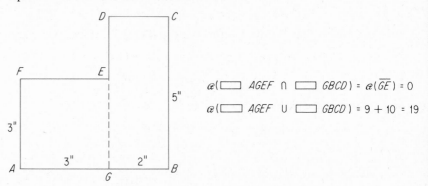

$\mathscr{C}(\square\ AGEF\ \cap\ \square\ GBCD) = \mathscr{C}(\overline{GE}) = 0$

$\mathscr{C}(\square\ AGEF\ \cup\ \square\ GBCD) = 9 + 10 = 19$

To find the area measures of triangular regions, we assume that congruent figures have equal area measures. Then since a diagonal of a rectangle divides the rectangular region into congruent triangular regions, the area measure of a right triangle with legs of length b and h is $\frac{1}{2}bh$, as indicated in the figure below.

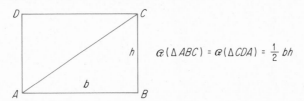

$\mathscr{C}(\triangle ABC) = \mathscr{C}(\triangle CDA) = \frac{1}{2}\,bh$

For any triangle ABC, a side, such as $\overline{AB}$, may be selected as base. Then the line that contains C and is perpendicular to $\overleftrightarrow{AB}$ intersects $\overleftrightarrow{AB}$ at a point D. There are three cases to consider according as D is an interior point of $\overline{AB}$, an endpoint of $\overline{AB}$, or an exterior point of $\overline{AB}$.

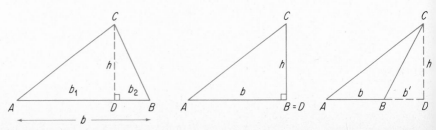

In the figure on the left, D is an interior point of $\overline{AB}$, $m(\overline{AB}) = b$, $m(\overline{AD}) = b_1$, $m(\overline{DB}) = b_2$, $b_1 + b_2 = b$, $m(\overline{CD}) = h$, $\mathcal{C}(\triangle ADC) = \frac{1}{2}b_1h$, $\mathcal{C}(\triangle DBC) = \frac{1}{2}b_2h$, and

$$\mathcal{C}(\triangle ABC) = \mathcal{C}(\triangle ADC) + \mathcal{C}(\triangle DBC)$$
$$= \tfrac{1}{2}b_1h + \tfrac{1}{2}b_2h$$
$$= \tfrac{1}{2}(b_1 + b_2)h = \tfrac{1}{2}bh$$

In the center figure, $B = D$, $m(\overline{AB}) = b$, $m(\overline{CD}) = h$, and $\mathcal{C}(\triangle ABC) = \frac{1}{2}bh$.

In the figure on the right, D is an exterior point of $\overline{AB}$, $m(\overline{AB}) = b$, $m(\overline{BD}) = b'$, $m(\overline{AD}) = b + b'$, $m(\overline{CD})\, h$, and

$$\mathcal{C}(\triangle ABC) = \mathcal{C}(\triangle ADC) - \mathcal{C}(\triangle BDC)$$
$$= \tfrac{1}{2}(b + b')h - \tfrac{1}{2}b'h$$
$$= \tfrac{1}{2}[(b + b') - b']h$$
$$= \tfrac{1}{2}[b + (b' - b')]h = \tfrac{1}{2}bh$$

In all cases and thus for any triangle ABC, $\mathcal{C}(\triangle ABC) = \frac{1}{2}bh$.

Areas of polygonal regions may be found, as in the supplementary exercises for this section, by subdividing the polygonal region into nonoverlapping rectangular and/or triangular regions, that is, regions with no common interior points. For example, any parallelogram $ABCD$ with base b and height h has area bh.

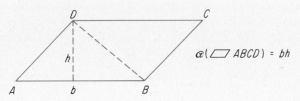

$$\mathcal{C}(\square\ ABCD) = bh$$

EXAMPLE

Find the area of the given region.

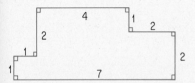

SOLUTION

We subdivide the region into rectangular regions. The three rectangular regions indicated in the next figure have area measures 1, 12, and 4. Thus the area measure of the given region is 17 and the area is 17 square units.

In the example notice that as for linear measure and length, the area measure is a number; the area is that number of square units.

Volume measure is a space measure and is used for cubic and other regions. As for area measure it is customary to speak of the volume of a cube or other polyhedron and to mean thereby the volume of the polyhedral region. The unit of volume measure is the measure of a cubic region having an edge with length 1 unit of linear measure. For example, 1 cubic inch, 1 cubic centimeter, and 1 cubic foot may each be used as units of volume measure.

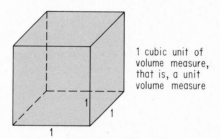

1 cubic unit of volume measure, that is, a unit volume measure

Volumes are considered in a manner very similar to areas. We define any rectangular parallelepiped with edges of linear measure a, b, and c on a common endpoint to have volume measure abc:

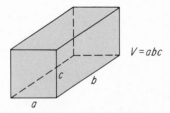

$V = abc$

Points, line segments, and plane regions have volume measure 0. Any pyramid with a base of area B square units and height to that base h linear units, can be proved in more advanced courses to have volume measure $\frac{1}{3}Bh$. See the figure at the top of page 337.

Volumes of many other regions may be obtained by subdividing the region into pyramids and/or rectangular parallelepipeds, or by using the prismoidal formula. (See the Explorations.) As in the case of length and area, the volume measure is a number; the volume is that number of cubic units.

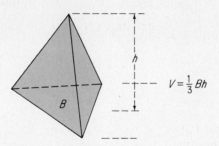

$$V = \frac{1}{3} Bh$$

exercises

Find the area measurement of each region.

1.

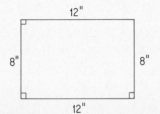

12"
8"   8"
12"

2.

12"
8"   6"   8"
12"

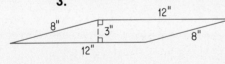

3.

8"   12"
      3"
12"   8"

4.

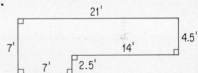

21'
7'   14'   4.5'
    7'   2.5'

5.

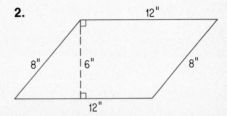

12"
            2"
6"
            4"
8"

6.

7'
5'
10'

7.

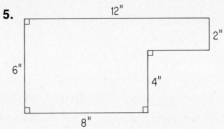

7"
6"
2"   14"
    16"

8.

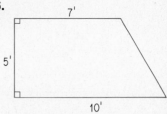

12'
7'
6'   7'
7'   7'   10'
    3'   5'

Find a formula for:

9. The surface area S of a cube with edge of linear measure e.

10. The surface area S of a rectangular parallelepiped with edges of linear measure b, d, and h.

Describe the change in the area of:

11. A square when the measure of each of its edges is doubled.

12. A square when the measure of each of its edges is multiplied by **(a)** 3; **(b)** $\frac{1}{2}$; **(c)** a constant k.

13. A rectangle when the measure of each of its edges is multiplied by **(a)** 2; **(b)** 3; **(c)** $\frac{1}{2}$; **(d)** a constant k.

14. A triangle when each of its edges is multiplied by **(a)** 2; **(b)** 3; **(c)** $\frac{1}{2}$; **(d)** a constant k.

15. A polygon when the measures of its angles are not changed and the measure of each of its edges is multiplied by **(a)** 2; **(b)** a constant k.

*Sketch and find **(a)** the surface area, and **(b)** the volume, of a cube with edge of the given length.*

16. 2 feet   **17.** 3 yards   **18.** 4 centimeters   **19.** 5 inches

*Find **(a)** the surface area, and **(b)** the volume, of each rectangular parallelepiped.*

20.

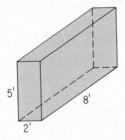

21.

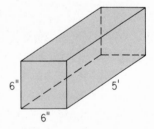

Find the volume of each pyramid.

22.

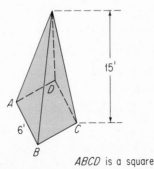

ABCD is a square

23.

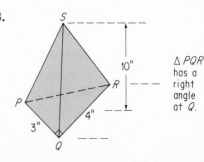

$\triangle PQR$ has a right angle at Q.

Describe the change in the volume of each figure.

24. A cube when the measure of each of its edges is doubled.

25. A cube when the measure of each of its edges is multiplied by **(a)** 3; **(b)** $\frac{1}{2}$; **(c)** a constant k.

26. A rectangular parallelepiped when the measure of each of its edges is multiplied by **(a)** 2; **(b)** 3; **(c)** $\frac{1}{2}$; **(d)** a constant k.

EXPLORATIONS

The most general common formula for volume measures is the **prismoidal formula**

$$V = \frac{h}{6}(B_1 + 4M + B_2)$$

This formula may be used, for example, for any prismatoid where B_1 and B_2 are the area measures of the bases, M is the area measure of the intersection of the prismatoid with the plane halfway between the bases, and the bases are h units apart; that is, the altitude to the bases has linear measure h. The plane section halfway between the bases is often called the **midsection.** Then M is the area measure of the midsection.

1. Show that for a cube with an edge of linear measure e, $B_1 = e^2$, $M = e^2$, $B_2 = e^2$, and the prismoidal formula gives the usual expression for the volume.

2. Think of a sphere as in the figure and use the prismoidal formula to find an expression for the volume of a sphere of radius r. (The area measure of a circle of radius r is πr^2.)

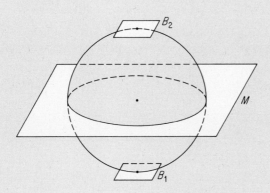

3. For a square pyramid with height h and a base with an edge of linear measure e, the midsection is a square with an edge $\frac{1}{2}e$. Use the prismoidal formula to find an expression for the volume of the square pyramid.

4. Use the prismoidal formula to find an expression for the volume of a circular cone with height h if the base has radius r and the midsection has radius $\frac{1}{2}r$.

8-8

Coordinates on a Plane

Any plane may be considered as a coordinate plane as in §7-7. Any two points $A: (x_1, y_1)$ and $B: (x_2, y_2)$ determine a line $\overleftrightarrow{AB}$.

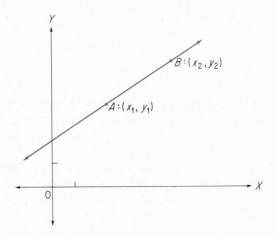

EXAMPLE 1

Graph $A: (2, 1)$ and $B: (2, 5)$. Draw the line $\overleftrightarrow{AB}$ and find the length of the line segment $\overline{AB}$.

SOLUTION   The given points may be graphed as in the figure. The line $\overleftrightarrow{AB}$ is parallel to the y-axis and has equation $x = 2$ since A and B both have x-coordinate 2.

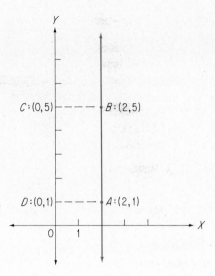

To find the length of $\overline{AB}$ we consider the two points D: $(0, 1)$ and C: $(0, 5)$ on the y-axis, note that the opposite sides $\overline{AB}$ and $\overline{CD}$ of the rectangle $ABCD$ are congruent (have the same length), and note that on the y-axis $m(\overline{CD}) = |5 - 1|$, that is, 4. Therefore, the length of $\overline{AB}$ is 4 units.

For any points A: (x_1, y_1) and B: (x_1, y_2) the line $\overleftrightarrow{AB}$ is parallel to the y-axis and the linear measure of the line segment $\overline{AB}$ is $|y_2 - y_1|$. Similarly for any points A: (x_1, y_1) and C: (x_2, y_1) the line $\overleftrightarrow{AC}$ is parallel to the x-axis and $m(\overline{AC}) = |x_2 - x_1|$. In general, the linear measure of any line segment is a number; the length of the line segment is that number of units of length.

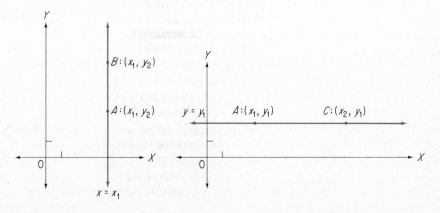

Consider the points P_x with coordinates x on a number line. We define the **directed distance** from P_r to P_s to be $s - r$. Then the directed distance from P_1 to P_4 is $4 - 1$, that is, 3; the directed distance from P_4 to P_1 is $1 - 4$, that is, -3; the directed distance from P_{-1} to P_1 is $1 - (-1)$, that is, 2. We use directed distances to develop a formula for the coordinate of the midpoint of a line segment. To find the midpoint M of a line segment $\overline{P_r P_s}$ we start at P_r and proceed halfway from P_r to P_s. Therefore, the coordinate of M is $r + \frac{1}{2}(s - r)$, that is, $\frac{1}{2}(r + s)$.

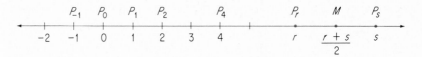

EXAMPLE 2

Find the linear measure and the coordinate of the midpoint of $\overline{AB}$ for $A : (1, 5)$ and $B : (7, 5)$.

SOLUTION

Consider the points $C : (1, 0)$ and $D : (7, 0)$ and the rectangle $CDBA$ as in the figure.

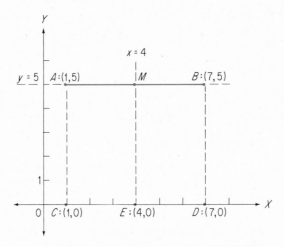

$$m(\overline{AB}) = m(\overline{CD}) = |7 - 1| = 6$$

The y-coordinate of the midpoint M of $\overline{AB}$ must be 5, since every point of $\overleftrightarrow{AB}$ has y-coordinate 5. The midpoint E of $\overline{DC}$ has x-coordinate $\frac{1}{2}(1 + 7)$, that is, 4. The lines $\overleftrightarrow{AC}$, $x = 4$, and $\overleftrightarrow{BD}$ are parallel and intercept (cut off) congruent segments $\overline{CE}$ and $\overline{ED}$ on the x-axis. Therefore these lines intercept congruent segments on $\overline{AB}$, M must be on the line $x = 4$, and M has coordinates $(4, 5)$.

In general, any two distinct points $A: (x_1, y_1)$ and $C: (x_2, y_1)$ on a line $y = y_1$ parallel to the x-axis are endpoints of a line segment $\overline{AC}$ with linear measure $|x_2 - x_1|$ and midpoint $(\frac{1}{2}(x_1 + x_2), y_1)$. Similarly, any two distinct points $A: (x_1, y_1)$ and $B: (x_1, y_2)$ on a line $x = x_1$ parallel to the y-axis are endpoints of a line segment $\overline{AB}$ with linear measure $|y_2 - y_1|$ and midpoint $(x_1, \frac{1}{2}(y_1 + y_2))$.

EXAMPLE 3

Find the coordinates of the midpoints of the line segments with endpoints (a) $(1, 0)$ and $(-3, 0)$; (b) $(2, 1)$ and $(2, 7)$.

SOLUTION

(a) $\left(\dfrac{1 + (-3)}{2}, 0 \right) = (-1, 0)$

(b) $\left(2, \dfrac{1 + 7}{2} \right) = (2, 4)$

For two points $A: (x_1, y_1)$ and $B: (x_2, y_2)$ such that $\overline{AB}$ is not parallel to a coordinate axis we consider also the point $C: (x_2, y_1)$. Then, as in the figure, the midpoints of $\overline{AC}$ and $\overline{AB}$ are on the line $x = \frac{1}{2}(x_1 + x_2)$, and

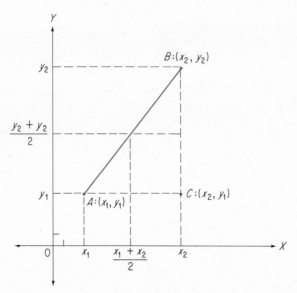

the midpoints of $\overline{CB}$ and $\overline{AB}$ are on the line $y = \frac{1}{2}(y_1 + y_2)$. Thus, as in the figure, the midpoint of $\overline{AB}$ has coordinates

$$\left(\dfrac{x_1 + x_2}{2}, \dfrac{y_1 + y_2}{2} \right)$$

This is the **midpoint formula** for points on a coordinate plane. Notice that the formula holds whether or not $\overline{AB}$ is parallel to a coordinate axis.

EXAMPLE 4

Find the midpoint of $\overline{AB}$ for (a) A: $(6, 1)$ and B: $(-2, 5)$; (b) A: $(-3, 4)$ and B: $(1, -2)$.

SOLUTION

(a) $\left(\dfrac{6 + (-2)}{2}, \dfrac{1 + 5}{2} \right) = (2, 3)$

(b) $\left(\dfrac{-3 + 1}{2}, \dfrac{4 + (-2)}{2} \right) = (-1, 1)$

exercises

1. Find the linear measure and the coordinates of the midpoint of the line segment with endpoints:
   (a) $(0, 3)$ and $(4, 3)$ **(b)** $(2, 1)$ and $(5, 1)$
   (c) $(2, 1)$ and $(2, 5)$ **(d)** $(5, -2)$ and $(5, 7)$
   (e) $(-3, 2)$ and $(-3, 7)$ **(f)** $(-3, 2)$ and $(-7, 2)$

2. Find formulas for the lengths of line segments with endpoints:
   (a) $(0, 0$ and $(x, 0)$ **(b)** $(0, 0$ and $(0, y)$
   (c) $(x_1, 0)$ and $(x_2, 0)$ **(d)** $(0, y_1)$ and $(0, y_2)$
   (e) (x_1, y_1) and (x_2, y_1) **(f)** (x_1, y_1) and (x_1, y_2)

   Find the coordinates of:

3. The midpoint of $\overline{AB}$ for **(a)** A: $(1, -6)$ and B: $(7, 2)$; **(b)** A: $(5, 11)$ and B: $(-1, 3)$; **(c)** A: $(-2, -7)$ and B: $(-4, -5)$.

4. The endpoint B if A: $(1, 3)$ and C: $(2, 5)$ is the midpoint of $\overline{AB}$.

5. The endpoint B if A: $(3, 0)$ and C: $(-1, 4)$ is the midpoint of $\overline{AB}$.

6. The endpoint A if B: $(-5, 6)$ and C: $(-3, -1)$ is the midpoint of $\overline{AB}$.

7. The vertex C of a square $ABCD$ with A: $(0, 0)$, B: $(a, 0)$, and D: $(0, a)$.

8. The vertex S of a rectangle $QRST$ with Q: $(0, 0)$, R: $(a, 0)$, and T: $(0, b)$.

9. The vertex S of an isosceles trapezoid $QRST$ with Q: $(0, 0)$, R: $(a, 0)$, and T: (b, c), where $2b < a$.

   Draw coordinate axes, sketch each set of points, and label each line with its equation.

10. The points 5 units from the y-axis.

11. The points 2 units from the x-axis.

12. The points 3 units from the line $x = 2$.

13. The points 4 units from the line $y = -1$.

Use properties of figures on a coordinate plane and prove:

14. If $A: (0, 0)$, $B: (a, 0)$, and $C: (b, c)$ are the vertices of a triangle, then the midpoints of $\overline{AC}$ and $\overline{BC}$ are endpoints of a line segment that is parallel to $\overline{AB}$ and half as long as $\overline{AB}$.

*15. The line segments with the midpoints of opposite sides of a quadrilateral as endpoints bisect each other. [*Hint:* Use vertices $A: (0, 0)$, $B: (a, 0)$, $C: (b, c)$, and $D: (d, e)$.]

EXPLORATIONS

Graph paper may be used for a wide variety of explorations.

1. (a) Use graph paper and cut out a square with 16 units on each side. Note that the area of this square is 256 square units.

(b) Mark and cut the square into four pieces as indicated in the figure.

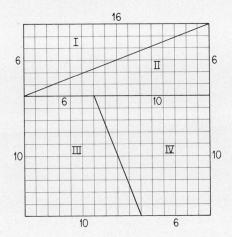

(c) Rearrange the pieces as in this figure.

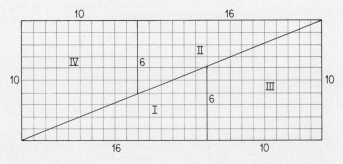

(d)   Note that a rectangular region 10 by 26 units has area 260 square units. How do you explain the increase of 4 square units?

The pegs of a pegboard may be considered as points with integral coordinates on a coordinate plane. Consider this figure showing an array of such points. Imagine stretching a rubber band

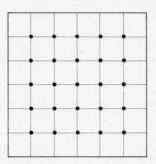

around a set of these dots. There is a formula that relates the area of the enclosed region to the number of points involved. See if you can discover the formula from the following explorations.

2.   Consider the given figures and complete the table.

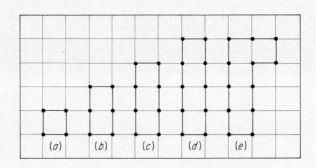

	(a)	(b)	(c)	(d)	(e)
Number (b) of boundary points	4	6	8	10	12
Areas (A)	1	2			

3.   Conjecture a formula that relates A and b as in the table for Exploration 2.

4. Each of the figures in Exploration 2 had no dots as interior points. Consider these figures and complete the table.

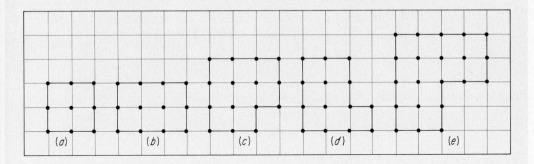

	(a)	(b)	(c)	(d)	(e)
Number (b) of boundary points	8	10	12		
Number (i) of interior points	1	2	3		
Area (A)	4	6			

5. Does the formula that you conjectured in Exploration 2 hold for the table in Exploration 4?

6. Consider $i = 0$ in Exploration 2 and conjecture a formula for the tables in Explorations 2 and 4.

7. Draw several figures of your own and test the formula conjectured in Exploration 6.

8-9

Distances on a Plane

The distance formula for points on a coordinate plane is based upon the *Pythagorean theorem* (§6-8):

For any right triangle the area of the square upon the hypotenuse is equal to the sum of the areas of the squares upon the other two sides.

This is essentially the form in which the early Greeks thought of the Pythagorean theorem before algebraic notations had been developed. We now

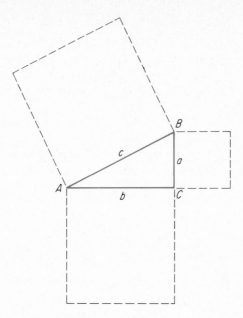

usually select the right angle at the vertex C of a right triangle ABC, let c be the linear measure of the hypotenuse (the side opposite the right angle), let a and b be the linear measures of the other two sides (legs), and write

$$c^2 = a^2 + b^2$$

This is an algebraic statement of the equivalence of the areas of the squares considered by the early Greeks.

The Pythagorean theorem may be used to obtain a general formula for the linear measure of any line segment on a coordinate plane. Let the

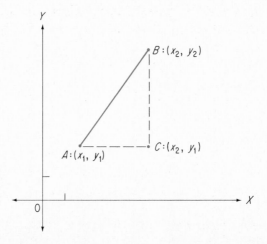

endpoints of the line segment be $A: (x_1, y_1)$ and $B: (x_2, y_2)$. As in §8-8, either the line segment $\overline{AB}$ is parallel to a coordinate axis, or a right triangle may be formed with the vertex of the right angle at $C: (x_2, y_1)$. The linear measures of the legs of the right triangle are $|x_2 - x_1|$ and $|y_2 - y_1|$. By the Pythagorean theorem, the linear measure of the hypotenuse is

$$m(\overline{AB}) = \sqrt{(x_2 - x_1)^2 + (y_2 - y_1)^2}$$

This is the **distance formula** on a plane. The formula holds whether or not the line segment is parallel to a coordinate axis.

EXAMPLE 1

Find the linear measure of the line segment with endpoints $A: (2, -3)$ and $B: (5, 1)$.

SOLUTION $m(\overline{AB}) = \sqrt{(5 - 2)^2 + [1 - (-3)]^2} = \sqrt{3^2 + 4^2} = 5$

The circle with center at $C: (h, k)$ and radius r is the set of points at a distance r from the point C. The distance of any point $P: (x, y)$ from $C: (h, k)$ is

$$\sqrt{(x - h)^2 + (y - k)^2}$$

Therefore, the circle with center $C: (h, k)$ and radius r has equation

$$(x - h)^2 + (y - k)^2 = r^2$$

Each point $P: (x, y)$ on the circle has coordinates that satisfy the equation. Each ordered pair (x, y) of numbers that satisfy the equation determines a point of the circle.

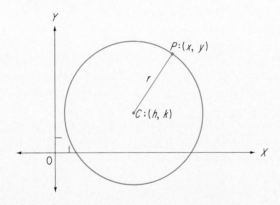

EXAMPLE 2

Describe the set of points (x, y) such that

(a) $x^2 + y^2 = 4$
(b) $(x - 1)^2 + (y - 3)^2 = 1$
(c) $(x + 2)^2 + (y + 1)^2 = 9$

SOLUTION   (a) Since $x = x - 0$ and $y = y - 0$, the given equation can be expressed as

$$(x - 0)^2 + (y - 0)^2 = 2^2$$

Each point of the set is 2 units from the origin. Thus the set of points is a circle with center $(0, 0)$ and radius 2. (b) The set of points is a circle with center $(1, 3)$ and radius 1. (c) Since $x + 2 = x - (-2)$ and $y + 1 = y - (-1)$, the given equation can be expressed as

$$[x - (-2)]^2 + [y - (-1)]^2 = 3^2$$

and the set of points is a circle with center $(-2, -1)$ and radius 3.

EXAMPLE 3

Give an algebraic representation for the points that are on a coordinate plane and are within 2 units of $(3, -5)$.

SOLUTION   The points form a **circular region** (the points of a circle and the interior points of the circle) with center $(3, -5)$ and radius 2. Thus we have

$$(x - 3)^2 + (y + 5)^2 \leq 4$$

exercises

Find the linear measures of the line segments with endpoints:
1. $A: (1, 2)$ and $B: (4, 6)$
2. $C: (-1, 6)$ and $D: (4, -6)$
3. $E: (2, 16)$ and $F: (9, -8)$
4. $G: (-2, 5)$ and $H: (7, 11)$
5. $I: (-1, -3)$ and $J: (2, 0)$
6. $L: (4, -5)$ and $M: (-7, -11)$

Describe the set of points (x, y) such that:

7. $x^2 + y^2 = 9$ **8.** $x^2 + y^2 \leq 36$

9. $x^2 + y^2 > 64$ **10.** $(x - 1)^2 + (y - 2)^2 = 4$

11. $(x - 1)^2 + (y - 2)^2 \leq 4$ **12.** $(x + 3)^2 + (y - 1)^2 < 4$

13. $(x + 1)^2 + (y + 2)^2 \leq 25$ **14.** $(x + 3)^2 + (y + 1)^2 > 1$

Consider the points on a coordinate plane and give an algebraic representation for:

15. The points of the circle with center $(2, 5)$ and radius 3.

16. The points of the circle with center $(-3, 4)$ and radius 2.

17. The points of the circular region with center $(-3, -1)$ and radius 4.

18. The interior points of the circle with center $(6, -4)$ and radius 5.

19. The exterior points of the circle with center $(-4, 3)$ and radius 4.

20. The exterior points of the circle with center $(-3, -2)$ and radius 6.

Graph the solution set of each sentence.

*21. $\{(x, y) | x^2 + y^2 \leq 9 \text{ and } x - y \geq 0\}$.

*22. $\{(x, y) | x^2 + y^2 \leq 4 \text{ and } x - y - < 0\}$.

EXPLORATIONS

Each of the following explorations involves the determination of a shortest path a spider has to crawl along the sides of a box to reach a fly F that is assumed to stay in one place. As in the figure, we consider a closed rectangular box 12 inches wide, 18 inches long, and 8 inches tall.

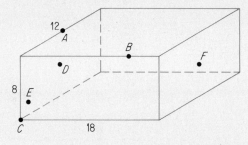

For each exploration assume that the fly is at the center F (intersection of the diagonals) of an end of the box. Copy the figure, draw the shortest path(s) for the spider, and find the length of the shortest path. If there are two or more paths of minimal length, draw them all.

1.   The spider is at position *A* in the center of the top edge of the opposite end of the box.

2.   The spider is at position *B* at the top of an adjoining side and 4 inches from the end containing the fly. (*Hint:* Think of the box as flattened out in some way.)

3.   The spider is in the corner *C* at the bottom of the opposite end of the box.

4.   The spider is at the middle *D* of the end of the box opposite the fly.

5.   The spider is at position *E* one inch from the vertical edge of the box over *C* and two inches from the bottom of the box.

chapter test

1.   Sketch **(a)** a simple curve that is not closed; **(b)** a convex polygon with five sides.

2.   Consider the six lines along the edges of a triangular pyramid *ABCD* as in the figure. Identify the lines that **(a)** intersect $\overleftrightarrow{AB}$ in a single point; **(b)** form skew lines with $\overleftrightarrow{AB}$.

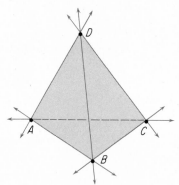

3.   Consider the given figure and identify each set of points:

(a) $\angle ABC \cap \angle CDE$ **(b)**   (Interior $\triangle ABC$) $\cap \overleftrightarrow{DE}$

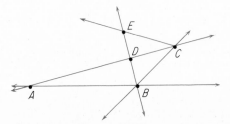

4. Sketch a cube, label its vertices, and name its edges.

5. Consider the given figure and state whether each network is traversable. If a network is traversable, identify the vertices that are possible
starting points.

(a)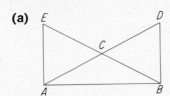

(b)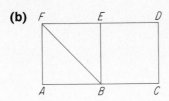

(c)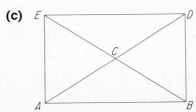

(d)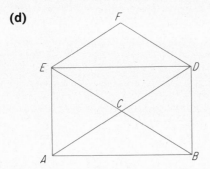

6. On a number line find the linear measure of: **(a)** P_2P_{-7}; **(b)** $P_{-8}P_{-3}$.

7. Find the surface area of a cube with an edge of length 3 inches.

On a coordinate plane:

8. Find the linear measure and the coordinates of the midpoint of the
line segment with endpoints $(2, -5)$ and $(7, 7)$.

9. Find the coordinates of the endpoint B of a line segment $\overline{AB}$ with
endpoint A: $(2, 6)$ and midpoint C: $(5, 1)$.

10. Describe the set of points such that:
    (a) $x^2 + y^2 \le 81$
    (b) $(x - 1)^2 + (y + 2)^2 > 4$

An Introduction to
PROBABILITY

We make frequent reference to probability in everyday language. For example, we say, "It probably will rain," "The odds are in his favor," and make similar comments about many things.

Ever since the fifteenth century, mathematicians have been exploring the topic of probability. Interestingly enough, the subject is said to have had its foundation in the realm of gambling and to have arisen from a discussion of the distribution of stakes in an unfinished game. Probabilities are still used to understand games of chance such as the tossing of coins, throwing of dice, or drawing of lottery tickets. Probabilities are also used extensively by persons who are concerned with the cost of insurance (based upon rates of mortality), the construction of various polls such as the Gallup Poll to appraise public opinion, and numerous other types of statistical studies. The topics in this chapter provide a basis for understanding probabilities and their applications to real life situations. As you read this chapter, look for other applications, if possible among your own activities, of the concepts considered.

Many problems depend for their solution upon an enumeration of all possible outcomes. Thus, the simple task of counting becomes an important one in the study of probability. To illustrate various problems in this chapter, we shall invent a fictitious club consisting of a set M of members:

$M = \{$Betty, Doris, Ellen, John, Tom$\}$

Let us form a committee that is to consist of one boy and one girl, each selected from the set M of club members. How many such committees are possible? One way to answer this question is by means of a *tree diagram*, which helps list each of the possibilities.

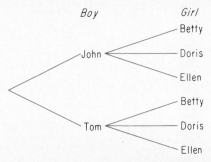

For each of the two possible choices of a boy, there are three possible choices of a girl. Thus the following six distinct possible committees can be formed and can be read from the tree diagram:

John–Betty	Tom–Betty
John–Doris	Tom–Doris
John–Ellen	Tom–Ellen

Suppose that we had selected a girl first. Then the tree diagram would be as shown in the next figure and there would still be six possibilities, the same six committees as before.

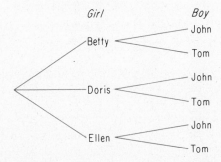

EXAMPLE 1

In how many different ways can two officers, a president and a vice-president, be elected from the set M of club members?

SOLUTION   Let us select the officers in two stages. There are five possible choices for the office of president. Each of these five selections may be paired with any one of the remaining four members. Thus there are 20 possible choices in all, which can be read from the following diagram.

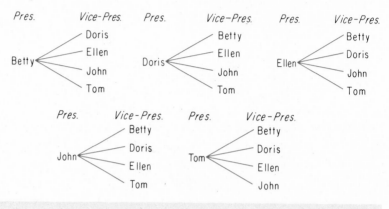

In general, if one task can be performed in m different ways and a second task can be performed in n different ways, then the first and second tasks together can be performed in $m \times n$ different ways. This **general principle of counting** can be extended if there are additional tasks:

$$m \times n \times r \times \cdots \times t$$

EXAMPLE 2

The club members M must send a delegate to a meeting tomorrow and also a delegate to a different meeting next week. In how many ways may these delegates be selected if any member of the club may serve as a delegate to each meeting?

SOLUTION   There are five possible choices of a delegate to the first meeting. Since no restriction is made, we assume that the same member may attend each of the two meetings. Thus, there are five choices for the delegate to next week's meeting. In all, there are 5×5, that is, 25 choices.

EXAMPLE 3

How many three-letter "words" may be formed from the set of vowels $V = \{a, e, i, o, u\}$ if no letter may be used more than once?

(A word in this sense is any arrangement of three letters, such as aeo, iou, etc.)

SOLUTION   There are five choices for the first letter, four for the second, and three for the third. In all, there are $5 \times 4 \times 3$, that is, 60 possible "words." Notice that 125 words would be possible if repetitions of letters were permitted.

exercises

1. How many three-letter "words" may be formed from the given set $R = \{m, a, t, h\}$ if **(a)** no letter may be used more than once? **(b)** repetitions of letters are permitted?

2. Repeat Exercise 1 for four-letter words.

3. Bob has four sport shirts and three pairs of slacks. Assuming that he can wear any combination of these, how many different outfits can he assemble?

4. Jane has five dresses, three hats, and four pairs of shoes. Assuming that she can wear any combination of these, how many different outfits can she assemble?

5. A baseball team has six pitchers and four catchers. How many different batteries consisting of a pitcher and a catcher can they form?

6. Show, by means of a tree diagram, the number of different routes from New York to Los Angeles, via Chicago, if you can go from New York to Chicago by one train or one plane, and from Chicago to Los Angeles by one train, one plane, or one bus.

7. How many different two-digit numbers may be formed from the set of digits $D = \{1, 2, 3, 4, 5\}$ if repetitions of digits are allowed? How many numbers of two different digits can be formed from the set D?

8. Repeat Exercise 7 for $D = \{1, 2, 3, 4, 5, 6, 7\}$.

9. The Jonesboro Swim Club has 12 members. How many different sets of officers consisting of a president, a vice-president, and a secretary-treasurer can it form? No person can hold more than one of these three offices.

10. The Portland Swim Club has 250 members with 200 of these members eligible to hold any office. Repeat Exercise 9 for the Portland Swim Club.

11. How many five-digit numbers can be formed using the ten decimal digits without repetitions if zero is not to be used as the first digit?

12. How many two-digit even numbers can be formed without repetitions from the set $I = \{1, 2, 3, \ldots, 9\}$?

13. Find the number of "words" of three different letters that may be formed from the set $V = \{a, e, i, o, u\}$ if **(a)** the first letter must be i; **(b)** the first letter must be e and the last letter must be i.

14. How many license plates can be made using a letter from our alphabet followed by three decimal digits if the first digit must not be zero? How many are possible if the first digit must not be zero and no digit may be used more than once?

*15. Find the number of three-digit numbers that may be formed from the set $W = \{0, 1, 2, \ldots, 9\}$ if zero is not an acceptable first digit and **(a)** the number must be even; **(b)** the number must be divisible by 5; **(c)** no digit may be used more than once; **(d)** the number must be odd and less than 500.

*16. In a certain so-called "combination" lock, there are 50 different positions. To open the lock you move to a certain number in one direction, then to a different number in the opposite direction, and finally to a third number in the original direction. What is the total number of such "combinations":
(a) If the first turn must be clockwise?
(b) If the first turn may be either clockwise or counterclockwise?

*17. Find the number of different possible license plates if each one is to consist of two letters of the alphabet followed by three decimal digits, the first digit may not be zero, and no repetitions of letters or numbers are permitted.

*18. Repeat Exercise 17 for license plates consisting of three consonants followed by three decimal digits.

EXPLORATIONS

Counting problems arise in many mathematical puzzles. Consider, for example, the following dart puzzles.

1. You are allowed to throw 4 darts, and we shall assume that there are no misses. In how many different ways can you obtain a score of 60?

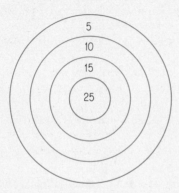

2. Again, use 4 darts and the next figure. Assume no misses and show two different ways of scoring 40.

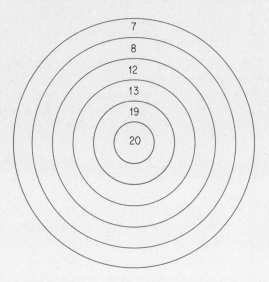

9-2
Definition of Probability

When an ordinary coin is tossed, we know that there are two distinct and **equally likely** ways in which it may land, heads or tails. We say that the probability of getting a head is one out of two, or simply $\frac{1}{2}$.

In rolling one of a pair of ordinary dice, there are six equally likely ways in which the dice may land. We say that the probability of rolling a 5 on one toss of a die is one out of six, or $\frac{1}{6}$.

In each of these two examples, the **events** that may occur are said to be **mutually exclusive.** That is, one and only one of the events can occur at any given time. When a coin is tossed, there are two possible events (heads and tails); one and only one of these may occur. When a single die is rolled, there are six events $(1, 2, 3, 4, 5, 6)$; one and only one of these may occur. Informally we define the probability of success as the ratio of the number of successes of an event to the number of possible outcomes of that event. More generally we define probability as follows:

If an event can occur in any one of n mutually exclusive and equally likely ways, and if m of these ways are considered favorable, then the probability $P(A)$ that a favorable event A will occur is given by the formula

$$P(A) = \frac{m}{n}$$

The probability m/n satisfies the relation $0 \le m/n \le 1$, since m and n are integers and $m \le n$. When success is inevitable, $m = n$ and the probability is 1; when an event cannot possibly succeed, $m = 0$ and the probability is 0. For example, the probability of getting either a head or a tail on a single toss of a coin is 1, assuming that the coin does not land on an edge. The probability of tossing a sum of 13 with a single toss of a pair of ordinary dice is 0. (Here, and in all future work, assume that normal dice are used unless otherwise instructed.)

The sum of the probability of an event's occurring and the probability of that same event's not occurring is 1.

If $P(A) = \dfrac{m}{n},$ then $P(\text{not } A) = 1 - \dfrac{m}{n}$

EXAMPLE 1

A single card is selected from a deck of 52 bridge cards. What is the probability that it is a spade? What is the probability that it is not a spade? What is the probability that it is an ace or a spade?

SOLUTION

Of the 52 cards, 13 are spades. Therefore, the probability of selecting a spade is $\frac{13}{52}$, that is, $\frac{1}{4}$. The probability that the card selected is not a spade is $1 - \frac{1}{4}$, that is, $\frac{3}{4}$. There are 4 aces and 12 spades besides the ace of spades. Therefore, the probability that the card selected is an ace or a spade is $(4 + 12)/52$; that is, $\frac{16}{52}$, which we express as $\frac{4}{13}$.

It is very important that only equally likely events be considered when the probability formula is applied; otherwise faulty reasoning can occur. Consider again the first question asked in Example 1. One might reason that any single card drawn from a deck of cards is either a spade or is not a spade. Thus there are two possible outcomes, and the probability of drawing a spade must therefore be $\frac{1}{2}$. Now it is correct to say that there are these two possible outcomes, but of course they are *not* equally likely since there are 13 spades in a deck of cards and 39 cards that are not spades.

EXAMPLE 2

A committee of two is to be selected from the set *and draw*
1st draw $+$ $=$ $\frac{3}{10}$
$\frac{3}{5}$ $\frac{1}{2}$

$M = \{$Betty, Doris, Ellen, John, Tom$\}$

by drawing names out of a hat. What is the probability that both
members of the committee will be girls?

SOLUTION   We solve this problem by first listing all of the possible committees
of two that can be formed from the set *M*.

Betty – Doris	Doris – John
Betty – Ellen	Doris – Tom
Betty – John	Ellen – John
Betty – Tom	Ellen – Tom
Doris – Ellen	John – Tom

Of the ten possible committees, there are three (those boxed) that
consist of two girls. Thus the probability that both members selected are girls is $\frac{3}{10}$. What is the probability of selecting a committee to consist of two boys?

exercises

What is the probability of tossing on a single toss of one die:

$\frac{1}{3}$ **1.** An even number?   **2.** An odd number? $\frac{1}{3}$

3. A number greater than 3? $\frac{1}{3}$ **4.** A number less than 5? $\frac{1}{4}$

5. A number different from 5? $\frac{1}{5}$ **6.** A number different from 7? $\frac{1}{6}$

7. The number 7? 0 **8.** A number less than 7? $\frac{1}{6}$

*In Exercises 9 through 12 what is the probability of drawing on a single
draw from a deck of 52 bridge cards:*

9. An ace? $\frac{4}{52} \frac{1}{13}$ **10.** A king? $\frac{1}{13}$

11. A spade? $\frac{1}{4}$ **12.** A red card? $\frac{1}{2}$

*In Exercises 13 through 16, we shall consider naming a committee of
two from the set $N = \{$Alice, Bob, Carol, Doug$\}$.*

13. How many different committees of two can be formed? 6

14. What is the probability that a committee of two will consist of two
boys? $\frac{1}{6}$

$\frac{1}{4}$

15. What is the probability that a committee of two will consist of two girls?

$\frac{1}{4}$

16. What is the probability that a committee of two will consist of one boy and one girl?

$\frac{1}{12}$

17. What is the probability that the first-named author of this text was born in the month of December?

$\frac{1}{8 \times 9 \times 10 \times 7}$

18. What is the probability that your instructor's telephone number has a 7 as its final digit?

$\frac{1}{8}$

19. The probability of obtaining all heads in a single toss of three coins is $\frac{1}{8}$. What is the probability that not all three coins are heads on such a toss?

$\frac{1}{6}$

20. What is the probability that the next person you meet was not born on a Sunday?

EXPLORATIONS

If you are enrolled in a class of 30 students, what would you guess is the probability that there are two members of the group that celebrate their birthdays on the same date? Normally one would reason that the probability is low, inasmuch as there are 365 different days of the year on which an individual's birth date could occur, and 30 is small relative to 365.

Although the mathematical explanation is beyond our scope at this time, it can be shown that the actual probability of at least two birth dates occurring on the same day of the year in a random group of 30 students is 0.71. That is, you can safely predict that 71% of the time such an occurrence will take place. Or, alternatively, if you bet that at least two members of the group have the same birth date, then you can expect to win such a bet approximately 71 times out of 100.

The following graph shows the probability of two common birth dates for people in groups of different sizes. Note that the probability in a group of 23 is approximately 0.50. That is, for a group of 23 individuals, there is a 50% chance of having such an occurrence take place. For a group of 60 the probability is 0.99, or almost certainty!

1. From the graph on page 363, what is the probability of two common birth dates for a group of 30 individuals?

2. Use the members of your class to test the results shown in the graph.

3. Refer to an encyclopedia, or other similar source and select the dates of birth of the first 30 individuals chosen at random.

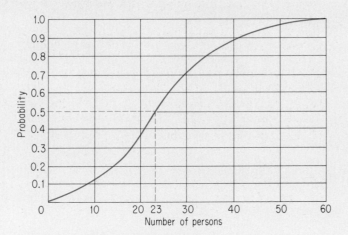

Number of persons

Repeat this experiment several times. Compare the frequency of repetitions of dates of birth with the probability given by the graph. Then repeat the experiment for a group of 40 individuals.

4. Make note of the last two digits of 20 license plates in any parking lot. Repeat this procedure for at least five sets of results and note the frequency with which you find a repetition of pairs of digits within a set of 20 numbers.

5. Repeat the preceding exploration by opening a telephone book at random and selecting 20 telephone numbers. Record the last two digits only, and note the frequency with which one finds a repetition of pairs of digits within a set of 20 two-digit numbers.

6. From the two preceding explorations, form a conjecture as to the probability of finding a pair of like digits in a random collection of 20 two-digit numbers. Does your conjecture, based upon experimentation, agree with your intuitive feeling of what the answer should be?

NANCY®   **By Ernie Bushmiller**

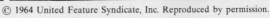

It is often convenient to solve problems of probability by making a list of all possible outcomes. Such a listing is called a **sample space**. Consider first the problem of tossing two coins. The sample space for this problem is given by the following set of all possible outcomes:

$\{HH, HT, TH, TT\}$

We may also summarize these data by means of a chart.

First coin	Second coin
H	H
H	T
T	H
T	T

Another convenient way to list all of the logical possibilities is by means of a tree diagram, where the branches represent the possible choices. The set of possible outcomes, $\{HH, HT, TH, TT\}$ may be read from the diagram. Note that there are four possible events: two heads occur in one event, one head and one tail occur in two events, no heads (that is, two tails) occur in one event. Thus we may list various probabilities regarding the tossing of two coins as follows:

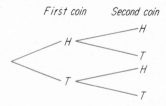

Event	Probabilty
2 heads	$\frac{1}{4}$
1 head	$\frac{2}{4}$
0 heads	$\frac{1}{4}$

Since all possibilities have been considered, the sum of the probabilities should be 1; $\frac{1}{4} + \frac{2}{4} + \frac{1}{4} = 1$. This provides a check on our computation. The list of probabilities is sometimes called a **probability distribution**.

For the case of three coins, the tree diagram and array on page 365 may be made. From the tree diagram, we may list the probabilities of specific numbers of heads by means of the distribution shown below the tree diagram.

	First coin	Second coin	Third coin	First coin	Second coin	Third coin
			H	H	H	H
		H	T	H	H	T
	H		H	H	T	H
		T	T	H	T	T
			H	T	H	H
		H	T	T	H	T
	T		H	T	T	H
		T	T	T	T	T

Event	Probabilty
0 heads	$\frac{1}{8}$
1 head	$\frac{3}{8}$
2 heads	$\frac{3}{8}$
3 heads	$\frac{1}{8}$

In each case, both for two coins and for three coins, the sum of the probabilities is 1; that is, all possible events have been listed and these events are mutually exclusive. Note also that for two coins there were four possible outcomes, for three coins there were eight possible outcomes, and, in general, for n coins there would be 2^n possible outcomes.

EXAMPLE 1

A box contains two red and three white balls. Two balls are drawn in succession without replacement. List a sample space for this experiment.

SOLUTION

To identify individual balls, we denote the red balls as R_1 and R_2; the white balls as W_1, W_2, W_3. Then the sample space is as follows:

$$
\begin{array}{lllll}
R_1 R_2 & R_2 R_1 & W_1 R_1 & W_2 R_1 & W_3 R_1 \\
R_1 W_1 & R_2 W_1 & W_1 R_2 & W_2 R_2 & W_3 R_2 \\
R_1 W_2 & R_2 W_2 & W_1 W_2 & W_2 W_1 & W_3 W_1 \\
R_1 W_3 & R_2 W_3 & W_1 W_3 & W_2 W_3 & W_3 W_2
\end{array}
$$

EXAMPLE 2

Use a tree diagram to show the possible selections for Example 1.

SOLUTION

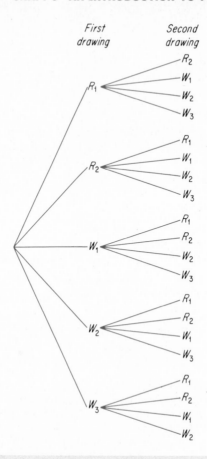

First
drawing

Second
drawing

exercises

Use the sample space of Example 1 to find the probability of each event.

1. Both balls are red.

2. Both balls are white.

3. The first ball is red.

4. The first ball is red and the second ball is white.

5. One ball is red and the other is white.

In Exercises 6 through 8 use the sample space for the outcomes when three coins are tossed to find the probability of each event.

6. All three coins are heads.

7. At least two coins are heads.

8. At most one coin is tails.

9. List a sample space for the outcomes when four coins are tossed.

In Exercises 10 through 12 use the sample space for the outcomes when four coins are tossed (Exercise 9) to find the probability of each event.

10. All four coins are heads.

11. At least three coins are heads.

12. At most two coins are tails.

13. List a sample space for the outcomes when a pair of dice is tossed. Represent each outcome by an ordered pair of numbers. For example, let $(1, 3)$ represent a 1 on the first die and a 3 on the second die.

In Exercises 14 through 22 use the sample space for the outcomes when a pair of dice is tossed (Exericse 13) to find the probability of each event.

14. The number on the first die is 2.

15. The number 2 is on both dice.

16. The same number is on both dice.

17. The sum of the numbers obtained is 11. $\frac{1}{18}$

18. The sum of the numbers obtained is not 11. $\frac{17}{18}$

19. The sum of the numbers obtained is 7. $\frac{1}{6}$

20. The sum of the numbers obtained is not 7. $\frac{5}{6}$

21. The number on one die is twice the number on the other die.

22. The number on one die is three more than the number on the other die.

23. A box contains two red balls R_1 and R_2 and two white balls W_1 and W_2. List a sample space for the outcomes when two balls are drawn in succession without replacement. Find the probability that both balls are red.

24. Repeat Exercise 23 for the case in which the first ball is replaced before the second ball is drawn.

25. Repeat Exercise 23 for a box that contains three red balls and two white balls.

EXPLORATIONS

Many interesting tree diagrams can be drawn that reflect problems of general interest. The following problem illustrates this point of view.

Complete each problem as instructed, and then attempt to find a similar one of current interest.

1. Consider a World Series baseball contest between the Dodgers D and the Yankees Y. Assume that each team is equally likely to win each game. Remember that four victories are needed to win the series, each game must be won by one of the teams, and thus at most seven games will be needed. Make a tree diagram for the entire series under the assumption that the Dodgers win the first two games. Circle the entry and assign a probability to each position in the diagram that represents a completion (winning) of the series. Find the probability of the Dodgers' winning the series on the fourth game, on the fifth game, on the sixth game, and on the seventh game. Find the probability of the Yankees' winning the series on each of these games. Check your work by being sure that the sum of the probabilities of someone winning the series is 1.

There are many interesting probability questions whose answers are not intuitively obvious. Here are two popular ones that are best solved by means of a sample space.

2.   Three cards are in a box. One is red on both sides, one is white on both sides, and one is red on one side and white on the other. A card is drawn at random and placed on a table. It has a red side showing. What is the probability that the side not showing is also red? (Contrary to popular belief, the answer is not $\frac{1}{2}$.)

As an aid to the solution of this problem, let us identify the three cards as in this diagram:

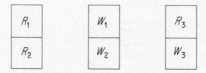

Now consider the set of possible outcomes when a card is drawn at random and placed on the table. In each of the following ordered pairs, the first side noted represents the one placed face up on the table whereas the second one indicates the face hidden from view.

$$R_1 R_2 \qquad R_2 R_1 \qquad R_3 W_3 \qquad W_1 W_2 \qquad W_2 W_1 \qquad W_3 R_3$$

Inasmuch as we are told that a red side is showing, we may narrow down the sample space to the first three pairs only. Of these three possibilities, if the first side is red can you now tell the probability that the second side is also red?

3.   One of the authors of this book has two children. One of these children is a boy. What is the probability that they are both boys? Believe it or not, the answer is *not* $\frac{1}{2}$!

9-4
Computation of Probabilities

If A and B represent two mutually exclusive events, then

$$P(A \text{ or } B) = P(A) + P(B)$$

That is, the probability that one event or the other will occur is the sum of the individual probabilities. Consider, for example, the probability of drawing an ace or a picture card (that is, a jack, queen, or king) from an ordinary deck of 52 bridge cards.

The probability of drawing an ace, $P(A)$, is $\frac{4}{52}$.
The probability of drawing a picture card, $P(B)$, is $\frac{12}{52}$.
Then $P(A \text{ or } B) = \frac{4}{52} + \frac{12}{52} = \frac{16}{52} = \frac{4}{13}$.

EXAMPLE 1

A bag contains three red, two black, and five yellow balls. Find the probability that a ball drawn at random will be red or black.

SOLUTION

The probability of drawing a red ball, $P(R)$, is $\frac{3}{10}$. The probability of drawing a black ball, $P(B)$, is $\frac{2}{10}$. Then

$$P(R \text{ or } B) = P(R) + P(B) = \frac{5}{10} = \frac{1}{2}$$

This process can be extended to find the probability of any finite number of mutually exclusive events.

$$P(A_1 \text{ or } A_2 \text{ or } A_3 \text{ or } \cdots A_n) = P(A_1) + P(A_2) + P(A_3) + \cdots + P(A_n)$$

EXAMPLE 2

A single die is tossed. What is the probability that either an odd number or a number greater than 3 appears?

SOLUTION

There are three odd numbers possible, $(1, 3, 5)$, so the probability of tossing an odd number is $\frac{3}{6}$. The probability of getting a number greater than 3 (that is, 4, 5, 6) is also $\frac{3}{6}$. Adding these probabilities gives $\frac{3}{6} + \frac{3}{6} = 1$. Something is obviously wrong, since a probability of 1 implies certainty and we can see that an outcome of 2 is neither odd nor greater than 3. The difficulty lies in the fact that the events are *not* mutually exclusive; a number may be both odd and also greater than 3 at the same time. In particular, 5 is both odd and greater than 3 at the same time. Thus $P(5)$ has been included twice. Since $P(5) = \frac{1}{6}$, our answer should be $\frac{3}{6} + \frac{3}{6} - \frac{1}{6}$, that is, $\frac{5}{6}$.

Example 2 can be illustrated by means of the following Venn diagram that lists all possible outcomes when a single die is tossed. Note that the outcome 5 is listed in the intersection of the two sets since it is both odd and greater than 3. On the other hand, the outcome 2 fits into neither of these descriptions and is thus placed outside the circles.

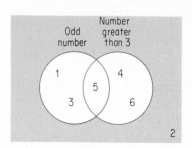

In situations like that of Example 2 we need to subtract the probability that both events occur at the same time. Thus, where A and B are not mutually exclusive events, we have

$$P(A \text{ or } B) = P(A) + P(B) - P(A \text{ and } B)$$

Note that in Example 2, we had the following probabilities:

$P(A)$, the probability of an odd number, is $\frac{3}{6}$.
$P(B)$, the probability of a number greater than 3, is $\frac{3}{6}$.
$P(A \text{ and } B)$, the probability that a number (in this case 5) is odd and greater than 3, is $\frac{1}{6}$.
$P(A \text{ or } B) = \frac{3}{6} + \frac{3}{6} - \frac{1}{6} = \frac{5}{6}$.

By an actual listing we can see that five of the six possible outcomes in Example 2 are either odd or greater than 3, namely 1, 3, 4, 5, and 6. The only "losing" number is 2. Thus the probability $P(A \text{ or } B)$ must be $\frac{5}{6}$.

In general, the two situations $P(A \text{ or } B)$ and $P(A \text{ and } B)$ that we have considered can be described by means of Euler diagrams, where the points of the circular regions represent probabilities of events.

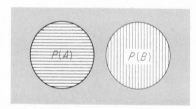   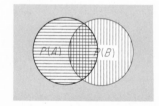

The first of these two diagrams shows mutually exclusive events:

$$P(A \text{ or } B) = P(A) + P(B)$$

In the second figure, the region that is shaded both horizontally and vertically represents $P(A \text{ and } B)$, $P(A \text{ and } B) \neq 0$, the events are **dependent events** (not mutually exclusive), and

$$P(A \text{ or } B) = P(A) + P(B) - P(A \text{ and } B)$$

We must subtract $P(A \text{ and } B)$, since we have counted it twice in the sum $P(A) + P(B)$.

Next we turn our attention to the probability that several events will occur, one after the other. Consider the probability of tossing a coin twice and obtaining heads on the first toss and tails on the second toss. From a sample space we see that the probability is $\frac{1}{4}$.

$$\{HH, \boxed{HT}, TH, TT\}$$

Furthermore, we see that the probability $P(A)$ that the first coin is heads is $\frac{1}{2}$. The probability $P(B)$ that the second coin is tails is $\frac{1}{2}$. Then $P(A$ and $B) = \frac{1}{2} \times \frac{1}{2} = \frac{1}{4}$. Note that these events are **independent**; that is, the outcome of the first toss does not affect the second toss.

In general, let the probability than an event A occurs be $P(A)$. Let the probability than an independent second event occurs, after A has occurred, be $P(B)$. Then

$$P(A \text{ and } B) = P(A) \times P(B)$$

This can be extended to any finite number of independent events.

$$P(A_1 \text{ and } A_2 \text{ and } A_3 \text{ and } \cdots A_n) = P(A_1) \times P(A_2) \times P(A_3) \times \cdots \times P(A_n)$$

EXAMPLE 3

Urn A contains three white and five red balls. Urn B contains four white and three red balls. One ball is drawn from each urn. What is the probability that they are both red?

SOLUTION

Let $P(A)$ be the probability of drawing a red ball from urn A and $P(B)$ be the probability of drawing a red ball from urn B. Then

$$P(A) = \tfrac{5}{8}, \qquad P(B) = \tfrac{3}{7}, \qquad P(A \text{ and } B) = \tfrac{5}{8} \cdot \tfrac{3}{7} = \tfrac{15}{56}$$

EXAMPLE 4

Two cards are selected in succession, without replacement, from an ordinary bridge deck of 52 cards. What is the probability that they are both aces?

SOLUTION   The probability that the first card is an ace is $\frac{4}{52}$. If it is an ace, then the probability that the second card is an ace is $\frac{3}{51}$. The probability that both cards are aces is $\frac{4}{52} \times \frac{3}{51}$, that is, $\frac{1}{221}$.

exercises

What is the probability of tossing on a single toss of one die:

1. An even number or a number greater than 3?
2. An odd number or a number less than 5?
3. An odd number or a number greater than 4?
4. An even number or a number greater than 6?
5. An odd number or a number greater than 10?
6. An odd number or a number less than 6?
7. An even number or a number less than 10?

A single card is drawn from a deck of 52 bridge cards. In Exercises 8 through 14 find the probability that the card selected is:

8. An ace or a king.
9. A spade or a heart.
10. A spade or a king.
11. A spade and a king.
12. A spade and a queen.
13. A heart or a king or a queen.
14. A club or an ace or a king.
15. Two cards are drawn in succession from a deck of 52 bridge cards without the first card's being replaced. Find the probability that **(a)** both cards are spades; **(b)** both cards are aces of spades; **(c)** the first card is a spade and the second card is a heart; **(d)** the first card is an ace and the second card is the king of hearts; **(e)** both cards are of the same suit.
16. Repeat Exercise 15 if the first card is replaced in the deck after the first drawing.
17. A coin is tossed five times. What is the probability that all five tosses are heads?
18. A coin is tossed five times. What is the probability that at least one head is obtained? (*Hint:* First find the probability of getting no heads.)
19. A coin is tossed and then a die is rolled. Find the probability of obtaining **(a)** a head and a 3; **(b)** a head and an even number; **(c)** a head or a 3; **(d)** a head or an even number.

20. A bag contains three red balls and seven white balls. **(a)** If one ball is drawn at random, what is the probability that it is white? **(b)** If two balls are drawn at random, what is the probability that they are both white?

21. Five cards are drawn at random from an ordinary bridge deck of 52 cards. Find the probability that all five cards drawn are spades.

22. A box contains three red, four white, and five green balls. Three balls are drawn in succession, without replacement. Find the probability that **(a)** all three are red; **(b)** the first is red, the second is white, and the third is green; **(c)** none are green; ***(d)** all three are of the same color.

23. Repeat Exercise 22 if each ball is replaced after it is drawn.

24. A die is tossed three times. Find the probability that **(a)** a 6 is tossed on the first toss; **(b)** a 6 is tossed on the first two tosses; **(c)** a 6 is tossed on all three tosses; **(d)** a 6 is tossed on the first toss and not tossed on the second or third toss.

25. A die is tossed three times. Find the probability that **(a)** an even number is tossed on all three tosses; **(b)** an even number is tossed on the first two tosses and an odd number on the third toss; **(c)** an odd number is tossed on the first toss and an even number on the second and third tosses; **(d)** exactly one even number is tossed.

*26. A die is tossed three times. Find the probability that **(a)** at least one 6 is tossed; **(b)** exactly one 6 is tossed.

EXPLORATIONS

Many interesting experiments can be performed that illustrate basic concepts of probability. Several such experiments are suggested here that you should attempt to complete. Also make a collection of any other such experiments that you can find.

1. Consider the network of streets shown in the following figure.

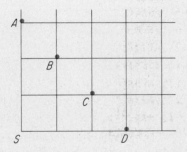

You are to start at the point S and move three "blocks." Each move is determined by tossing a coin. If the coin lands "tails," move one block to the right; for "heads," move one block up. Your terminal point will be at A, B, C, or D. Try to predict the number of times you will land at each point if the experiment is to be repeated 16 times. Then complete 16 trials, keeping a tally of the number of times you land at each point. Compare your actual results with your predictions as well as with the results obtained by your classmates. Finally, on the basis of your experimentation, revise your predictions if necessary.

2. Repeat Exploration 1 for the following network. This time you are to make four moves and will land at point A, B, C, D, or E.

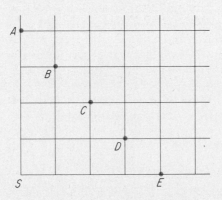

3. If a pair of normal dice is tossed repeatedly, one can theoretically expect a sum of 7 to occur in one out of six tosses. Thus if a pair of dice is tossed 36 times, on the average six of the tosses will give a sum of 7. Toss a pair of dice 36 times and count the frequency with which a sum of 7 appears. Compare your actual results with the theoretical probability of $\frac{1}{6}$. Collect these data for each member of your class and compare actual results with expected results.

4. Repeat Exploration 3 for 72 tosses of a pair of dice. Record each toss by placing an x over the numeral representing the sum, using graph paper. The final result should show a bar graph that is fairly "normal" in shape. The figure at the top of page 375 shows the results obtained after one experiment of 18 tosses.

5. By the "probability" of an event, such as obtaining heads on the toss of a single coin, we always mean "probability in the long run." That is, the fact that the probability of obtaining a head on one toss of a coin is $\frac{1}{2}$ does not imply that you will obtain exactly

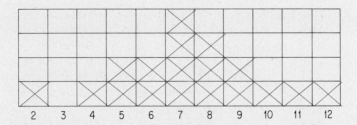

4 heads in 8 tosses. Rather, it means that the longer you continue to toss a coin, the closer you can expect to come to having 50% of the tosses produce a head. This can be demonstrated by actually tossing a coin and recording the results. For example, the results obtained from 10 tosses of a single coin are summarized in the next chart. Note that the first two tosses were heads, the third was a tail, the fourth was a head, and so forth.

Tosses	1	2	3	4	5	6	7	8	9	10
Head	x	x		x			x		x	x
Tail			x		x	x		x		
Ratio of heads	$\frac{1}{1}$	$\frac{2}{2}$	$\frac{2}{3}$	$\frac{3}{4}$	$\frac{3}{5}$	$\frac{3}{6}$	$\frac{4}{7}$	$\frac{4}{8}$	$\frac{5}{9}$	$\frac{6}{10}$
Ratio of tails	$\frac{0}{1}$	$\frac{0}{2}$	$\frac{1}{3}$	$\frac{1}{4}$	$\frac{2}{5}$	$\frac{3}{6}$	$\frac{3}{7}$	$\frac{4}{8}$	$\frac{4}{9}$	$\frac{4}{10}$
Per cent of heads	100	100	67	75	60	50	57	50	56	60
Per cent of tails	0	0	33	25	40	50	43	50	44	40

The graphical presentation of the results obtained from these 10 tosses gives a visual interpretation of the meaning of probability. Despite minor fluctuations, in the long run the graphs of the per cent of heads and tails will come closer and closer to the 50% line.

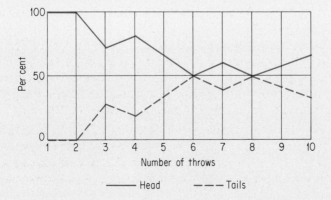

That is, in the long run we expect to come close to having one-half of our tosses produce heads and one-half produce tails.

Repeat the experiment described here for 20 tosses of a coin.

6. Have each member of your class write down in advance the predicted outcome for five tosses of a coin. For example, one guess might be *H, T, T, H, T.* Then actually complete five tosses of a coin and count how many members of your class guessed correctly. Can you predict in advance how many members of the class can be expected to have perfect scores?

9-5

Odds and Mathematical Expectation

One often reads statements in the sports section of the daily newspaper concerning "odds" in favor of, or against, a particular team or individual's winning or losing some encounter. For example, we may read that the odds in favor of the Cardinals' winning the pennant are "4 to 1." In this section we shall attempt to discover just what such statements really mean.

Consider the problem of finding the odds against obtaining a 3 in one toss of a die. Since the probability of obtaining a 3 is known to be $\frac{1}{6}$, most people would say that the odds are therefore 6 to 1 against rolling a 3. This is not correct—because out of every six tosses of the die, in the long run, one expects to toss one 3. The other five tosses are not expected to be 3's. Therefore, the correct odds against rolling a 3 in one toss of a die are 5 to 1. The odds in favor of tossing a 3 are 1 to 5. Formally we define odds as follows:

> The **odds in favor** of an event are defined as the ratio of the probability that an event will occur to the probability that the event will not occur. The reciprocal of this ratio gives the **odds against** the occurrence of the event.

Thus the odds in favor of an event that may occur in several equally likely ways is the ratio of the number of favorable ways to the number of unfavorable ways.

Notice that odds and probabilities are very closely related. Indeed, if either the odds for or the probability of an event is known, then the other can be found. For example, if the odds for an event are 1 to 2, then we have the ratio

$$\frac{\text{number of favorable ways}}{\text{number of unfavorable ways}} = \frac{1}{2}$$

and the probability is the ratio

$$\frac{\text{number of favorable ways}}{\text{total number of all ways (favorable and unfavorable)}} = \frac{1}{3}$$

Similarly, if the probability is $\frac{2}{5}$, then the odds are 2 to 3.

EXAMPLE 1

Find the odds in favor of drawing a spade from an ordinary deck of 52 bridge cards.

SOLUTION

Since there are 13 spades in a deck of cards, the probability of drawing a spade is $\frac{13}{52} = \frac{1}{4}$. The probability of failing to draw a spade is $\frac{3}{4}$. The odds in favor of obtaining a spade are $\frac{1}{4} \div \frac{3}{4}$; that is, $\frac{1}{3}$.

The odds in favor of drawing a spade are stated as $\frac{1}{3}$. They may also be stated as "1 to 3" or as "1:3." Similarly, the odds against drawing a spade are $\frac{3}{1}$, which may be written as "3 to 1" or "3:1."

Mathematical expectation is closely related to odds and is defined as the product of the probability that an event will occur and the amount to be received upon such occurrence. Suppose that you are to receive $2.00 each time you obtain two heads on a single toss of two coins. You do not receive anything for any other outcome. Then your mathematical expectation will be one-fourth of $2.00; that is, $0.50. This means that you should be willing to pay $0.50 each time you toss the coins if the game is to be a fair one. In the long run, both you and the person who is running the game would break even. For example, if you played the game four times it would cost you 4 × $0.50, that is, $2.00. You expect to win, in the long run, once out of every four games played. Assuming that you do so, you will win $2.00 once every four times and thus be even.

If an event has several possible outcomes that occur with probabilities p_1, p_2, p_3, and so forth, and for each of these outcomes one may expect the amounts m_1, m_2, m_3, and so on, then the mathematical expectation E may be defined as

$$E = m_1 p_1 + m_2 p_2 + m_3 p_3 + \cdots$$

Whenever this formula is used, it is worthwhile to check that all possible outcomes have been considered by checking that the sum of the probabilities is 1.

EXAMPLE 2

Suppose that you play a game wherein you are to toss a coin twice and are to receive 10 cents if two heads are obtained, 5 cents if one head is obtained, and nothing if both tosses produce tails. What is your expected value in this game?

SOLUTION

The probabilities of obtaining two, one, and no heads, respectively, are $\frac{1}{4}$, $\frac{1}{2}$, and $\frac{1}{4}$. Therefore, the expected value E, in cents, is found to be

$$E = (10)(\tfrac{1}{4}) + (5)(\tfrac{1}{2}) + (0)(\tfrac{1}{4}) = 5$$

This solution may be interpreted in several ways. For one thing, it is the price you should be willing to pay for the privilege of playing this game. It may also be interpreted as the average amount of winnings per game that one may expect when one is playing a large number of games.

exercises

What are the odds in favor of obtaining:

1. Two heads in a single toss of two coins?
2. At least two heads in a single toss of three coins?
3. Two heads when a single coin is tossed twice?
4. At least two heads when a single coin is tossed three times?

In Exercises 5 through 8 what are the odds against obtaining:

5. Two heads in a single toss of two coins?
6. An ace in a single draw from a deck of 52 bridge cards?
7. An ace or a king in a single draw from a deck of 52 bridge cards?
8. A 7 or an 11 in a single toss of a pair of dice?

9. What are the odds in favor of rolling a 7 or an 11 in a single toss of a pair of dice?
10. One hundred tickets are sold for a lottery. The grand prize is $1000. What is a fair price to pay for a ticket if there is only one prize?
11. Repeat Exercise 10 for the case in which 250 tickets are sold.

12. What is your mathematical expectation when you buy one of 300 tickets for a single prize worth $750?

13. What is your mathematical expectation in a game in which you will receive $10 if you toss a "double" (the same number on both dice) on a single toss of a pair of dice?

14. A box contains three dimes and two quarters. You are to reach in and select one coin, which you may then keep. Assuming that you are not able to determine which coin is which by its size, what would be a fair price for the privilege of playing this game?

15. There are three identical boxes on a table. One contains a five-dollar bill, one contains a one-dollar bill, and the third is empty. A man is permitted to select one of these boxes and to keep its contents. What is his expectation?

16. Three coins are tossed. What is the expected number of heads?

17. Two bills are to be drawn from a purse that contains three five-dollar bills and two ten-dollar bills. What is the mathematical expectation for this drawing?

18. If there are two pennies, a nickel, a dime, a quarter, and a half-dollar in a hat, what is the mathematical expectation of the value of a random selection of a coin from the hat?

EXPLORATIONS

Someone announces that he has just tossed a coin 19 times in a row and has obtained 19 heads. He is about to toss the coin for the twentieth time. You are given an opportunity to place a bet on the outcome of this toss.

1. Would you bet on heads? on tails? or would you say that it makes no difference? (That is, you might decide that both heads and tails alike have an equally likely chance of occurrence.)

2. What is the probability of tossing 20 heads in a row?

3. What are the odds against tossing 20 heads in a row?

9-6
Permutations

Suppose that three people, Ruth, Joan, and Debbie, are waiting to play singles at a tennis court. Two of the three can be selected in six different ways if the order in which they are named is significant, for example, if the first person named is to serve first. We may identify these six ways from a tree diagram.

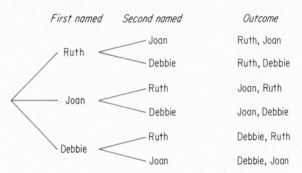

We say that there are 3×2 (that is, 6) *permutations* of the set of three people selected two at a time. In each case the *order* in which the two people are named is significant. A **permutation** of a set of elements is an arrangement of certain of these elements in a specified order. In the problem just discussed, the number of permutations of three things taken two at a time is 6. In symbols we write

$$_3P_2 = 6$$

(read as "the permutation of 3 things taken 2 at a time.")

In general, we wish to find a formula for $_nP_r$, that is, the permutations of n things taken r at a time. To do this, we note that we can fill the first of the r positions in any one of n different ways. Then the second position can be filled in $n - 1$ different ways, and so on.

Position:	1	2	3	4	$\cdots$	r
	$\downarrow$	$\downarrow$	$\downarrow$	$\downarrow$		$\downarrow$
Number of choices:	n	$n-1$	$n-2$	$n-3$	$\cdots$	$n-(r-1)$
						(i.e., $n-r+1$)

The product of these r factors gives the number of different ways of arranging r elements selected from a set of n elements, that is, the permutation of n things taken r at a time

$$_nP_r = (n)(n-1)(n-2) \cdots (n-r+1)$$

where n and r are integers and $n \geq r$.

EXAMPLE 1

Find $_8P_4$.

SOLUTION

Here $n = 8$, $r = 4$, and $n - r + 1 = 5$. Thus

$$_8P_4 = 8 \times 7 \times 6 \times 5 = 1680$$

Note that there are r, in this case 4, factors in the product.

EXAMPLE 2

How many different three-letter "words" can be formed from the 26 letters of the alphabet if each letter may be used at most once?

SOLUTION   We wish to find the number of permutations of 26 things taken three at a time.

$$_{26}P_3 = 26 \times 25 \times 24 = 15,600$$

A special case of the permutation formula occurs when we consider the permutations of n things taken n at a time. For example, let us see in how many different ways we may arrange the five members of set M in a row. Here we have the permutations of five things taken five at a time:

$$_5P_5 = 5 \times 4 \times 3 \times 2 \times 1$$

In general, for n things n at a time, $n = n$, $r = n$, and $n - r + 1 = 1$;

$$_nP_n = (n)(n - 1)(n - 2) \cdots (3)(2)(1)$$

We use a special symbol, $n!$, read "n factorial," for this product of integers from 1 through n. The following examples should illustrate the use of the new symbol:

$1! = 1$ $5! = 5 \times 4 \times 3 \times 2 \times 1$
$2! = 2 \times 1$ $6! = 6 \times 5 \times 4 \times 3 \times 2 \times 1$
$3! = 3 \times 2 \times 1$ $7! = 7 \times 6 \times 5 \times 4 \times 3 \times 2 \times 1$
$4! = 4 \times 3 \times 2 \times 1$ $8! = 8 \times 7 \times 6 \times 5 \times 4 \times 3 \times 2 \times 1$

Also, we *define* $0! = 1$ so that $(n - r)!$ may be used when $r = n$.

Using this **factorial notation**, we are now able to provide a different, but equivalent, formula for $_nP_r$:

$$_nP_r = n(n - 1)(n - 2) \cdots (n - r + 1)$$
$$\times \frac{(n - r)(n - r - 1)(n - r - 2) \cdots (3)(2)(1)}{(n - r)(n - r - 1)(n - r - 2) \cdots (3)(2)(1)} = \frac{n!}{(n - r)!}$$

EXAMPLE 3

Evaluate $_7P_3$ in two different ways.

SOLUTION (a) $_7P_3 = 7 \times 6 \times 5$

(b) $_7P_3 = \dfrac{7!}{4!} = \dfrac{7 \times 6 \times 5 \times 4 \times 3 \times 2 \times 1}{4 \times 3 \times 2 \times 1} = 7 \times 6 \times 5$

EXAMPLE 4

A certain class consists of 10 boys and 12 girls. They wish to elect officers in such a way that the president and treasurer are boys and the vice-president and secretary are girls. In how many ways can this be done?

SOLUTION The number of different ways of selecting the president and treasurer is $_{10}P_2$. The number of ways of selecting the vice-president and secretary is $_{12}P_2$. The total number of ways of choosing officers is

$$(_{10}P_2) \times (_{12}P_2) = (10 \times 9) \times (12 \times 11) = 11,880$$

exercises

Evaluate.

1. 5!   **2.** 6!

3. $\dfrac{8!}{6!}$ **4.** $\dfrac{11!}{7!}$

5. $_7P_2$ **6.** $_7P_3$

7. $_{10}P_1$ **8.** $_{10}P_{10}$

9. $_{12}P_{12}$ **10.** $_{12}P_3$

11. $_6P_2$ **12.** $_6P_4$

13. $_{10}P_3$ **14.** $_{10}P_7$

In Exercises 15 through 18 solve for n.

15. $_nP_1 = 6$ **16.** $_nP_2 = 6$

17. $_nP_2 = 20$ **18.** $_nP_3 = 24$

19. Find the number of different arrangements of the set of letters $V = \{a, e, i, o, u\}$ if they are taken **(a)** two at a time; **(b)** five at a time.

20. Find the number of four-digit numbers that can be formed using the digits $1, 2, 3, 4, 5$ if no digit may be used more than once in a number.

How many of these numbers will be even? What is the probability that such a four-digit number will be even? odd?

21. Find the number of different signals that can be formed by running up three flags on a flagpole, one above the other, if seven different flags are available.

22. Find the number of different ways that a disc jockey can arrange a musical program of seven selections. How many of the arrangements will have a particular one of the seven selections first? What is the probability that a particular one of the seven selections will be first on the program?

23. Find the number of different ways that a manager of a nine-man baseball team can arrange his batting order.

Find for any positive integer n a formula for:

24. $_nP_r \times (n - r)!$ **25.** $_nP_0$

EXPLORATIONS

Permutations are assumed to be in a row (along a line) unless otherwise specified. Suppose that we consider the permutations of seating people at a circular table. The various places at the table are assumed to be indistinguishable and only the relative positions of the people are considered. For such permutations (circular permutations) one person is seated as a reference point and then the various arrangements of the others with reference to this person are considered.

1. In how many ways can four people be seated at a circular table?

2. In how many ways can five people be seated at a circular table?

3. Explain the appropriateness of the expression $\dfrac{_nP_n}{n}$ for the number of permutations of n people in n seats around a circular table.

4. Explain the appropriateness of the expression $\dfrac{_nP_r}{r}$ for the number of permutations of r people from a set of n people in r seats around a circular table.

Unlike the seating of people around a table, a key ring can be turned over. Consider the effect of this phenomenon on the number of distinguishable permutations of keys on a circular key ring.

5. In how many ways can five keys be arranged on a circular key ring?

6. In how many ways can seven keys be arranged on a circular key ring?

7. Give an expression for the number of ways in which n keys can be arranged on a circular key ring.

9-7
Combinations

A fictitious club with five members was considered in §9-1. The set of members was {Betty, Doris, Ellen, John, Tom}. The number of ways in which a president and a vice-president could be selected was essentially the number $_5P_2$ of permutations of five things taken two at a time. Here order is important in that Betty as President and Doris as vice-president is a different set of officers than Doris as president and Betty as vice-president.

Now, suppose we wish to select a committee of two members from set M without attaching any meaning to the order in which the members are selected. Then the committee consisting of Betty and Doris is certainly the same as the one consisting of Doris and Betty. In this case, we see that order is not important, and we call such an arrangement a **combination**. One way to determine the number of possible committees of two to be formed from the set M is by enumeration. We find that there are 10 possible committees, as follows:

Betty–Doris	Doris–John
Betty–Ellen	Doris–Tom
Betty–John	Ellen–John
Betty–Tom	Ellen–Tom
Doris–Ellen	John–Tom

We summarize this discussion by saying that the number of combinations of five things, taken two at a time, is 10. In symbols we write

$$_5C_2 = 10$$

(read as "the combinations of 5 things taken 2 at a time.")

In general, we wish to find a formula for $_nC_r$; that is, the combinations of n things taken r at a time. This is written in symbols in the form

$$_nC_r \quad \text{or} \quad \binom{n}{r}$$

To find a formula for $_nC_r$, let us first consider a specific problem, that of selecting committees of three from the set M. There are 10 such possibilities, and we list them using the first initial of each name only:

$$
\begin{array}{lll}
B, D, E & B, J, T & M = \{B, D, E, J, T\} \\
B, D, J & D, E, J & _5C_3 = 10 \\
B, D, T & D, E, T & \\
B, E, J & D, J, T & \\
B, E, T & E, J, T &
\end{array}
$$

Note that selecting committees of three is equivalent to selecting groups of two to be omitted. That is, omitting J and T is the same as selecting B, D, and E. Therefore, we find that $_5C_3 = {_5C_2} = 10$.

Inasmuch as we wanted only committees, and assigned no particular jobs to the members of each committee, we see that order is not important. However, suppose that each committee is now to elect a chairman, secretary, and historian. In how many ways can this be done within each committee? This is clearly a problem in which order is important, involving permutations. The number of such possible arrangements within each committee is $_3P_3$; that is, 3!. For example, the committee consisting of B, D, and E can rearrange themselves as chairman, secretary, and historian, respectively, as follows:

$$B, D, E; \quad B, E, D; \quad D, E, B; \quad D, B, E; \quad E, B, D; \quad E, D, B$$

All six of these permutations constitute only one combination.

We know that $_5C_3 = 10$. If each of these combinations is multiplied by 3! we then will have the total number of permutations of five things taken three at a time:

$$_5C_3 \times 3! = {_5P_3} \quad \text{and} \quad _5C_3 = \frac{_5P_3}{3!}$$

In general, consider $_nC_r$. Each of these combinations, consisting of r elements each, may be used to form $r!$ permutations. Thus the number of combinations of n things taken r at a time is given by the formula

$$_nC_r \times r! = {_nP_r} \quad \text{and} \quad _nC_r = \frac{_nP_r}{r!}$$

Since

$$_nP_r = (n)(n-1)(n-2)\cdots(n-r+1) = \frac{n!}{(n-r)!}$$

we also have the formula

$$_nC_r = \frac{(n)(n-1)(n-2)\cdots(n-r+1)}{r!} = \frac{n!}{r!(n-r)!}$$

EXAMPLE 1

Evaluate $_7C_2$ in two ways.

SOLUTION

(a) $_7C_2 = \dfrac{_7P_2}{2!} = \dfrac{7 \times 6}{2 \times 1} = 21$

(b) $_7C_2 = \dfrac{7!}{2!5!} = \dfrac{7 \times 6 \times 5 \times 4 \times 3 \times 2 \times 1}{2 \times 1 \times 5 \times 4 \times 3 \times 2 \times 1} = 21$

EXAMPLE 2

In how many different ways can a hand of 5 cards be dealt from a deck of 52 cards? What is the probability that a particular hand contains four aces and the king of hearts?

SOLUTION

The order of the five cards is unimportant, so this is a problem involving combinations.

$$_{52}C_5 = \frac{52!}{5!47!} = \frac{52 \times 51 \times 50 \times 49 \times 48 \times (47!)}{5! \qquad\qquad (47!)} = 2{,}598{,}960$$

The probability of obtaining any one particular hand, such as that containing the four aces and the king of hearts, is $\dfrac{1}{2{,}598{,}960}$.

Many problems in probability are most conveniently solved through the use of the concepts of combinations presented here. For example, let us consider again Example 4 of §9-4. There we were asked to find the probability that both cards would be aces if two cards are drawn from a deck of 52 cards. This problem can be solved by noting that $_{52}C_2$ is the total number of ways of selecting two cards from a deck of 52 cards. Also $_4C_2$ is the total number of ways of selecting two aces from the four aces in a deck. The required probability is then given as

$$\frac{_4C_2}{_{52}C_2} = \frac{\dfrac{4!}{2!2!}}{\dfrac{52!}{2!50!}} = \frac{4 \times 3}{52 \times 51} = \frac{1}{221}$$

exercises

In Exercises 1 through 5 evaluate each expression.

1. (a) $\dfrac{6!}{4!2!}$ (b) $\dfrac{8!}{5!3!}$

2. (a) $\dfrac{10!}{4!6!}$ (b) $\dfrac{12!}{7!5!}$

3. (a) $_7C_2$ (b) $_7C_3$
4. (a) $_8C_3$ (b) $_9C_4$
5. (a) $_{11}C_2$ (b) $_{15}C_3$

6. List the $_3P_2$ permutations of the elements of the set $\{a, b, c\}$. Then identify the permutations that represent the same combinations and find $_3C_2$.

7. List the $_4P_3$ permutations of the elements of the set $\{p, q, r, s\}$. Then identify the permutations that represent the same combinations and find $_4C_3$.

8. List the elements of each combination of $_4C_3$ for the set $\{w, x, y, z\}$. Then match each of these combinations with a combination of $_4C_1$ and thereby illustrate the fact that $_4C_3 = _4C_1$.

9. Find a formula for $_nC_n$ for any positive integer n.

10. Find the value and give an interpretation of $_nC_0$ for any positive integer n.

11. Evaluate $_3C_0, _3C_1, _3C_2, _3C_3$ and check that the sum of these combinations is 2^3, the number of possible subsets that can be formed from a set of three elements.

12. Evaluate $_5C_0, _5C_1, _5C_2, _5C_3, _5C_4, _5C_5$ and check the sum as in Exercise 11.

13. Use the results obtained in Exercises 11 and 12 and conjecture a formula for any positive integer n for

$$_nC_0 + _nC_1 + _nC_2 + _nC_3 + \cdots + _nC_{n-1} + _nC_n$$

14. How many sums of money (include the case of no money) can be selected from a set of coins consisting of a penny, a nickel, a dime, a quarter, and a half-dollar?

15. A man has a penny, a nickel, a dime, a quarter, and a half-dollar in his pocket. In how many different ways can he give a tip if he wishes to use exactly two coins?

16. A class consists of 10 boys and 12 girls. How many different committees of four can be selected from the class if each committee is to consist of two boys and two girls?

17. In how many ways can a hand of 13 cards be selected from a bridge deck of 52 cards?

18. In how many ways can one choose three books to read from a set of seven books?

19. Explain why a so-called "combination" lock should really be called a permutation lock.

20. Urn A contains five balls and urn B contains ten balls. In how many ways can ten balls be selected if three are to be drawn from urn A and seven from urn B?

21. An urn contains seven black and three white balls. In how many ways can four balls be selected from this urn? How many of these selections will include exactly three black balls?

22. Repeat Exercise 21 for an urn that contains ten black balls and five white balls.

23. A group of six girls live in the same house. Three of the girls have blue eyes. If two of the girls are chosen at random, what is the probability that both will have blue eyes?

24. Three arbiters are to be chosen by lot from a panel of ten. What is the probability that a certain individual on the panel will be one of those chosen?

25. Five cards are drawn at random from an ordinary bridge deck of 52 cards. Find the probability that all four aces are among the cards drawn.

State whether each question involves a permutation, a combination, or a permutation and a combination. Then answer each question.

26. The 20 members of the Rochester Tennis Club are to play on a certain Saturday evening. In how many ways can pairs be selected for playing "singles"?

27. In how many different ways can a disc jockey with ten records: **(a)** select a set of four records? **(b)** present a program in which four records are played?

28. Students taking a certain examination are required to answer four out of eight questions. In how many ways can a student select the four questions that he tries to answer?

29. In how many ways can seven people line up at a single theater ticket window?

30. How many lines are determined by ten points if no three points are collinear?

31. In how many ways can the 15 boys in a certain class be selected in a group of 9 to play baseball?

32. How many different hands can be dealt from a deck of 52 bridge cards if each hand contains: **(a)** 4 cards? **(b)** 7 cards?

33. In how many different ways may five guests be seated in the five seats of a six-passenger car after the driver is seated?

34. In how many ways can four pictures be arranged with one in each of four given places on the walls of a room?

35. A class is to be divided into two committees. In how many different ways can some or all of eight of the students be assigned to one of the committees?

36. In how many different ways can eight students be divided into two groups of four students each?

37. In how many different ways can eight students be divided into two groups of students?

EXPLORATIONS

Recall that the notation $_nC_r$ may also be written in the form $\binom{n}{r}$.

Consider the following array for sets of n elements where $n = 1$, $2, 3, \ldots$:

$$n = 1: \qquad \binom{1}{0} \quad \binom{1}{1}$$

$$n = 2: \qquad \binom{2}{0} \quad \binom{2}{1} \quad \binom{2}{2}$$

$$n = 3: \qquad \binom{3}{0} \quad \binom{3}{1} \quad \binom{3}{2} \quad \binom{3}{3}$$

$$n = 4: \qquad \binom{4}{0} \quad \binom{4}{1} \quad \binom{4}{2} \quad \binom{4}{3} \quad \binom{4}{4}$$

$$n = 5: \quad \binom{5}{0} \quad \binom{5}{1} \quad \binom{5}{2} \quad \binom{5}{3} \quad \binom{5}{4} \quad \binom{5}{5}$$

. . . .

If we replace each symbol by its equivalent number, we may write the following array, known as **Pascal's triangle**. Generally ascribed to the French mathematician Blaise Pascal (1623–1662), this array of numbers is said to have been known to the Chinese in the early fourteenth century.

```
n = 1:                    1     1
n = 2:                 1     2     1
n = 3:              1     3     3     1
n = 4:           1     4     6     4     1
n = 5:        1     5    10    10     5     1
```
. . . .

We read each row in this array by noting that the first entry in the nth row is $\binom{n}{0}$, the second is $\binom{n}{1}$, the third is $\binom{n}{2}$, and so on until the last entry, which is $\binom{n}{n}$. Since $\binom{n}{0} = \binom{n}{n} = 1$, each row begins and ends with 1.

There is a simple way to continue the array with very little computation. In each row the first number is 1 and the last number is 1. Each of the other numbers may be obtained as the sum of the two numbers appearing in the preceding row to the right and left of the position to be filled. Thus, to obtain the sixth row, begin with 1. Then fill the next position by adding 1 and 5 from the fifth row. Then add 5 and 10 to obtain 15, add 10 and 10 to obtain 20, and so forth as in this diagram:

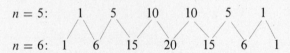

```
n = 5:      1     5    10    10     5     1

n = 6:   1     6    15    20    15     6     1
```

Pascal's triangle may be used mechanically to compute probabilities as follows. The elements of the second row are the numerators for the probabilities when two coins are tossed; the elements of the third row are the numerators when three coins are tossed; and so on. The denominator in each case is found as the sum of the elements in the row used. For example, when three coins are tossed, we examine the third row $(1, 3, 3, 1)$. The sum is 8. The probabilities of 0, 1, 2 and 3 heads are then given as

$$\frac{1}{8}, \quad \frac{3}{8}, \quad \frac{3}{8}, \quad \frac{1}{8}$$

Note that the sum of the entries in the second row is 4, the sum in the third row is 2^3 or 8, the sum in the fourth row is 2^4 or 16, and, in general, the sum in the nth row will be 2^n.

1. Use the fourth row of Pascal's triangle to find the probability of 0, 1, 2, 3, 4 heads in a single toss of four coins.

2. Construct Pascal's triangle for $n = 1, 2, 3, \ldots, 10$.

3. List the entries in the eleventh row of Pascal's triangle, using the notation

$$\binom{n}{r}$$

4. Repeat Exploration 3 for the twelfth row.

5. See how many different patterns of numbers you can find in Pascal's triangle. For example, do you see the sequence 1, 2, 3, 4, 5, . . . ?

6. A recognizable pattern may be found by considering the number of ways in which each of the numbers 2, 3, 4, . . . can be expressed as sums of counting numbers one at a time, two at a time, three at a time, and so forth. For example, 3 can be expressed in one way (3) as a single number, in two ways $(1 + 2$ and $2 + 1)$ as a sum of two counting numbers, and in one way $(1 + 1 + 1)$ as a sum of three counting numbers. Consider also numbers 4, 5, 6, . . . until you recognize the pattern. Then state the number of ways in which 75 can be expressed as a sum of two numbers, and the number of ways in which 79 can be expressed as a sum of 78 numbers.

chapter test

1. How many four-letter "words" can be formed from the set $B = \{p, e, n, c, i, l\}$ if **(a)** no letter may be used more than once? **(b)** repetitions of letters are permitted?

2. A single card is selected from a deck of 52 bridge cards. What is the probability that it is **(a)** not a spade? **(b)** a spade or a heart?

3. A certain florist has five different kinds of roses. In how many ways can one each of three different kinds be selected?

4. A man has a nickel, a dime, a quarter, and a half-dollar. In how many ways can he leave a tip of exactly two coins?

5. Use a sample space to represent the possible outcomes described in Exercise 4 and find the probability that the tip is at least 30 cents when the selection of the two coins is at random.

6. What is the probability of tossing on a single toss of two dice **(a)** a 7? **(b)** a number greater than 7?

7. What are the odds in favor of obtaining an ace when 1 card is drawn from a deck of 52 bridge cards?

8.   What is your mathematical expectation when you buy 3 tickets out of a total of 6000 tickets for a single prize worth $1000?

9.   A bag contains five red balls and three green balls. Two balls are drawn in succession, without replacement. Find the probability that **(a)** both balls are red; **(b)** the first ball is red and the second one is green.

10.   A disc jockey has 25 records. In how many ways can he **(a)** select a set of 5 records? **(b)** arrange a program of 5 records?

10

An Introduction to
STATISTICS

The average consumer is besieged daily with statistical data that are presented over radio and television, in newspapers, and in various other media. He is urged to buy a certain commodity because of statistical evidence presented to show its superiority over other brands. He is cautioned *not* to consume a particular item because of some other statistical study carried out to show its danger. He is told to watch certain programs, read certain magazines, see certain movies, and eat certain foods because of evidence produced to indicate the desirability of these acts as based upon data gathered concerning the habits of others.

Almost every issue of the daily newspaper presents data in graphical form to help persuade the consumer to follow certain courses of action. Thus we are told what the "average" citizen eats, what he reads, what he earns, and even what he does with his leisure time. Unfortunately, too many of us are impressed by statistical data regardless of their source.

Because of the widespread use of statistics for the consumer, we find that even elementary school children are ready to quote facts that they read or hear, often without real understanding. Unfortunately, this is equally true for the adult. Therefore in this chapter we shall make an effort to acquaint the reader with a sufficient number of basic concepts so that he may better understand and interpret statistical data.

10-1

Uses and Misuses of Statistics

Statistical data are often presented incorrectly to impress or mislead the reader. We shall consider several imaginary misuses as samples of the sort of abuses of statistics that occur daily. Consider the statement:

> Brush your teeth with GLUB and you will have fewer cavities.

The basic question that this statement fails to answer is: "Fewer cavities than what?" Regardless of the merits of GLUB, it is probably safe to say the following of every brand of toothpaste: "Brush your teeth with XXXX and you will have fewer cavities than *if you never brush your teeth at all.*" In other words, the example given is misleading in that it implies the superiority of GLUB without presenting sufficient data to warrant comparisons. Here is another example:

> There is a high degree of correlation between expenditures on education and alcohol. Education leads one to drink.

It may well be that some students face examinations in school with such dread that alcohol becomes their only salvation. Hopefully, this book will not drive anyone to drink. There is no doubt that expenditures on education and alcohol are correlated; that is, both items have seen increases in recent years. This does not mean, however, that the two items are related or that one necessarily affects the other. Thus the conclusion drawn from the given fact is unwarranted and misleading.

Many types of statements must be examined very carefully to separate their statistical implications from the impressions they are intended to make. Consider this statement:

> More people are killed in automobile accidents than in wars. Therefore, it is safer to be on the battlefield.

Although the first statement may be true, the second one does not necessarily follow. Inasmuch as there are generally many more people at home driving cars than there are in combat, the *per cent* of deaths on the battlefield is actually much higher than in automobiles. Consider this advertisement:

95% of mothers who use BOON milk for their babies do so on advice of their doctors.

This statement is made to impress the consumer. It is hoped that he will then conclude, incorrectly of course, that 95% of all doctors recommend BOON milk for children. Actually the claim being made, if true, is still not very impressive. Most modern mothers follow doctor's recommendations concerning their babies' formula. Therefore, this statement could undoubtedly be made not only of BOON milk, but also of every other milk used for babies.

Frequently graphs are used to mislead readers:

Why not subscribe to MATH magazine? The next graph shows our growth of new subscribers over the past five years.

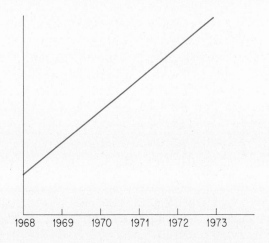

1968   1969   1970   1971   1972   1973

The visual impression given by the graph is that of tremendous growth over a five-year period of time. However, without any scale on the vertical axis it is not possible to draw any conclusion other than that there has been a change in the circulation. If the vertical scale is oriented upward and there has been an increase, this increase may be quite small, since we have no assurance that the vertical axis begins with 0. For example, the actual facts may be as in the next graph, where there has been an increase, but only a very modest one of 400 readers (an increase of 0.4%). In this first graph on page 396, the scale on the vertical axis does not begin at 0, as indicated by the broken line near the base. The second graph shows what the same data would look like if the horizontal scale were set to begin at 0. This time the graph shows the correct picture, namely a very gradual and almost imperceptible increase over the five-year period.

At times data are presented graphically purposefully to mislead a reader. In the graph on page 397, let us assume that the data presented

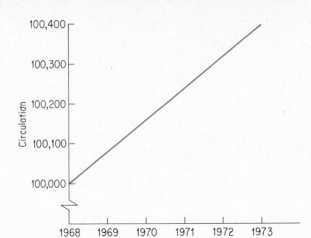

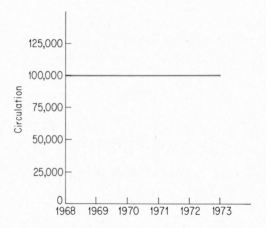

are correct. Note, however, how the picture graph distorts the data by showing the box representing KORNIES to be not only three times as high as the OATIES picture, but also three times as wide. Thus the area of the KORNIES box is nine times as large as the OATIES box, instead of three times as large as indicated by the data. With the numerical facts absent, as is often the case, the picture would be completely dishonest.

The appearance of a contrast between the sales of KORNIES and OATIES is even more striking when three-dimensional boxes are pictured. (See page 397.) Then the volumes of the boxes are visualized and the volume of the KORNIES box is 27 times that of the OATIES box.

The collecting of actual examples of uses of statistics to mislead readers can be a very interesting activity. In particular, an analysis of advertising claims can be rewarding in the insight it brings to an understanding of basic statistical concepts.

Number of boxes of leading cereals sold last year

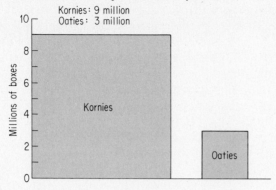

Kornies: 9 million
Oaties: 3 million

Number of boxes of leading cereals sold last year

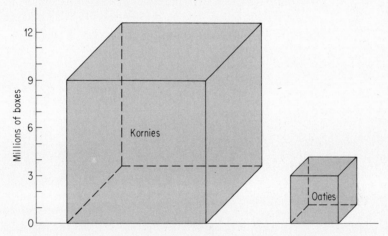

The intelligent consumer needs to ask basic questions about statistical statements that he hears or reads. Unfortunately too many people tend to accept as true any statement that contains numerical data, probably because such statements sound impressive and authentic. Consider, for example, this statement:

Two out of three adults no longer smoke.

Before accepting and repeating such a statement, one should ask such questions as the following:

Does the statement sound reasonable?
What is the source of these data?
Can the facts be verified?
Do the facts appear to conform with your own observations?

The reader will undoubtedly think of other questions to raise. The important thing is that one raises these questions, and does not accept all statistical facts as true in general. Furthermore this is not to imply that all such facts are biased or untrue. Rather, as we have seen and shall see later in this chapter, even true statements can be presented in many ways so as to provide various impressions and interpretations.

Finally, one should examine all statements to search for ambiguous words or meaningless comparisons. Thus, consider this assertion:

If you take vitamin C daily, your health will be better.

In addition to the questions one might raise about the scientific basis for this statement, note that the word "better" is ambiguous. We need to ask the question: "Better than what?" That is, we need to know the basis for comparison. As an extreme case, we certainly can say that your health will be better than one who is desperately ill! The exercises that follow will provide the reader with an opportunity to test his powers of reasoning and questioning.

exercises

Discuss each of the following examples and tell what possible misuse or misinterpretation of statistics each one involves.

1. Most automobile accidents occur near home. Therefore one is safer taking long trips than short ones.

2. Over 95% of the doctors interviewed endorse SMOOTHIES as a safe cigarette to smoke. Therefore it is safe to smoke SMOOTHIES.

3. More college students are now studying mathematics than ever before. Therefore mathematics must be a very popular subject.

4. Professor *X* gave out more A's last semester than Professor *Y*. Therefore, one should try to enroll in *X*'s class next semester rather than in *Y*'s class.

5. At a certain college 100% of the students bought *Elements of Mathematics* last year. Therefore this must be a very popular book at that school.

6. Most accidents occur in the home. Therefore it is safer to be out of the house as much as possible.

7. Over 75% of the people surveyed favor the Democratic candidate. Therefore he is almost certain to win the election.

8. A psychologist reported that the American girl kisses an average of 79 men before getting married.

9. In July, 1960 a magazine reported the presence of 200,000 stray cats in New York City.

10. A recent report cited the statistic of the presence of 9,000,000 rats in New York City.

11. Arizona has the highest death rate in the nation for asthma. Therefore if you have asthma, you should not go to Arizona.

12. During World War II more people died at home than on the battlefield. Therefore it was safer to be on the battlefield than to be at home.

13. In a pre-election poll, 60% of the people interviewed were registered Democrats. Thus the Democratic candidate will surely win the election.

14. Over 90% of the passengers that fly to a certain city do so with Airline *X*. It follows that most people prefer Airline *X* to other airlines.

15. A newspaper exposé reported that 50% of the children in the local school system were below average in arithmetic skills.

Assume that each of the following statements are made. For each one, list two or three questions that you would wish to raise before accepting the statement as true. Identify words, if any, that are ambiguous or misleading.

16. Most people can swim.

17. Short men are more aggressive than tall men.

18. If you walk three miles a day you will live longer.

19. Brand *A* aspirin is twice as effective as Brand *B*.

20. Teen-agers with long hair are happier individuals.

21. People who swim have fewer heart attacks.

22. If you sleep eight hours the night before an exam you will do better.

23. Most students have success with this textbook.

24. Teachers tend to be less conscious about social ills.

25. About three out of four college students marry within one year of graduation.

Use the following graph for Exercises 26–27.

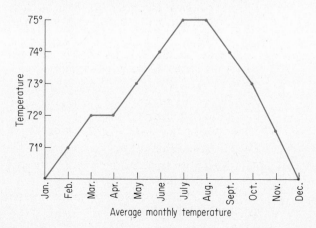

Average monthly temperature

26. Disregarding the scale, what visual impression do you get from this graph about the fluctuation of temperatures during the year?

27. In what ways, if any, is the graph misleading?

Use the following bar graph to answer Exercises 28–30.

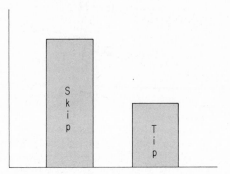

Circulation of two leading magazines

28. Can one deduce from the graph that the circulation of SKIP is greater than the circulation of TIP?

29. Can one deduce from the graph that the circulation of SKIP is approximately twice as great as the circulation of TIP?

30. What, if anything, could possibly be misleading or deceiving about the graph?

EXPLORATIONS

1. Read *How to Lie With Statistics* by Darrell Huff, New York: W. W. Norton and Co., Inc., 1954.

Begin a collection from newspapers, magazines, and other media of examples of misuses of statistics. In particular, look for examples with one or more of the following:

2. Scales that have been cut off to mislead the reader.

3. Areas or volumes used to exaggerate comparisons.

4. Parts of the figure magnified or enlarged to provide undue emphasis.

Discuss the following classic quotes:

5. Facts are facts.

6. Figures don't lie but liars figure.

"Looks better this way."

Cartoon by Lurie, LIFE Magazine, © 1971 Time Inc. Reproduced by permission.

10-2

Collecting and Presenting Data

Although the typical consumer needs to be able to interpret rather than to present data, a discussion of methods of presentation should serve to help develop skills of interpretation. As an illustration of such methods, let us assume that four coins have been tossed 32 times and that each time the number of heads showing has been recorded. These **raw data** are listed in the order of the occurrence of the events.

Number of Heads

1	2	3	2	4	2	2	1
0	2	4	2	2	1	3	3
2	3	1	1	2	2	3	0
4	1	3	2	4	2	1	2

The information is somewhat more meaningful when the number of heads are tallied and summarized in tabular form as a **frequency distribution.** From the frequency distribution on page 402, the original raw data can be treated graphically in a number of different ways.

Number of heads	Tally	Frequency
0	||	2
1	~~||||~~ ||	7
2	~~||||~~ ~~||||~~ |||	13
3	~~||||~~ |	6
4	||||	4

One common form of presenting data is by means of a **histogram,** a bar graph without spaces between the bars.

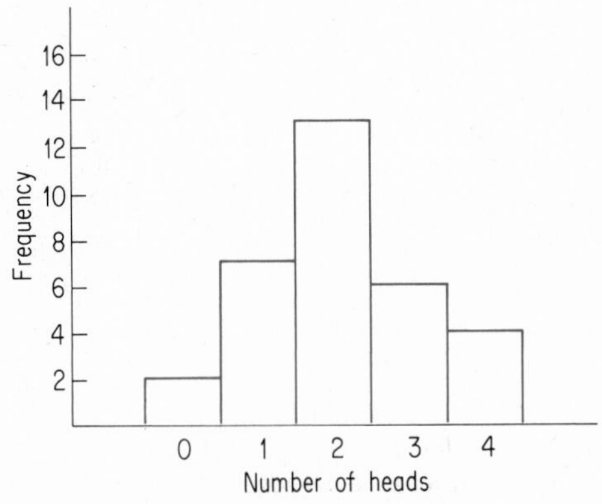

Frequently a histogram is used to construct a **frequency polygon** or **line graph.** Thus the histogram can be approximated by a line graph by connecting the midpoints of the tops of each bar. The graph is then extended to the base line as in the figure.

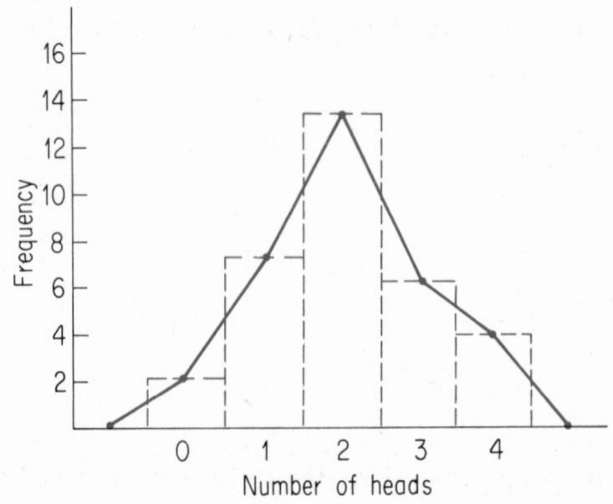

Line graphs appear to associate values with all points on a continuous interval rather than just with points with integers as coordinates. Thus such line graphs are more appropriate for **continuous data** (all intermediate values have meaning) rather than with **discrete data** (only isolated values have meaning). For example, consider a graph of the temperatures in a particular city. Although it was 40° at 5 P.M. and 35° at 6 P.M., the temperature must have passed through *all* possible values between 40 and 35° in one hour. This, then, is an example of continuous data. Temperatures may be plotted and connected to produce a line graph as in the figure. Since gradual changes are expected, we "smooth out" the graph as a curve without "sharp turns."

Hour	6 A.M.	7	8	9	10	11	12	1	2	3	4	5	6
Temp.	32	35	35	38	40	42	45	46	45	42	41	40	35

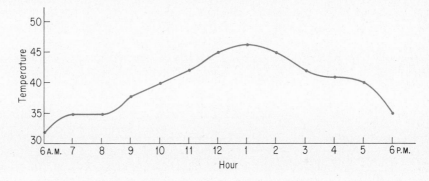

Many different types of graphs are used to present statistical data. One very popular type is the **circle graph,** which is especially effective when one wishes to show how an entire quantity is divided into parts. Local and federal government documents frequently use this type of graph to show the distribution of tax money, and various budget distributions. As an example, consider a family that plans to distribute their income according to the following guide.

Food:	40%
Household:	25%
Recreation:	5%
Savings:	10%
Miscellaneous:	20%
Total:	100%

To draw a circle graph for this data, we first recognize that there are 360° in a circle. Then we find each of the given per cents of 360, and with the aid of a protractor construct central angles of the appropriate sizes as in the next figure.

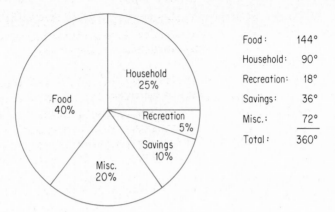

Food :	144°
Household:	90°
Recreation:	18°
Savings :	36°
Misc.:	72°
Total :	360°

The same data can also be presented in the form of a divided bar graph. Here we select an arbitrary unit of length to represent 100%, and divide this in accordance with the given per cents.

Food 40%	Household 25%	Rec. 5%	Savings 10%	Misc. 20%

There is an almost endless supply of graphs that one could use to illustrate this method of presenting data. Indeed, the reader will note from observation that he is constantly subjected to graphical presentations in the daily newspapers and in other periodicals as well. Other examples of graphs are given in the exercises that follow.

exercises

1. Toss four coins simultaneously and record the number of heads obtained. Repeat this for a total of 32 tosses of four coins. Present your data in the form of a frequency distribution.
2. Present the data for Exercise 1 in the form of a bar graph.
3. Present the data for Exercise 1 in the form of a line graph.
4. Repeat Exercise 1 for 32 tosses of a set of five coins.
5. Present the data for Exercise 4 in the form of a bar graph.
6. Present the data for Exercise 4 in the form of a line graph.
7. Here is the theoretical distribution of heads when four coins are tossed for a total of 64 times. Present these data in the form of a bar graph.

Number of Heads	0	1	2	3	4
Frequency	4	16	24	16	4

8. Here is the theoretical distribution of heads when five coins are tossed for a total of 64 times.

Number of Heads	Frequency
0	2
1	10
2	20
3	20
4	10
5	2

Present these data in the form of a bar graph.

9. Construct a circle graph to show this distribution of time spent by one student.

Sleep:	8 hours
School:	6 hours
Homework:	4 hours
Eating:	2 hours
Recreation:	4 hours
Total:	24 hours

10. Construct a divided bar graph for the data of Exercise 9.

Use the following graph to answer Exercises 11–17.

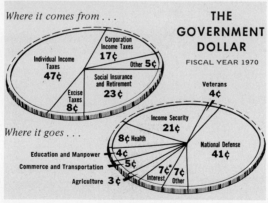

Executive Office of the President/Office of Management and Budget

*Excludes Interest Paid to Trust Funds

11. What per cent of the government dollar comes from individual income taxes?

12. What per cent of the government dollar goes to national defense?

13. Can you tell from the graphs **(a)** what per cent of the government expenditures were for health matters? **(b)** how many dollars were spent on health matters? Explain your answers.

14. What is the size of the central angle in the graph that shows the amount spent for commerce and transportation?

15. To the nearest degree, what is the size of the central angle that shows the amount obtained from individual income taxes?

16. Of every million dollars collected, how many dollars were collected from excise taxes?

17. Of every million dollars spent, how many dollars were spent for health?

Insurance companies compile data on deaths, and use this information to make predictions of life expectancy and to determine insurance rates. Use the following two graphs to answer Exercises 18–22.

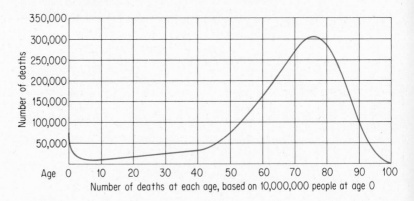

Number of deaths at each age, based on 10,000,000 people at age 0

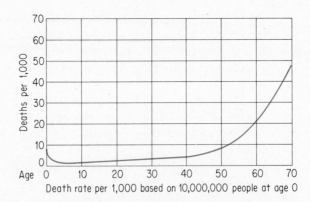

Death rate per 1,000 based on 10,000,000 people at age 0

18. At approximately what age do the largest number of deaths occur?

19. At approximately what ages are there 150,000 deaths?

20. Explain why both graphs show an initial decrease before both start to rise again.

21. Both graphs have the same shape at first. The latter portions, however, are quite different. Explain why this is so.

22. The graph of the death rate per 1000 stops at age 70. What do you think happens to this curve after age 70?

Use the following graph for Exercises 23–30.

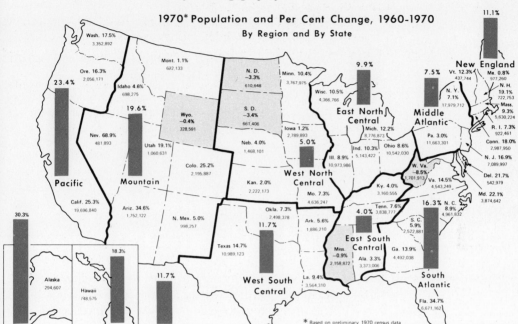

1970* Population and Per Cent Change, 1960-1970
By Region and By State

© 1970 the National Industrial Conference Board. Reproduced by permission.

23. Which states showed a decrease in population during the past decade?
24. Which state showed the largest per cent increase in population?
25. Which state has the largest population?
26. Which state showed the greatest per cent decrease in population?
27. Alaska shows a greater per cent increase in population than Hawaii. Yet the population of Hawaii is more than twice as great as that of Alaska. How can you explain this apparent paradox?
28. Within the continental U.S., which region showed the greatest per cent increase?
29. Within the continental U.S., which region showed the greatest per cent decrease?
30. What general conclusions can you form on the basis of this graph?

EXPLORATIONS

1. Select a textbook from another course at your college or university. (A psychology text or an economics text would be par-

ticularly appropriate.) Determine the amount of attention given to the construction and interpretation of graphs. Ascertain the number of different types of graphs presented.

2. Make a collection of graphs found in newspapers and magazines. See how many different types of graphs you can find.

3. Write for a copy of the booklet *Sets, Probability, and Statistics* by Clifford, Keiffer, and Sobel. This can be obtained free of charge from The Institute of Life Insurance, 277 Park Avenue, New York, New York, 10017.

10-3
Measures of Central Tendency

Most of us have neither the ability nor the desire to digest large quantities of statistical data. Rather, we prefer to see a graphical presentation of such data, or some figure cited as representative of the entire collection. Thus we are often faced by such statements as:

> On the average, 9 out of 10 doctors recommend H_2O.
> The average family earns $9800 per year and has 2.3 children.
> The average college teacher gives out 21% A's each year.

The word "average" is used loosely and may have a number of different meanings. However, in each case it is used in an effort to provide a capsule summary of a collection of data by means of a single number.

Consider the following set of data, which represents the number of dates that each of 10 girls in Alpha Beta Gamma sorority had last semester:

6, 7, 8, 9, 10, 12, 15, 15, 20, 28

Each of the following statements is now correct:

1. *The average sorority member had 13 dates last semester.*

In this case the average is computed by finding the sum of the given numbers and dividing the sum by 10, the number of girls. This is the most commonly used type of average, is referred to as the **arithmetic mean,** or simply the **mean,** of a distribution, and is generally denoted by the symbol $\bar{X}$.

6
7
8
9     $\overline{X} = \frac{130}{10} = 13$
10
12   Arithmetic mean: 13
15
15
20
28
___
130

2. *The average sorority member had 11 dates last semester.*

In this case the word *average* is used to denote the number that divides the data so that half of the scores are above this number and half are below it.

6 ⎫
7 ⎪
8 ⎬ Five scores below 11
9 ⎪
10 ⎭
—   Median: 11
12 ⎫
15 ⎪
15 ⎬ Five scores above 11
20 ⎪
28 ⎭

The number that divides a set of scores in this way is called the **median** of the distribution. In this case it is determined as the number midway between 10 and 12. Note that the data have to be arranged in order of size before the median can be found.

3. *The average sorority member had 15 dates last semester.*

In this case, the average is determined as the number that appears most frequently in the distribution. That is, it is correct to say that more girls had 15 dates last semester than any other number of dates. Such an average is referred to as the **mode** of a set of data. Obviously, in this case, the mode is the best average to use if the sorority wishes to impress others with the success of their members. Actually, however, the median is more representative of the number of dates of the typical member of the sorority.

Each of the three types of averages that we have discussed is known as a **measure of central tendency**. That is, each average is an attempt to describe a set of data by means of a single representative number. Note, however, that not every distribution has a mode, whereas some may have more than one mode. The set of scores $\{8, 12, 15, 17, 20\}$ has no mode. The set $\{8, 10, 10, 12, 15, 15, 17\}$ is **bimodal** and has both 10 and 15 as modes.

EXAMPLE 1

Find (a) the mean; (b) the median; and (c) the mode for this set of test scores:

$$72, 80, 80, 82, 88, 90, 96$$

SOLUTION   (a) The sum of the seven scores is 588.

$$\overline{X} = \tfrac{588}{7} = 84$$

(b) The median is 82. This is the middle score; there are three scores below 82 and three above it.
(c) The mode is 80, since this score appears more frequently than any other one.

EXAMPLE 2

Repeat Example 1 for these scores:

$$30, 80, 80, 82, 88, 90, 96$$

SOLUTION   (a) $\overline{X} = \tfrac{546}{7} = 78$. (b) The median is still 82. (c) The mode is still 80.

A comparison of Examples 1 and 2 indicates that the test scores are the same except for the first one. Note that the arithmetic mean is the only one of the three averages that is affected by the one low score. In general, the mean is affected by extreme scores, whereas the median and mode are not. Thus the arithmetic mean should be used as a representative of a set of data when extreme (high or low) scores should be reflected in the average. Otherwise, the arithmetic mean should not be used to describe a set of data that contains extreme scores unless deception is one's major objective.

Consider this example of the earnings of the employees of a small business run by a foreman and three other employees. The foreman earns $19,000 per year. The others earn $4800, $4500, and $4500, respectively. To impress the union, the owner claims that he pays his employees an average salary of $8200. He selects the arithmetic mean as the representative salary.

$$19,000 + 4800 + 4500 + 4500 = 32,800$$
$$32,800 \div 4 = 8200$$

Actually, the median, $4650, would present a fairer picture of the average, or typical, salary.

As a more extreme example, consider a group of 49 people with a mean income of $5000. Let us see what happens to the average income of the group when an additional person with an income of $100,000 joins the group.

$$49 \times 5000 = 245,000 \quad \text{(total income of the 49 people)}$$
$$1 \times 100,000 = 100,000$$
$$\text{Sum} = 345,000 \quad \text{(total income of the 50 people)}$$
$$\bar{X} = \frac{345,000}{50} = 6900$$

The average (arithmetic mean) salary is now $6900. The addition of this one extreme score has raised the mean $1900, whereas the median and the mode of the salaries have not been affected.

In reading or hearing advertisements, the consumer must always raise the question of the type of average being used. Furthermore, he should question the source and plausibility of the data. For example, one might wonder how a psychologist can say that the "average" American girl kisses 79 men before getting married!

exercises

Find **(a)** *the mean;* **(b)** *the median; and* **(c)** *the mode for each set of data.*

1. 73, 79, 80, 82, 85, 85, 97
2. 62, 63, 67, 67, 72, 75, 75, 81, 86
3. 7, 12, 15, 16, 18, 21, 23, 24
4. 6, 7, 10, 10, 10, 13, 14
5. 87, 63, 70, 75, 93, 70, 95

In Exercises 6 through 8 tell which one, if any, of the three measures of central tendency seems most appropriate to represent the data described.

6. The average salary in a shop staffed by the owner and five employees.
7. The average salary of the workers in a factory that employs 100 people.
8. The average number of cups of coffee ordered by individual diners in a restaurant.

9. What relationship does the mode have to the use of this word in everyday language?

10. The mean score on a set of 15 tests is 74. What is the sum of the 15 test scores?

11. The mean score on a set of 30 tests is 82. What is the sum of the test scores?

12. The mean score on a set of 40 tests is 76. What is the sum of the test scores?

13. The mean score on nine of a set of ten tests is 80. The tenth score is 50. What is the sum of the test scores?

14. The mean score on 25 of a set of 27 tests is 78. The other two scores are 30 and 35. What is the sum of the scores?

15. Two sections of a course took the same test. In one section the mean score on the 25 tests was 80. In the other section the mean score on the 20 tests was 70. **(a)** What is the sum of the 45 test scores? **(b)** To the nearest integer what is the mean of the 45 test scores?

16. The mean score on five tests is 85. The mean on three other tests is 75. What is the mean score on all eight tests? (*Hint:* The answer is *not* 80.)

17. A student has a mean score of 85 on seven tests taken to date. What score must she achieve on the eighth test in order to have a mean score of 90 on all eight tests? Comment on your answer.

*18. An interesting property of the arithmetic mean is that the sum of the deviations (considered as signed numbers) of each score from the mean is 0. Show that this is true for the set of scores: 9, 10, 13, 17, 21.

EXPLORATIONS

The term **percentile rank** is frequently found in educational literature, especially as it applies to scores on standardized tests. For example, when we say that someone's test score has a percentile rank of 80 we mean that 80% of the scores on this test fall below this particular grade. Of course this also means that 20% of all the scores lie above the given grade. A score with a percentile rank of 80 is also said to be at the eightieth percentile.

1. What is the percentile rank of the median of a distribution?

2. A student scores a grade of 127 on an aptitude test. What can you say about his aptitude as a result of this test score? Suppose you are then given the additional information that the score of 127 is at the ninety-fifth percentile in a distribution of scores for all students who have taken this particular test. What can you conclude then about his aptitude?

A measure of central tendency describes a set of data through the use of a single number. However, as in the examples considered in §10-3, a single number without other information can be misleading. Some information can be obtained by comparing the arithmetic mean and the median, since the mean is affected by extreme scores and the median is relatively stable. In this section we consider other ways of obtaining information about sets of data without considering all of the elements of the data individually.

A **measure of dispersion** is a number that provides some information on the variability of a set of data. The simplest such measure to use for a set of numbers is the **range**, the difference between the largest and the smallest number in the set.

Consider the test scores of these two students

Betty: $\{68, 69, 70, 71, 72\}$, $\quad \bar{X} = 70$
Jane: $\{40, 42, 70, 98, 100\}$, $\quad \bar{X} = 70$

The range for Betty's test scores is $72 - 68$, or 4; the range for Jane's test scores is $100 - 40$, or 60. The averages and ranges provide a more meaningful picture of the two students' test results than the averages alone:

Betty: $\bar{X} = 70$, $\quad$ range, 4
Jane: $\bar{X} = 70$, $\quad$ range, 60

Although the range is an easy measure of dispersion to use, it has the disadvantage of relying on only two extreme scores—the lowest and the highest scores in a distribution. For example, consider Bob's test scores:

$\{40, 90, 95, 95, 100\}$

Bob is quite consistent in his work, and there may well be some good explanation for his one low grade. However, in summarizing his test scores one would report a range of 60, since $100 - 40 = 60$.

There is another measure of dispersion that is widely used in describing statistical data. This measure is known as the **standard deviation,** and is denoted by the Greek letter σ (sigma). The standard deviation is relatively difficult to compute but can be very informative. Fortunately, the average consumer of statistics needs to be able to understand and interpret, rather than to compute standard deviations. This need is particularly acute for teachers since they need to interpret the performance of their students on standardized tests.

To understand the significance of standard deviations as measures of dispersion, we first turn our attention to a discussion of **normal distributions.**

The line graphs of these distributions are the familiar bell-shaped curves that are used to describe distributions for so many physical phenomena. For example, the distribution of intelligence quotient (I.Q.) scores in the entire population of the United States can be pictured by a **normal curve.** The area under the curve represents the entire population.

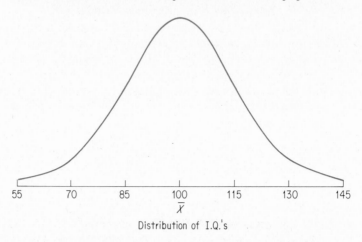

Distribution of I.Q.'s

According to psychologists, a score over 130 is considered to be very superior. At the other end of the scale, a score below 70 generally indicates some degree of retardation.

A normal-appearing curve should be expected only for data that involve a large number of elements and are based upon a general population. For example, the I.Q. scores of the honor society members in a school should not be expected to fit the distribution for the school as a whole.

In a normal distribution the mean, the median, and the mode all have the same value. This common value is associated with the axis of symmetry of the normal curve.

If three standard deviations are added to and subtracted from the mean of a normal distribution, practically all (99.7%) of the data will fall on the interval from $\bar{X} - 3\sigma$ to $\bar{X} + 3\sigma$. If an interval of two standard deviations from the mean is considered, approximately 95% of all data is included. An interval of one standard deviation about the mean includes approximately 68% of all data in a normal distribution. We may summarize these statements as follows

$$\bar{X} \pm 1\sigma \approx 68\%$$
$$\bar{X} \pm 2\sigma \approx 95\%$$
$$\bar{X} \pm 3\sigma \approx 100\%$$

and as in the following figure:

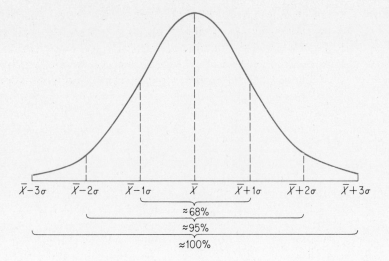

$\approx 68\%$

$\approx 95\%$

$\approx 100\%$

Let us now return to the graph of I.Q. scores. Suppose we are told that for this distribution $\sigma = 15$. We may then show these standard deviations on the base line of the graph. According to our prior discussion, we may now say that approximately 68% of the population have I.Q. scores between 85 and 115, that is, on the interval $100 \pm 1\sigma$. Approximately 95% of the population have I.Q. scores between 70 and 130, that is, on the interval $100 \pm 2\sigma$. Finally, almost everyone has an I.Q. score between 55 and 145, that is, on the interval $100 \pm 3\sigma$. "Almost everyone" really means 99.7%; that is, we might expect 0.3% (3 in 1000) of the population to have scores below 55 or above 145.

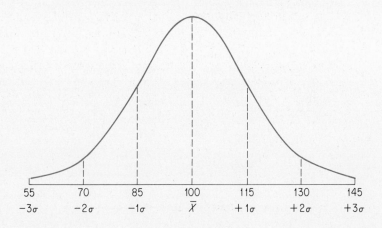

EXAMPLE 1

What per cent of the population have I.Q. scores less than 115?

SOLUTION   We know that 50% of the population have I.Q. scores below 100, the mean. Also, 68% have scores on the interval $\bar{X} \pm 1\sigma$. Then by the symmetry of the normal curve, 34% have scores between 100 and 115, and, as in the figure, 84% have scores less than 115.

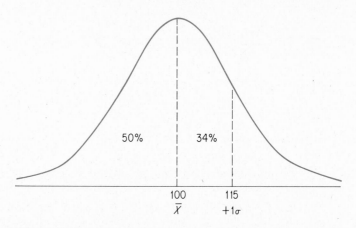

Notice that when considering questions such as that in Example 1 we do not concern ourselves with the number of people who make any particular score such as 100 or 115. Also the scale $-3, -2, -1, 0, 2, 3$ in standard deviations enables us to use the properties of a normal distribution, such as the 68% within one standard deviation of the mean, for a wide variety of situations. For example, we found in Example 1 that 84% of the population have I.Q. scores less than 115, that is, less than one standard deviation above the mean. This same result can be used in many other situations. Scholastic Aptitude Test (S.A.T.) scores have a mean of 500 and a standard deviation of 100. Thus 84% of the S.A.T. scores on a particular test are below 600. College Entrance Examination Board (C.E.E.B.) scores also have a mean of 500, a standard deviation of 100, and 84% of the scores below 600.

EXAMPLE 2

What per cent of the population have I.Q. scores greater than 130?

SOLUTION   The interval $\bar{X} \pm 2\sigma$ includes the scores of 95% of the population, so 47.5% of the population have scores on the interval from 100 to 130. Thus 97.5% of the population have scores below 130, and 2.5% of the population have I.Q. scores above 130.

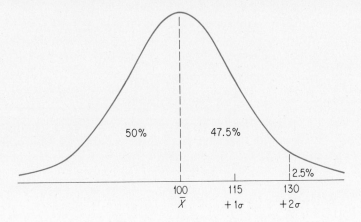

50%   47.5%   2.5%

100   115   130
$\overline{X}$ $+1\sigma$ $+2\sigma$

The result obtained in Example 2 as applied to S.A.T. and C.E.E.B. scores shows that 97.5% of the scores are below 700 and 2.5% of the scores are over 700.

Note that we have discussed only normal distributions. However, the standard deviation can be computed for other distributions as well. In general, many sets of data tend to be approximately normal, and the standard deviation is a very useful measure of dispersion.

As stated earlier, the standard deviation is relatively difficult to compute, especially for data grouped in a table. However, it may be instructive to note the computation of σ for a small set of scores. For such data, we generally use the formula

$$\sigma = \sqrt{\frac{\Sigma\, d^2}{n}}$$

where $\Sigma\, d^2$ represents the sum of the squares of the deviations of each score from the mean, and n is the number of scores.

Consider the set of scores

$\{8, 10, 12, 16, 19\}$

We first compute the mean as 13. In the column headed d we find the differences when the mean $\overline{X}$ is subtracted from each score. The sum of these deviations is 0 as noted in §10-3, Exercise 18. Thus to obtain a meaningful average of these deviations we may use either absolute values or the squares of the deviations. The standard deviation is based upon the squares of the deviations, as in the column headed d^2.

Scores	d	d^2
8	-5	25
10	-3	9
12	-1	1
16	3	9
19	6	36
5 $\overline{65}$		$\overline{80}$

$\sigma = \sqrt{\frac{80}{5}} = \sqrt{16}$

$\sigma = 4$

$\overline{X} = 13$

Recall that σ is a measure of dispersion. In Example 3 we have a set of scores with the same mean as the preceding set, but with a much smaller range. Note that the standard deviation is correspondingly smaller.

EXAMPLE 3

Compute σ: $\{10, 11, 13, 14, 17\}$.

SOLUTION

Scores	d	d^2
10	-3	9
11	-2	4
13	0	0
14	1	1
17	4	16
5 $\overline{65}$		$\overline{30}$

$\sigma = \sqrt{\frac{30}{5}} = \sqrt{6}$

$\sigma \approx 2.5$

$\overline{X} = 13$

These two sets of five scores for which standard deviations have been computed are much too small to expect a close match with a normal distribution. In each of these two cases all scores are within one and a half standard deviations of the mean.

exercises

1. Find the range for the set of scores:

   $\{55, 60, 80, 85, 90, 92, 98\}$

2. Give a set of seven scores with the same mean as in Exercise 1 but with a smaller range.

3. Repeat Exercise 2 for a larger range.

4. Give a set of seven scores with the same range as in Exercise 1 but with a smaller mean.

5. Repeat Exercise 4 for a larger mean.

In Exercises 6 through 8 tell which of the distributions can be expected to be approximately normal in shape.

6. The scores of graduating high school seniors on a particular college board examination.

7. The weights of all college freshmen boys.

8. The number of heads obtained if 100 coins are tossed by each college graduate in the country.

9. For a normal distribution of 10,000 test scores the mean is found to be 500 and the standard deviation is 100.
    (a) What per cent of the scores will be above 600?
    (b) What per cent of the scores will be below 300?
    (c) About how many scores will be above 700?
    (d) About how many scores will be below 400?
    (e) About how many scores will be between 400 and 700?

10. If 100 coins are tossed repeatedly, the distribution of the number of heads is a normal one with a mean of 50 and a standard deviation of 5. What per cent of the number of heads will be: (a) greater than 60? (b) less than 45? (c) between 40 and 60?

11. Compute the standard deviation for the set of scores:

    $\{7, 9, 10, 11, 13\}$

12. Compute the standard deviation for the set of scores:

    $\{11, 12, 13, 15, 20, 20, 21\}$

EXPLORATIONS

In reading educational literature one frequently comes across the term **coefficient of correlation,** usually denoted by the letter r. The coefficient of correlation is given as a decimal from -1.00 to $+1.00$ and provides an indication of how two variables are related.

A perfect positive correlation of 1.00 indicates that two sets of data are related so that as one increases, so does the other, and does so uniformly. For example, here is a set of ages and weights for a group of six individuals. Note that for each increase of 5 years of age there is a corresponding increase of 10 pounds of weight.

Age	Weight	
20	130	
25	140	
30	150	The coefficient of
35	160	correlation is 1.00.
40	170	
45	180	

Now consider the following table for six other individuals. For each increase of 5 years of age, there is a corresponding decrease of 10 pounds of weight.

Age	Weight	
20	180	
25	170	
30	160	The coefficient of
35	150	correlation is -1.00.
40	140	
45	130	

A coefficient of correlation of 0 indicates no uniform change of either variable with respect to the other. In general practice one would not expect to have these extreme cases occur. Furthermore one has to read the literature accompanying any particular test or research study to determine whether any particular coefficient of correlation can be considered as significant.

Finally, it is important to note that correlation does not imply causation. For example, there is a positive correlation between size of shoe that one wears and handwriting ability. Now this does not mean that big feet improves one's handwriting. Rather the explanation is probably that as one grows older, one's feet tend to get larger and also at the same time one tends to write more legibly.

1. Obtain a copy of a standardized test and read the instructions to see what references, if any, are made to correlation.

2. There has been a high positive correlation between expenditures for alcohol and for higher education in recent years. Does this mean that drinking alcohol provides one with the thirst for knowledge? Does it mean that education leads one to drink? How can you explain this high correlation?

3. Estimate whether there is a high or a low correlation between the following pairs of data, and attempt to justify your answer:
(a) I.Q. scores and scores on college entrance examinations.
(b) Scores made by elementary school students in reading and in arithmetic.
(c) Age and physical ability.

10-5
Binomial Distributions

Any distribution of data for which there are only *two* possible events is a **binomial distribution**. The distribution of heads in tossing coins is a binomial distribution since each coin is assumed to have exactly one of two possible outcomes, heads or tails. A "fair" coin can be expected to land heads about half the time. Accordingly, we say that the *probability* (§9-2) of heads is $\frac{1}{2}$ on each toss of the coin; that is, the ratio of the number of heads to the total number of tosses is expected to be about $\frac{1}{2}$. If the experiment of tossing a coin n times is repeated over and over, the expected mean of the distribution of heads on these trials of the experiment is $\frac{1}{2}n$. The result of tossing one coin n times is the same as that of tossing n coins once. For example, if one coin is tossed 100 times or 100 coins are tossed once, we have probability $p = \frac{1}{2}$ and $n = 100$. Therefore, the mean or average number of heads expected is $(\frac{1}{2})(100) = 50$.

The standard deviation of a binomial distribution is easily found by the formula

$$\sigma = \sqrt{p \times (1 - p) \times n}$$

where p is the probability of success and n is the number of trials. For heads on 100 coins, we have $p = \frac{1}{2}$, $1 - p = \frac{1}{2}$, $n = 100$, and

$$\sigma = \sqrt{(\tfrac{1}{2})(\tfrac{1}{2})(100)} = \sqrt{25} = 5$$

It has been shown for a normal distribution that if three standard deviations are added to and subtracted from the mean, we have the limits within which almost all of the data will fall. For the example of 100 coins, the mean is 50, and three times the standard deviation is 15, thus giving the limits 35 to 65. This is frequently stated in terms of **confidence limits:** we may say with "*almost* 100%" confidence that the number of heads will be between 35 and 65. Furthermore, we may say with approximately 95% confidence that the number of heads will be between 40 and 60, that is, within two standard deviations of the mean. We may say with approximately 68% confidence that the number of heads will be between 45 and 55, that is, within one standard deviation of the mean.

A **control chart** is a graph on which lines are drawn to represent the limits within which all data are expected to fall a given per cent of the time. In industry, for example, a chart like the following may be drawn on which data from samples are plotted.

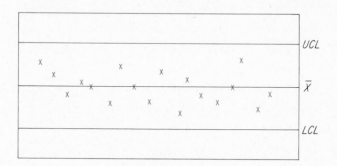

The middle horizontal line represents the **mean** ($\bar{X}$) or average around which these data are expected to lie. The other horizontal lines, called the **upper control limit** (UCL) and **lower control limit** (LCL), represent the limits within which all the data are expected to fall most of the time. Occasionally data may fall outside of these limits; as long as most of the samples tested produce data which lie between these lines, the process is said to be "in control."

For the case of 100 coins, the theoretical average is 50. The lower control limit would be $\bar{X} - 3\sigma$, or 35. The upper control limit would be $\bar{X} + 3\sigma$, or 65. We would suspect that something was wrong if a toss of 100 coins were to produce fewer than 35 or more than 65 heads. Such an event *can* take place just by chance, but it would be very unusual, occurring less than 3 times in 1000.

EXAMPLE 1

What is the mean number of fives to be expected in 180 tosses of a single die?

SOLUTION   When dice (singular die) are tossed, the possible results are

$$1, \quad 2, \quad 3, \quad 4, \quad 5, \quad 6$$

For a "fair" die these six possibilities are equally likely and we say that the probability of a 5 is 1 out of 6, that is, $\frac{1}{6}$. Thus we have $p = \frac{1}{6}$ and $n = 180$. Therefore, the average number of fives expected is $(\frac{1}{6})(180) = 30$. Theoretically we can expect 1 out of every 6 tosses to produce a 5 and expect 30 fives in 180 tosses.

EXAMPLE 2

What is the standard deviation for the distribution of Example 1?

SOLUTION   The probability of tossing a five is $\frac{1}{6}$. Thus $p = \frac{1}{6}$, $(1 - p) = \frac{5}{6}$, $n = 180$, and $\sigma = \sqrt{(\frac{1}{6})(\frac{5}{6})(180)} = \sqrt{25} = 5$.

The results of Examples 1 and 2 can be shown in terms of a graph as follows:

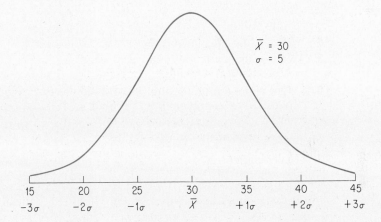

$$\overline{X} = 30$$
$$\sigma = 5$$

| 15 | 20 | 25 | 30 | 35 | 40 | 45 |
| -3σ | -2σ | -1σ | $\overline{X}$ | $+1\sigma$ | $+2\sigma$ | $+3\sigma$ |

Thus, in a set of 180 tosses of a die, we would expect with almost 100% confidence to have between 15 and 45 fives. With 95% confidence we would expect to have between 20 and 40 fives; and with 68% confidence we would expect between 25 and 35 fives.

These examples should serve to show the reader that fluctuation is normal. That is, although the mean of the distribution is 30, we expect there to be variability. Not every set of 180 tosses will produce exactly 30 fives—in fact, very few will. However, on the other hand, we expect this fluctuation to be within limits that can be described by our knowledge of the behavior of normal distributions. Someone could conceivably toss a die 180 times and produce 100 fives. However, an understanding of the principles of this chapter would dictate that you challenge the tosser of the die and examine the die quite carefully for evidence of foul play. In other words, although it is possible to obtain 100 fives, it is far from probable that this would happen.

exercises

In Exercises 1 through 5 consider a single die that is tossed 720 times.

1. What is the mean number of sixes to be expected?

2. What is the standard deviation for the distribution of sixes?

3. What is the maximum number of sixes we can expect to obtain with almost 100% confidence?

4. What is the minimum number of sixes we can expect to obtain with almost 100% confidence?

5. Within what limits can we expect the number of sixes to fall approximately 95% of the time?

6. Toss 100 coins 10 times and record the results on a control chart.

7. A coin is tossed 64 times. Within what limits can we say, with 95% confidence, that we will find the total number of heads?

In Exercises 8 through 11 consider a single die that is tossed 180 times.

8. What is the mean number of threes to be expected?

9. What is the maximum number of threes we can expect to obtain with almost 100% confidence?

10. What is the minimum number of threes we can expect to obtain with almost 100% confidence?

11. Within what limits can we expect the number of threes to fall approximately 95% of the time?

12. Every student in a class is asked to toss a coin 64 times and record the number of heads obtained. One student claims to have tossed 45 heads. React to his claim, basing your reaction on the statistical concepts presented in this section.

EXPLORATIONS

1. Use the concepts of this section to establish the limits within which the total number of heads can be expected to fall when a coin is tossed 64 times, using the 95% confidence limits. Then test these results by tossing a coin 64 times, and asking as many other people as possible to repeat this experiment. (Note that instead of tossing a single coin for 64 tosses, you may also toss four coins for 16 tosses, sixteen coins for 4 tosses, etc.)

2. Repeat Exercise 1 for 180 tosses of a single die, counting the number of sixes that appear.

chapter test

1. Tell what possible misuse or misinterpretation of statistics each of the following statements involves:

   (a) Mathville's population increased by 100% last year. This is a greater per cent increase than for any other city in the state. Therefore the population of Mathville must be the largest of any city in that state.

   (b) Statistics show that college students drink more.

2. Draw a circle graph to show the distribution of grades given by Professor X last semester.

Grade	A	B	C	D	F
Per cent	20	25	40	10	5

3. Draw a histogram for the following data of the number of heads obtained on 32 tosses of five coins.

Number of heads	0	1	2	3	4	5
Frequency		1	4	11	9	5	2

   Use the following graph to answer Exercises 4–6.

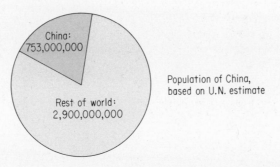

China: 753,000,000

Rest of world: 2,900,000,000

Population of China, based on U.N. estimate

4. The population of China is approximately what per cent (nearest whole number) of the population of the entire world? (Note that the population of the entire world includes the population of China as well.)

5. What is the size of the central angle (nearest degree) that represents the population of China?

6. Assume you wish to draw a bar graph to show the same facts as does the circle graph. If the population of China is represented by a bar that is one inch high, approximately how long a bar (nearest inch) would you need to show the population of the rest of the world?

7. Find **(a)** the mean; **(b)** the median; **(c)** the mode for this set of scores:

55, 62, 70, 74, 74, 79

8. Find the range for each of the following sets of scores:
    (a) 75, 82, 64, 98, 79   **(b)** 100, 45, 79, 82, 96, 68

9. A collection of 1000 scores forms a normal distribution with a mean of 50 and a standard deviation of 10.
    (a) What per cent of the scores will be above 70?
    (b) What per cent will be below 40?

10. A coin is tossed 64 times.
    (a) Within what limits can we say, with almost 100% confidence, that we shall find the total number of heads?
    (b) If the experiment is repeated many times, what per cent of the times should be expected to produce fewer than 36 heads?

Epilogue

Twenty-five hundred years ago arithmetic (the theory of numbers), geometry, music, and spherics (astronomy) were the basic subjects in the liberal arts program of study of the Pythagoreans. Grammar, logic, and rhetoric were added in the Middle Ages and these seven liberal arts were considered essential for an educated person. Mathematics is still an important part of the background of any educated person.

Mathematics today includes many topics such as arithmetic, number theory, algebra, geometry, logic, probability, statistics, and linear programming that we have considered in this introduction to mathematical concepts. The concepts of sets provide a unifying theme for all of these topics.

The study of separate courses in arithmetic, algebra, and geometry is misleading since it makes them appear as separate subjects rather than as parts of the same subject, mathematics. We have seen that algebra may be used in the study of geometry as, for example, in the use of coordinates. Also geometry may be used in the study of algebra as, for example, in the use of the number line. The very close relationship between algebra and geometry may be observed by noticing that the algebra of the real numbers and the geometry of the points on a Euclidean line are essentially the same. Indeed, the early Greeks represented numbers by line segments before more convenient notations were developed. Thus algebra provides one approach or point of view for the study of mathematics and geometry provides another approach or point of view.

Mathematics is a living and rapidly growing subject. More new mathematics has been developed during the last fifty years than in all previous time of recorded history. Also mathematics is permeating the scientific advances of our time so that we must either understand the basic mathematical concepts or live in fear of the scientific advances. In this book we have tried to help you remove your fears of mathematics and develop an understanding that will enable you to proceed with confidence in your chosen careers.

Answers to Odd-Numbered Exercises

1

Explorations with
MATHEMATICS

1-1 *Explorations with Number Patterns*

1. The text shows the diagrams for 3×9, 7×9, and 4×9 respectively. Here are the others:

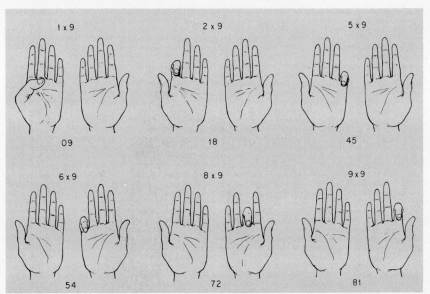

3. $6^2 = 1 + 2 + 3 + 4 + 5 + 6 + 5$
$\qquad + 4 + 3 + 2 + 1$
$7^2 = 1 + 2 + 3 + 4 + 5 + 6 + 7$
$\qquad + 6 + 5 + 4 + 3 + 2 + 1$
$8^2 = 1 + 2 + 3 + 4 + 5 + 6 + 7$
$\qquad + 8 + 7 + 6 + 5 + 4 + 3$
$\qquad\qquad\qquad\qquad + 2 + 1$
$9^2 = 1 + 2 + 3 + 4 + 5 + 6 + 7$
$\qquad + 8 + 9 + 8 + 7 + 6 + 5$
$\qquad\qquad + 4 + 3 + 2 + 1$

7. **(a)** $9 \times 47 = 423$

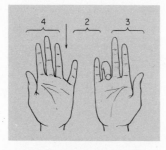

(b) $9 \times 39 = 351$

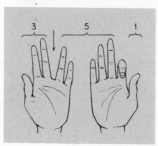

(c) $9 \times 18 = 162$

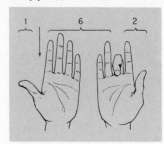

(d) $9 \times 27 = 243$

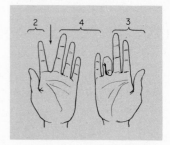

9. **(a)** 40×81, that is, 3240;
(b) 100×201, that is, 20,100;
(c) $\frac{25}{2} \times 50$, that is, 625;
(d) 50×200, that is, 10,000;
(e) 100×402, that is, 40,200.

11. **(a)** $\frac{2}{3}$; $\frac{3}{4}$.
(b) The last fraction in each row is of the form $1/[n(n + 1)]$ and the sum is $n/(n + 1)$. Thus, since the last fraction in the original series is $1/(9 \times 10)$, the sum is $\frac{9}{10}$.
(c) $\frac{4}{5}$.
(d) $\frac{99}{100}$.

13. **(a)** $\frac{3}{4}$;
(b) $\frac{7}{8}$;
(c) $\frac{15}{16}$. In general: $\dfrac{2^n - 1}{2^n}$.

1-2 *Explorations with Geometric Patterns*

1. There will be 4 holes produced by 4 folds, and 2^{n-2} holes with n folds.

3. $V + R = A + 2$: the sum of the number of vertices and the number of regions is 2 more than the number of arcs.

5.

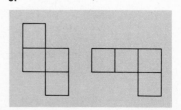

7. (a) One possible tracing:

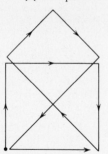

(b) Impossible.

9. Dotted lines show segments removed.

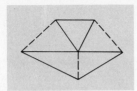

11. Dotted lines show original outline of the lake.

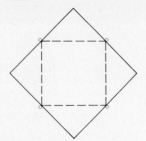

1-3 *Explorations with Mathematical Recreations*

1. (a) 12.
   (b) Only one, if it's long enough.
   (c) Only halfway; then you start walking out.
   (d) *One* of them is not a nickel, but the other one is.
   (e) There is no dirt in a hole.
   (f) Brother-sister.

3. There are eleven trips needed. First one Indian and one missionary go over; the missionary returns. Then two Indians go over and one returns. Next two missionaries go over; one missionary and one Indian return. Two missionaries go over next and one Indian returns. Then two Indians go over and one of them returns. Finally, the last two Indians go over.

5. After 27 days the cat still has 3 feet to go. It does this the next day and is at the top after 28 days.

7.

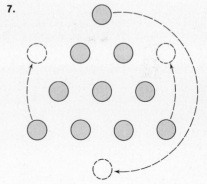

9. Use the matchsticks to form a triangular pyramid.

11. Both *A*'s and *B*'s will say they are *B*'s. Therefore, when the second man said that the first man said he was a *B*, the second man was telling the truth. Thus the first two men told the truth and the third man lied.

13. If the penny is in the left hand and the dime in the right hand, the computation will give $3 + 60 = 63$, an odd number. If the coins are reversed, we have $30 + 6 = 36$, an even number.

15. There really is no missing dollar. The computation may be done in one of two ways: $(30 - 3) - 2 = 25$, or $25 + 2 + 3 = 30$. In the problem the arithmetic was done in a manner that is not legitimate; that is, $(30 - 3) + 2$.

17. Eight moves are needed. The coins are identified in the following diagram as well as the squares which may be used. The moves are as follows, where the first numeral indicates the position of the coin and the second one tells you where to move it: D_1: 4–3; P_2: 2–4; P_1: 1–2; D_1: 3–1; D_2: 5–3; P_2: 4–5; P_1: 2–4; D_2: 3–2.

1 P_1	2 P_2	3	4 D_1	5 D_2

19. E; N. These are the first letters of the names (one, two, three, etc.) of the counting numbers.

21. Let H represent the half-dollar, Q the quarter, and N the nickel. The moves are then as follows, in the order given: N to C, Q to B, N to B, H to C, N to A, Q to C, and N to C. If a fourth coin P (penny) is added, the 15 moves required are as follows: P to B, N to C, P to C, Q to B, P to A, N to B, P to B, H to C, P to C, N to A, P to A, Q to C, P to B, N to C, and P to C. For a discussion of the number of moves required for 64 discs, see *Mathematics and the Imagination* by Kasner and Newman (page 171). They estimate that it would take more than 58,454,204,609 centuries to complete the task!

23. Note that the sum of the values on two opposite sides of a die is always 7.

1-4   *Explorations with Flow Charts*

1. 56

3. 8

5. 23

7. 29

9. 37

11. $5(n + 6)$; 90

13. $\frac{1}{3}(n + 6)$; 6.

15.

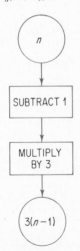

17.

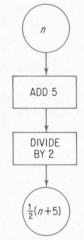

19. 12, 19, 26, 33, 40, 47, 54, 61.

2

An Introduction to

SETS

2-1 Set Notation

1. $\in$

3. $\notin$

5. Well-defined.

7. Not well-defined.

9. Equal.

11. Equal.

13. {January, February, March, . . . , December}

15. $\{1, 2, 3, \ldots, 9\}$

17. The set of counting numbers 1 through 6.

19. The set of counting numbers greater than 50.

21. The set of multiples of 10 from 10 through 150.

*23. The set of numbers that can be expressed in the form $n(n - 1)$, where n ranges from 1 through 10.

2-2 Subsets

1. Among the many correct answers are: the set of odd integers divisible by 2; the set of integers between 4 and 5; the set of even integers between 6 and 8; the human beings who traveled to Venus prior to 1968; the hens that lay glass eggs; the people born with ten left feet.

3. $\{c\}$

5. $\{c, a, r, t\}$

7. $\{f, s\}$

9. $\{f\}$

11. $\{2, 4, 6, 8\}$

13. $\{2, 3, 5\}$

15. $\{a\}, \varnothing.$

17. $\{a, b, c\}, \varnothing, \{a\}, \{b, c\}, \{b\}, \{a, c\}, \{c\}, \{a, b\}.$

19. $\{a, b, c, d, e\}, \varnothing, \{a\}, \{b, c, d, e\}, \{b\}, \{a, c, d, e\}, \{c\}, \{a, b, d, e\}, \{d\}, \{a, b, c, e\}, \{e\}, \{a, b, c, d\}, \{a, b\}, \{c, d, e\}, \{a, c\}, \{b, d, e\}, \{a, d\}, \{b, c, e\}, \{a, e\}, \{b, c, d\}, \{b, c\}, \{a, d, e\}, \{b, d\}, \{a, c, e\}, \{b, e\}, \{a, c, d\}, \{c, d\}, \{a, b, e\}, \{c, e\}, \{a, b, d\}, \{d, e\}, \{a, b, c\}.$

21. $\varnothing$

23. $\varnothing, \{a\}, \{b, c\}, \{b\}, \{a, c\}, \{c\}, \{a, b\}.$

25. $\varnothing, \{a\}, \{b, c, d, e\}, \{b\}, \{a, c, d, e\}, \{c\}, \{a, b, d, e\}, \{d\}, \{a, b, c, e\}, \{e\}, \{a, b, c, d\}, \{a, b\}, \{c, d, e\}, \{a, c\}, \{b, d, e\}, \{a, d\}, \{b, c, e\}, \{a, e\}, \{b, c, d\}, \{b, c\}, \{a, d, e\}, \{b, d\}, \{a, c, e\}, \{b, e\}, \{a, c, d\}, \{c, d\}, \{a, b, e\}, \{c, e\}, \{a, b, d\}, \{d, e\}, \{a, b, c\}.$

2-3 Equivalent Sets

1. $\{1, 2\}$ $\{1, 2\}$
   $\updownarrow\updownarrow$ $\updownarrow\updownarrow$
   $\{p, q\}$ $\{q, p\}$

3. $\{1, 2, 3, 4\}$ $\{1, 2, 3, 4\}$
   $\updownarrow\updownarrow\updownarrow\updownarrow$ $\updownarrow\updownarrow\updownarrow\updownarrow$
   $\{r, e, s, t\}$ $\{r, e, t, s\}$

$\{1, 2, 3, 4\}$ $\{1, 2, 3, 4\}$
↕ ↕ ↕ ↕   ↕ ↕ ↕ ↕
$\{r, s, e, t\}$ $\{r, s, t, e\}$

$\{1, 2, 3, 4\}$ $\{1, 2, 3, 4\}$
↕ ↕ ↕ ↕   ↕ ↕ ↕ ↕
$\{r, t, e, s\}$ $\{r, t, s, e\}$

$\{1, 2, 3, 4\}$ $\{1, 2, 3, 4\}$
↕ ↕ ↕ ↕   ↕ ↕ ↕ ↕
$\{e, r, s, t\}$ $\{e, r, t, s\}$

$\{1, 2, 3, 4\}$ $\{1, 2, 3, 4\}$
↕ ↕ ↕ ↕   ↕ ↕ ↕ ↕
$\{e, s, r, t\}$ $\{e, s, t, r\}$

$\{1, 2, 3, 4\}$ $\{1, 2, 3, 4\}$
↕ ↕ ↕ ↕   ↕ ↕ ↕ ↕
$\{e, t, r, s\}$ $\{e, t, s, r\}$

$\{1, 2, 3, 4\}$ $\{1, 2, 3, 4\}$
↕ ↕ ↕ ↕   ↕ ↕ ↕ ↕
$\{s, r, e, t\}$ $\{s, r, t, e\}$

$\{1, 2, 3, 4\}$ $\{1, 2, 3, 4\}$
↕ ↕ ↕ ↕   ↕ ↕ ↕ ↕
$\{s, e, r, t\}$ $\{s, e, t, r\}$

$\{1, 2, 3, 4\}$ $\{1, 2, 3, 4\}$
↕ ↕ ↕ ↕   ↕ ↕ ↕ ↕
$\{s, t, r, e\}$ $\{s, t, e, r\}$

$\{1, 2, 3, 4\}$ $\{1, 2, 3, 4\}$
↕ ↕ ↕ ↕   ↕ ↕ ↕ ↕
$\{t, r, e, s\}$ $\{t, r, s, e\}$

$\{1, 2, 3, 4\}$ $\{1, 2, 3, 4\}$
↕ ↕ ↕ ↕   ↕ ↕ ↕ ↕
$\{t, e, r, s\}$ $\{t, e, s, r\}$

$\{1, 2, 3, 4\}$ $\{1, 2, 3, 4\}$
↕ ↕ ↕ ↕   ↕ ↕ ↕ ↕
$\{t, s, r, e\}$ $\{t, s, e, r\}$

5. Among the many correct answers is: the set $\{n, u, m, b, e, r\}$ of letters in "number."

7. Among the many correct answers is: the empty set.

9. 3

11. 8

13. 0

15. Among the many correct answers is: the set of counting numbers.

17. **(a)** Yes. Any two sets that consist of the same elements have the same number of elements.
(b) No. Two sets may have the same number of elements without having the same elements.

19. Among the many correct answers is:

$$\{a, e, i, o, u\}$$
↕ ↕ ↕ ↕ ↕
$$\{2, 4, 6, 8, 10\}$$

*21. $\{1, 3, 5, 7, 9, \ldots, 2n - 1, \ldots\}$
↕ ↕ ↕ ↕ ↕   ↕
$\{5, 10, 15, 20, 25, \ldots, 5n, \ldots\}$

2-4   Sets of Numbers

1. Ordinal.

3. Ordinal.

5. Ordinal.

7. Identification.

9. Ordinal.

11. $\{1, 2, 3, 4, 5, \ldots, n, \ldots\}$
↕ ↕ ↕ ↕ ↕   ↕
$\{1, 3, 5, 7, 9, \ldots, 2n - 1, \ldots\}$

13. $\{2, 4, 6, 8, 10, \ldots, 2n, \ldots\}$
↕ ↕ ↕ ↕ ↕   ↕
$\{1, 3, 5, 7, 9, \ldots, 2n - 1, \ldots\}$

15.
$\{101, 103, 105, 107, 109, \ldots, n, \ldots\}$
↕ ↕ ↕ ↕ ↕   ↕
$\{102, 104, 106, 108, 110, \ldots, n + 1, \ldots\}$

17. Infinite.

19. Finite.

21. Finite.

23. Infinite.

25. The set $\{1, 2, 3, \ldots, n, \ldots\}$ has the transfinite cardinal number $\aleph_0$; the set $\{\triangle, \square, 0, 1, 2, 3, \ldots, n, \ldots\}$ has cardinal number $\aleph_0 + 3$, which is the same as $\aleph_0$. The equivalence of the two sets may be shown as follows:

$$\{\, 1, \quad 2, \quad 3, 4, 5, \ldots, \quad n, \quad \ldots\},$$
$$\updownarrow \ \updownarrow \ \updownarrow \updownarrow \updownarrow \qquad\quad \updownarrow$$
$$\{\triangle, \square, 0, 1, 2, \ldots, n - 3, \ldots\}$$

27. The set $\{1, 2, 3, \ldots, n, \ldots\}$ has the transfinite cardinal number $\aleph_0$; the set with the number 1 removed has cardinal number $\aleph_0 - 1$, which is the same as $\aleph_0$. The equivalence of the two sets may be shown as follows:

$$\{1, 2, 3, \ldots, \quad n, \quad \ldots\}$$
$$\updownarrow \updownarrow \updownarrow \qquad\ \updownarrow$$
$$\{2, 3, 4, \ldots, n + 1, \ldots\}$$

2-5 *Intersection and Union*

1. (a) $\{1, 2, 3, 5, 7\}$;
  (b) $\{1, 3\}$.

3. (a) $\{1, 3, 5, 7, 9\}$;
  (b) $\{7\}$.

5. (a) $\{1, 3, 5, 7, 8, 10\}$;
  (b) $\varnothing$

7. (a) $\{1, 2, 3, \ldots\}$;
  (b) $\varnothing$

9. (a) $\{3, 4, 5\}$;
  (b) $\{2, 4\}$;
  (c) $\{2, 3, 4, 5\}$;
  (d) $\{4\}$

11. (a) $\{2, 4, 6, \ldots\}$;
  (b) $\{1, 3, 5, \ldots\}$;
  (c) $\{1, 2, 3, \ldots\}$;
  (d) $\varnothing$.

13. (a) $\{2, 3\}$;
  (b) $\{1, 2\}$;
  (c) $\{1, 2, 3\}$;
  (d) $\{2\}$.

15. $\{6, 8\}$

17. $\{2, 4, 5, 6, 8, 9, 10\}$

19. $\{4, 5, 9\}$

21. (a) A;
  (b) B;
  (c) $\varnothing$;
  (d) U.

23. $U = \{1, 2, 3, \ldots\}$

25. $\varnothing$

27. $A = \{1, 3, 5, \ldots\}$

2-6 *Sets of Points*

1. (a) 2;
  (b) 9;
  (c) 4;
  (d) 13.

3.

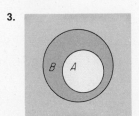

5.

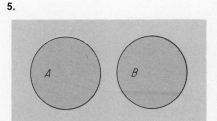

7.

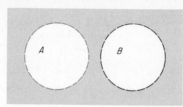

9. A' is shaded with horizontal lines; B is shaded with vertical lines. The union of these two sets is the subset of U that is shaded with lines in either or both directions.

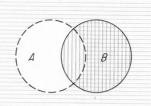

11. A is shaded with vertical lines; B' is shaded with horizontal lines. The intersection of these two sets is the subset of U that is shaded with lines in both directions.

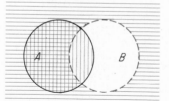

13. $A \cup B = B \cup A$

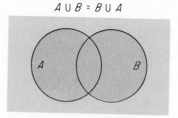

15. $A \cap B \subseteq A$

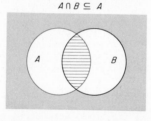

17. In the following pair of diagrams, the final result is the same, showing the equivalence of the statements given.

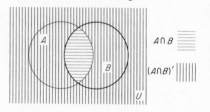

The set $(A \cap B)$ is shaded with horizontal lines. Its complement, $(A \cap B)'$, is the remaining portion of U shaded with vertical lines.

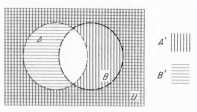

The set A' is shaded with vertical lines; the set B' is shaded with horizontal lines. Their union, $A' \cup B'$, is the portion of U shaded with lines in either or both directions.

19. **(a)** 1;
    (b) 5;
    (c) 4;
    (d) 12;
    (e) 21;
    (f) 25.

21. **(a)** 25;
    (b) 7;
    (c) 7;
    (d) 29;
    (e) 37;
    (f) 6.

23. There are 11 students who are not taking any of the three subjects, 10 taking only chemistry, and 2 taking physics and chemistry but not biology. The given data are shown in the following Venn diagram on page 437.

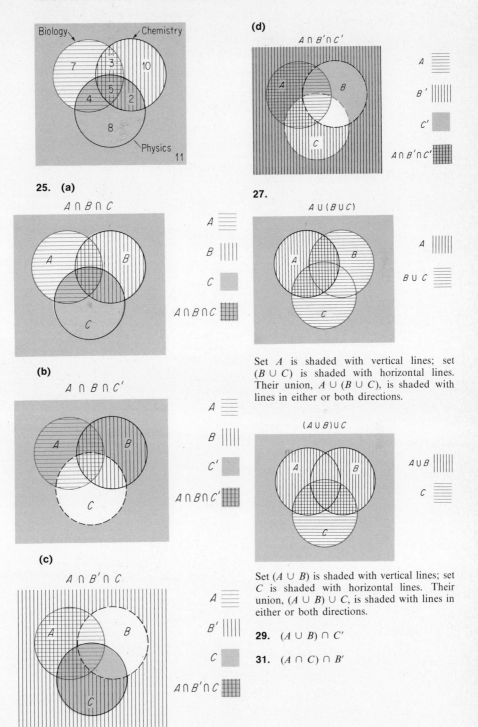

25. (a)

$A \cap B \cap C$

(b)

$A \cap B \cap C'$

(c)

$A \cap B' \cap C$

(d)

$A \cap B' \cap C'$

27.

$A \cup (B \cup C)$

Set A is shaded with vertical lines; set $(B \cup C)$ is shaded with horizontal lines. Their union, $A \cup (B \cup C)$, is shaded with lines in either or both directions.

$(A \cup B) \cup C$

Set $(A \cup B)$ is shaded with vertical lines; set C is shaded with horizontal lines. Their union, $(A \cup B) \cup C$, is shaded with lines in either or both directions.

29. $(A \cup B) \cap C'$

31. $(A \cap C) \cap B'$

3

Concepts of
LOGIC

3-1 Sets of Statements

1. **(a)** $(\sim p) \wedge (\sim q)$;
   (b) $(\sim p) \wedge q$;
   (c) $(\sim p) \wedge q$;
   (d) $\sim[p \wedge (\sim q)]$;
   (e) $p \vee (\sim q)$.

3. **(a)** $p \wedge (\sim q)$;
   (b) $p \vee (\sim q)$;
   (c) $(\sim p) \wedge q$;
   (d) $\sim[(\sim p) \wedge q]$;
   (e) $\sim[(\sim p) \wedge q]$.

5. **(a)** I like this book and I like mathematics.

(b) I do not like mathematics.
(c) I do not like this book.
(d) I do not like this book and I do not like mathematics.

7. 5(a), 6(b), 6(d).

*9. **(a)** True;
   (b) true;
   (c) true;
   (d) true.

3-2 Truth Values of Statements

1.

p	q	$(\sim p) \wedge q$
T	T	F
T	F	F
F	T	T
F	F	F

3.

p	q	$(\sim p) \vee (\sim q)$
T	T	F
T	F	T
F	T	T
F	F	T

5.

p	q	$\sim(p \wedge q)$
T	T	F
T	F	T
F	T	T
F	F	T

7.

p	q	$\sim$	$[p$	$\vee$	$(\sim q)]$
T	T	F	T	T	F
T	F	F	T	T	T
F	T	T	F	F	F
F	F	F	F	T	T

(d) (a) (c) (b)

9.

p	q	$\sim$	$[(\sim p)$	$\wedge$	$(\sim q)]$
T	T	T	F	F	F
T	F	T	F	F	T
F	T	T	T	F	F
F	F	F	T	T	T

(d) (a) (c) (b)

11.

p	q	$p \underline{\vee} q$
T	T	F
T	F	T
F	T	T
F	F	F

13. $\sim(p \wedge q)$

3-3 Conditional Statements

1. If you will study hard, then you will get an A.

3. If you do not study hard, then you will not get an A.

5. (a) If a triangle is equilateral, then the triangle is isosceles.
(b) If a triangle is isosceles, then the triangle is equilateral.
(c) If a triangle is not equilateral, then the triangle is not isosceles.
(d) If a triangle is not isosceles, then the triangle is not equilateral.

7. (a) T;
(b) T;
(c) T.

9.

p	q	$p \rightarrow q$	$(\sim p)$	$\vee$	q
T	T	T	F	T	T
T	F	F	F	F	F
F	T	T	T	T	T
F	F	T	T	T	F

    (a)   (b) (d) (c)

Note that the truth values in columns (a) and (d) are the same.

11.

p	q	$(p \rightarrow q)$	$\wedge$	$(q \rightarrow p)$
T	T	T	T	T
T	F	F	F	T
F	T	T	F	F
F	F	T	T	T

    (a)   (c)   (b)

13.

p	q	$[(\sim p)$	$\wedge$	$q]$	$\rightarrow$	$(p \vee q)$
T	T	F	F	T	T	T
T	F	F	F	F	T	T
F	T	T	T	T	T	T
F	F	T	F	F	T	F

    (a)   (c) (b) (e)   (d)

15. {7}

*17. The set of all real numbers.

*19. The set of all real numbers different from 2.

3-4 Equivalent Statements

1.

p	q	$[\sim (p \wedge q)]$	$\leftrightarrow$	$[(\sim p)$	$\vee$	$(\sim q)]$
T	T	F T	T	F	F	F
T	F	T F	T	F	T	T
F	T	T F	T	T	T	F
F	F	T F	T	T	T	T

    (b) (a)   (f)   (c) (e) (d)

3.

p	q	$(p \rightarrow q)$	$\leftrightarrow$	$[q$	$\vee$	$(\sim p)]$
T	T	T	T	T T	F	
T	F	F	T	F F	F	
F	T	T	T	T T	T	
F	F	T	T	F T	T	

    (a)   (e) (b) (d) (c)

5. *Converse:* If we buy a new car, then we can afford it.
    Inverse: If we cannot afford it, then we do not buy a new car.
    Contrapositive: If we do not buy a new car, then we cannot afford it.

7. *Converse:* If the triangles are congruent, then two sides and the included angle of one are congruent to two sides and the included angle of the other.

Inverse: If two sides and the included angle of one triangle are not congruent to two sides and the included angle of another triangle, then the triangles are not congruent.

Contrapositive: If two triangles are not congruent, then two sides and the included angle of one are not congruent to two sides and the included angle of the other.

9. *Converse:* If $x = 1$, then $x(x - 1) = 0$.
Inverse: If $x(x - 1) \neq 0$, then $x \neq 1$.
Contrapositive: If $x \neq 1$, then
$$x(x - 1) \neq 0.$$

11. The statement is always true in Exercises 7 and 8.

13. The inverse is always true in Exercises 7 and 9.

3-5 *Forms of Statements*

1. If it is a duck, then it is a bird.

3. If two angles are complements of the same angle, then they are congruent.

5. If two lines are parallel, then they are coplanar.

7. If a geometric figure is a circle, then it is round.

9. If a person is a teacher, then he is boring.

11. If you like this book, then you like mathematics.

13. If you like mathematics, then you like this book.

15. If you like mathematics, then you will like this book.

17. $q \to p$

19. $q \to p$

21. $p \to (\sim q)$

23. $p \leftrightarrow q$

25. $p \to q$

27. $p \leftrightarrow q$

29. If $9 + 3 < 10$, then $11 - 3 > 8$; true.

31. If $5 + 3 = 8$, then $7 \times 4 = 25$; false.

33. If $7 \times 6 = 42$, then $8 \times 5 \neq 40$; false.

35. The assertion is equivalent to the statement: "If you give me $10,000, then I will marry your daughter." If he received the money, then he should have married the girl. Thus he should be sued for breach of promise.

3-6 *The Nature of Proof*

1. A tautology.

p	p	$\vee$	$(\sim p)$
T	T	T	F
F	F	T	T
	(a)	(c)	(b)

3. A tautology.

p	q	$[(p \to q)$	$\wedge$	$(\sim q)]$	$\to$	$(\sim p)$
T	T	T	F	F	T	F
T	F	F	F	T	T	F
F	T	T	F	F	T	T
F	F	T	T	T	T	T
		(a)	(c)	(b)	(e)	(d)

5. Not a tautology.

p	q	$[(p \rightarrow q)$	$\wedge$	$(\sim p)]$	$\rightarrow$	$(\sim q)$
T	T	T	F	F	T	F
T	F	F	F	F	T	T
F	T	T	T	T	F	F
F	F	T	T	T	T	T
		(a)	(c)	(b)	(e)	(d)

7. A tautology.

p	q	$[(p \vee q)$	$\wedge$	$(\sim p)]$	$\rightarrow$	q
T	T	T	F	F	T	T
T	F	T	F	F	T	F
F	T	T	T	T	T	T
F	F	F	F	T	T	F
		(a)	(c)	(b)	(e)	(d)

9. A tautology.

p	q	r	$[(p \rightarrow q)$	$\wedge$	$(q \rightarrow r)]$	$\rightarrow$	$(p \rightarrow r)$
T	T	T	T	T	T	T	T
T	T	F	T	F	F	T	F
T	F	T	F	F	T	T	T
T	F	F	F	F	T	T	F
F	T	T	T	T	T	T	T
F	T	F	T	F	F	T	T
F	F	T	T	T	T	T	T
F	F	F	T	T	T	T	T
			(a)	(c)	(b)	(e)	(d)

3-7 *Valid Arguments*

1. For

p: Elliot is a freshman.
q: Elliot takes mathematics.

the argument has the form

$[(p \rightarrow q) \wedge p] \rightarrow q$.

This statement is a tautology (§3-6, Example), and the argument is valid.

3. For

p: The Yanks win the game.
q: The Yanks win the pennant.

the argument has the form

$[(p \rightarrow q) \wedge (\sim q)] \rightarrow (\sim p)$.

This statement is a tautology (§3-6, Exercise 3), and the argument is valid.

5. For

p: You work hard.
q: You are a success.

the argument has the form

$[(p \rightarrow q) \wedge (\sim q)] \rightarrow (\sim p)$.

This statement is a tautology (§3-6, Exercise 3), and the argument is valid.

7. For

p: You are reading this book.
q: You like mathematics.

the argument has the form

$[(p \rightarrow q) \wedge (\sim p)] \rightarrow (\sim q)$.

This statement is not a tautology (§3-6, Exercise 5) and the argument is not valid.

9. This argument is of the form $[(p \rightarrow q) \wedge (q \rightarrow r)] \rightarrow (r \rightarrow p)$ and is not valid.

11. You do not drink milk.

13. If you like to fish, then you are a mathematician.

15. If you like this book, then you will become a mathematician.

3-8   *Euler Diagrams*

1. All juniors are intelligent.
All undergraduates are intelligent.
Some juniors are undergraduates.
Some intelligent people are juniors.
Some intelligent people are undergraduates.
Some undergraduates are juniors.

3. Valid.

5. Not valid.

7. (a) Valid;
(b) valid;
(c) valid;

(d) not valid.

9. (a) Valid;
(b) not valid;
(c) not valid;
(d) not valid;
(e) not valid;
(f) not valid;
(g) not valid.

11. (a) Valid;
(b) valid;
(c) valid;
(d) not valid.

3-9   *Quantifiers*

1. 9

3. 6

5. (1) 7, or any value different from 9.
(2) 7, or any value different from 6.
(3) 5, or any value different from 6.
(4) 5, or any value different from 3 or -3.

7. In each case use $\forall_x$.

9. Not all numbers are whole numbers; that is, there exists at least one number that is not a whole number.

11. Not all fractions represent rational numbers; that is, there exists at least one fraction that does not represent a rational number.

13. All real numbers are rational numbers.

15. No rational numbers are integers; that is, all rational numbers are not integers.

4

Systems of
NUMERATION

4-1   *Egyptian Numeration*

1. ∩ ∩ | | | | |

3. 9 9 9 9 ∩ ∩ | | | | | |

5. ⟋ ⌓ ⌓ 9 9 9 9 | | | | | | |

7. 22

9. 1102

11. 1324

13.

$$\cap\cap\cap\cap\,|\,|$$
$$+\qquad\cap\cap\,|$$
$$\overline{\cap\cap\cap\cap\cap\cap\,|\,|\,|}$$

15.

$$9\,9\cap\cap\cap\,|\,|\,|\,|\,|\,|\,|\,|$$
$$+\quad 9\cap\cap\cap\,|\,|\,|\,|\,|$$
$$\overline{9\,9\,9\cap\cap\cap\cap\cap\cap\cap\,|\,|\,|}$$

17.

$$\overset{\curvearrowright}{\underset{\curlywedge}{9}}\,9\,9\cap\cap\cap\cap\,|\,|\,|$$
$$-\qquad 9\cap\cap\cap\quad|\,|\,|\,|\,|\,|\,|$$
$$\rule{5cm}{0.4pt}$$

$$|\,|\,|\,|\,|\,|\,|\,|\,|\,|$$
$$\overset{\curvearrowright}{\underset{\curlywedge}{9}}\,9\cap\cap\cap\,|\,|\,|$$
$$-\qquad 9\cap\cap\cap\,|\,|\,|\,|\,|\,|\,|$$
$$\rule{5cm}{0.4pt}$$
$$\overset{\curvearrowright}{\underset{\curlywedge}{9}}\qquad|\,|\,|\,|\,|\,|$$

19. ① $\times\,45 =$ ㊺
2 $\times\,45 =$ 90
4 $\times\,45 =$ 180
8 $\times\,45 =$ 360
⑯ $\times\,45 =$ ⟨720⟩
$17 = 1 + 16$
$17 \times 45 = (1 + 16) \times 45$
$\qquad\qquad = 45 + 720 = 765$

21. ① $\times\,31 =$ ㉛
② $\times\,31 =$ ㉒
4 $\times\,31 =$ 124
⑧ $\times\,31 =$ ⟨248⟩
⑯ $\times\,31 =$ ⟨496⟩
$27 = 1 + 2 + 8 + 16$
$27 \times 31 = (1 + 2 + 8 + 16) \times 31$
$\qquad\qquad = 31 + 62 + 248 + 496$
$\qquad\qquad = 837$

23. $19 \rightarrow$ ㉝
$9 \rightarrow$ 66
$4\qquad 132$
$2\qquad 264$
$1 \rightarrow$ ⟨528⟩
$19 \times 33 = 33 + 66 + 528 = 627$

25. $21 \rightarrow$ 52
$10\qquad 104$
$5 \rightarrow$ ⟨208⟩
$2\qquad 416$
$1 \rightarrow$ ⟨832⟩
$21 \times 52 = 52 + 208 + 832 = 1092$

27. We need a symbol for zero because the value of a numeral in our decimal system depends upon both the value and the position of the digits in the numeral. For example, 405 and 450 represent two entirely different numbers, yet the same digits are used in the two numerals. The ancient Egyptians were not dependent upon the position of a symbol in determining its value, so zero as a "place-holder" was not needed.

4-2 *Other Methods of Computation*

1.
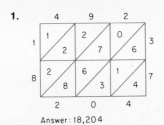
Answer: 18,204

3.
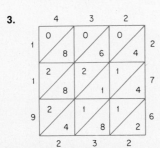
Answer: 119,232

5.
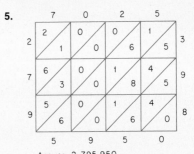
Answer: 2,795,950

7.
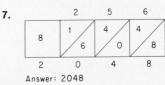
Answer: 2048

9.

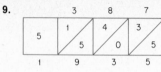

Answer: 1935

11.

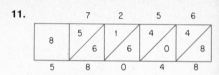

Answer: 58,048

4-3 *Decimal Notation*

1. $(2 \times 10^2) + (3 \times 10) + (5 \times 10^0)$

3. $(1 \times 10^3) + (4 \times 10^2) + (9 \times 10) + (2 \times 10^0)$

5. $(1 \times 10^4) + (7 \times 10^3) + (2 \times 10^2) + (5 \times 10) + (9 \times 10^0)$

7. $(1 \times 10^3) + (0 \times 10^2) + (7 \times 10) + (5 \times 10^0)$

9. $(1 \times 10^7) + (0 \times 10^6) + (6 \times 10^5) + (5 \times 10^4) + (0 \times 10^3) + (0 \times 10^2) + (0 \times 10) + (2 \times 10^0)$

11. One thousand seven hundred twenty.

13. One hundred seven thousand twenty.

15. Seventeen million two.

17. 25,017

19. 5,600,000

21. 3253

23. 403,205

25. 9,000,007

4-4 *Other Systems of Numeration*

1. 22_{five}

3. 30_{four}

5.

7.

9.

⬭ ✳ ✳ ✳ ✳ ⬭
⬭ ✳ ✳ ✳ ✳ ⬭
✳ ✳ ✳

11. 23

13. 26

15. 19

17. 32_{five}

19. 32_{six}

4-5 *Base Five Notation*

1. 113

3. 124

5. 86

7. 141

9. 209

11. 3012_5

13. 11112_5

15. 10000_5

17. 1022_5

19. 4012_5

*21. $(4 \times 8^2) + (3 \times 8) + (7 \times 8^0)$; 287.

*23. $(1 \times 2^3) + (0 \times 2^2) + (1 \times 2) +$ (1×2^0); 11.

*25. $(1 \times 20^2) + (3 \times 20) + (2 \times 20^0)$; 462.

*27. $(2 \times 15^2) + (1 \times 15) + (4 \times 15^0)$; 469.

4-6 *Computation in Base Five Notation*

1. 111_5

3. 432_5

5. 11312_5

7. 112_5

9. 134_5

11. 413_5

13. 222_5

15. 3023_5

17. 3401_5

19. 1423_5

21. 21022_5

23. 22_5

25. 224_5 R 2

27. 13_5

29. 41_5

31.

X	0	1	2	3	4
0	0	0	0	0	0
1	0	1	2	3	4
2	0	2	4	11_5	13_5
3	0	3	11_5	14_5	22_5
4	0	4	13_5	22_5	31_5

4-7 *Other Number Bases*

1. 314

3. 1076

5. 50

7. 662

9. 2590

11. 3598_{12}

13. $87e0_{12}$

15. 11010_4

17. 1551_6

19. 11143_5

21. 655_9

23. 187_{12}

*25. $1e7_{12}$

*27. 101111_2

4-8 *Binary Notation*

1. 11100_2

3. 10011_2

5. 10011000_2

7. 15

9. 55

11. 38

13. 11000_2

15. 101_2

17. 100111_2

19. (a) 1000010_2;
   (b) 1000100_2;
   (c) 1000111_2

21. $214 = 326_8 = 11010110_2$. When placed in groups of three, starting from the units digit, the binary representation can be translated into the octal system, and conversely.

Thus

$\underbrace{011}_{3}$ $\underbrace{010}_{2}$ $\underbrace{110_2}_{6}$ $= 326_8$.

23. 3531_8

25. $22{,}533_8$

27. $1{,}531{,}046_8$

29. $110{,}000{,}010{,}011_2$

31. $11{,}110{,}010{,}100_2$

33. $111{,}110{,}101{,}100{,}011_2$

5

Mathematical
SYSTEMS

5-1 *An Abstract System*

1. #

3. Σ

5. Commutative property for multition.

7. Identity element (*).

9. Associative property for multition.

11. Inverse element.

13. $\triangle$

15. Yes.

17. Yes.

19. No; it must hold for all possible choices of three elements.

21. Yes, the inverse of $\triangle$ is Q, the inverse of $\square$ is $\square$, the inverse of Q is $\triangle$.

23. One counterexample is sufficient; for example, $8 \div 3$ is not a counting number.

25. 1; $8 \times 1 = 8$, $13 \times 1 = 13$.

5-2 *The Distributive Property*

1. 8, 8, 3, 7.

3. 7, 7.

5. 3, 3.

7. No.

9. $7 \times 79 = 7 \times (80 - 1)$
$= 560 - 7 = 553.$

11. 3

13. 5

15. 7

17. 7

19. True for all replacements of n.

5-3 *Clock Arithmetic*

1. (a) Yes;
(b) yes.

3. The identity with respect to addition is 12; the identity with respect to multiplication is 1.

5. 3

7. 9

9. 3

11. 5

13. 8

15. 6

17. 1, 5, or 9.

19. 4

*21. An impossible equation; that is, there is no value of t for which this equation is true.

*23. An identity; that is, this equation is true for all possible replacements of t.

25. (a) 6; **(b)** 16_{12}.

27. (a) 9; **(b)** 39_{12}.

29. (a) 12; **(b)** 30_{12}.

5-4 *Modular Arithmetic*

1.

X	0	1	2	3	4
0	0	0	0	0	0
1	0	1	2	3	4
2	0	2	4	1	3
3	0	3	1	4	2
4	0	4	3	2	1

3. Two specific cases are $(3 \cdot 2) \cdot 4 = 3 \cdot (2 \cdot 4) = 4$ and $(4 \cdot 3) \cdot 4 = 4 \cdot (3 \cdot 4) = 3$. There many others.

5. The inverse of 1 is 1, of 2 is 3, of 3 is 2, and of 4 is 4. Note that 0 does not have an inverse with respect to multiplication.

7. 4

9. 4

11. 3

13. 3

15. 4

17. 4

19. 3

21. 4

23. An impossible equation; that is, there is no value of x for which this equation is true.

5-5 *Two by Two Matrices*

1. (a) $\begin{bmatrix} 2 & 2 \\ 4 & 1 \end{bmatrix}$;

(b) $\begin{bmatrix} 1 & 2 \\ 4 & 2 \end{bmatrix}$;

(c) $\begin{bmatrix} 3 & 2 \\ 3 & 0 \end{bmatrix}$.

3. (a) $\begin{bmatrix} z & y & 1+x \\ w & 1+v & u \\ 1+t & s & r \end{bmatrix}$;

(b) $\begin{bmatrix} t & s & r \\ w & v & u \\ z & y & x \end{bmatrix}$;

(c) $\begin{bmatrix} x & y & z \\ u & v & w \\ r & s & t \end{bmatrix}$.

5. For any two by two matrix

$$\begin{bmatrix} a & b \\ c & d \end{bmatrix}$$

$$\begin{bmatrix} a & b \\ c & d \end{bmatrix} + \begin{bmatrix} 0 & 0 \\ 0 & 0 \end{bmatrix} = \begin{bmatrix} a & b \\ c & d \end{bmatrix} = \begin{bmatrix} 0 & 0 \\ 0 & 0 \end{bmatrix} + \begin{bmatrix} a & b \\ c & d \end{bmatrix}.$$

7. For any two matrices of S the sums

$$\begin{bmatrix} a & b \\ c & d \end{bmatrix} + \begin{bmatrix} r & s \\ t & u \end{bmatrix} = \begin{bmatrix} a+r & b+s \\ c+t & d+u \end{bmatrix}$$

$$\begin{bmatrix} r & s \\ t & u \end{bmatrix} + \begin{bmatrix} a & b \\ c & d \end{bmatrix} = \begin{bmatrix} r+a & s+b \\ t+c & u+c \end{bmatrix}$$

are equal since addition of integers is commutative.

9. The product of any two matrices of S

$$\begin{bmatrix} a & b \\ c & d \end{bmatrix}\begin{bmatrix} r & s \\ t & u \end{bmatrix} = \begin{bmatrix} ar+bt & as+bu \\ cr+dt & cs+du \end{bmatrix}$$

is an element of S since the set of integers is closed under multiplication and addition.

11. $\begin{bmatrix} 1 & 1 \\ 0 & 1 \end{bmatrix}\begin{bmatrix} 0 & 1 \\ 1 & 0 \end{bmatrix} = \begin{bmatrix} 1 & 1 \\ 1 & 0 \end{bmatrix} \neq$

$\begin{bmatrix} 0 & 1 \\ 1 & 1 \end{bmatrix} = \begin{bmatrix} 0 & 1 \\ 1 & 0 \end{bmatrix}\begin{bmatrix} 1 & 1 \\ 0 & 1 \end{bmatrix}$

13. $\begin{bmatrix} a & b \\ c & d \end{bmatrix}\begin{bmatrix} 0 & 1 \\ 1 & 0 \end{bmatrix} = \begin{bmatrix} b & a \\ d & c \end{bmatrix}$

15. The set is closed (Exercise 9), associative (assumed), contains an identity element (Exercise 10), and each element of the set T has an inverse (assumption and Exercise 14).

6

Sets of

NUMBERS

6-1 The Set of Counting Numbers

1. $\{(1, 1), (1, 2), (1, 3), (1, 4), (2, 1), (2, 2), (2, 3), (2, 4)\}$.

3. Use any counter example such as $3 - 2 \neq 2 - 3$.

5. $8 - (3 - 2) = 8 - 1 = 7$, $(8 - 3) - 2 = 3$; no, no.

7. Use any counter example such as $2 + (3 \times 5) \neq (2 + 3) \times (2 + 5)$; $17 \neq 5 \times 7$.

9. Commutative, $+$.

11. Associative, $\times$.

13. Distributive, $\times$ with respect to $+$.

15. Commutative, $+$.

17. Distributive, $\times$ with respect to $+$.

19. (a) Associative, $+$;
(b) commutative, $+$;
(c) associative, $+$;
(d) associative, $+$;
(e) distributive, $\times$ with respect to $+$.

21. $92 + (50 + 8)$
$= 92 + (8 + 50)$ Commutative, $+$.
$= (92 + 8) + 50$ Associative, $+$.

23. $37 \times (1 + 100)$
$= 37 \times (100 \times 1)$ Commutative, $+$.
$= 37 \times 100 + 37 \times 1$ Distributive,
$\times$ w.r.t. $+$.
$= 37 \times 100 + 37$ Identity, $\times$.
$= 3700 + 37$ Number system.

25. $(73 + 19) + (7 + 1)$
$= [(73 + 19) + 7] + 1$ Associative, $+$.
$= [73 + (19 + 7)] + 1$ Associative, $+$.
$= [73 + (7 + 19)] + 1$ Commutative, $+$.
$= [(73 + 7) + 19] + 1$ Associative, $+$.
$= (73 + 7) + (19 + 1)$ Associative, $+$.

27. No.

6-2 *Prime Numbers*

1. Any number that is divisible by 15 is also divisible by 3.

3. The set of numbers divisible by 3 and 5 is the set of numbers divisible by 15.

5. Any number that is divisible by 12 is also divisible by 3.

7. $\{21, 22, 24, 25, 26, 27, 28, 30, 32, 33, 34, 35, 36, 38, 39\}$

9. No. No.

11. There are other possible answers in many cases. $4 = 2 + 2$, and

$6 = 3 + 3$;	$8 = 3 + 5$;
$10 = 3 + 7$;	$12 = 5 + 7$;
$14 = 7 + 7$;	$16 = 3 + 13$;
$18 = 5 + 13$;	$20 = 7 + 13$;
$22 = 5 + 17$;	$24 = 7 + 17$;
$26 = 3 + 23$;	$28 = 5 + 23$;
$30 = 7 + 23$;	$32 = 3 + 29$;
$34 = 5 + 29$;	$36 = 7 + 39$;
$38 = 7 + 31$;	$40 = 3 + 37$.

13. Any three consecutive odd numbers includes 3 or a multiple of 3. Each multiple of 3 is composite, and 1 is by definition not a prime. Therefore 3, 5, 7 is a set of three consecutive odd numbers that are all prime numbers and, since any other such set contains a composite number that is a multiple of 3, this is the only prime triplet.

*15. (a) 13;
      (b) 19;
      (c) 31.

17. 1×15; 3×5.

19. 1×20; 2×10; 4×5.

21. 1×29

23. $2^2 \times 17$

25. 3×41

27. $3 \times 5^2 \times 19$

29. 2×409

6-3 *Applications of Prime Factorizations*

1. $\{1, 2, 4, 5, 10, 20\}$

3. $\{1, 3, 9\}$

5. $\{1, 2, 4, 7, 14, 28\}$

7. $\{1, 2, 3, 4, 5, 6, 10, 12, 15, 20, 30, 60\}$

9. $68 = 2^2 \times 17$; $76 = 2^2 \times 19$; G.C.F. $= 4$.

11. $76 = 2^2 \times 19$; $1425 = 3 \times 5^2 \times 19$; G.C.F. $= 19$.

13. $215 = 5 \times 43$; $1425 = 3 \times 5^2 \times 19$; G.C.F. $= 5$.

15. $12 = 2^2 \times 3$; $15 = 3 \times 5$; $20 = 2^2 \times 5$; G.C.F. $= 1$.

17. $12 = 2^2 \times 3$; $18 = 2 \times 3^2$; $30 = 2 \times 3 \times 5$; G.C.F. $= 2 \times 3 = 6$.

19. $\{7, 14, 21, 28, 35\}$

21. $\{15, 30, 45, 60, 75\}$

23. See Exercise 9; L.C.M. $= 2^2 \times 17 \times 19 = 1292$.

25. See Exercise 11; L.C.M. $= 2^2 \times 3 \times 5^2 \times 19 = 5700$.

27. See Exercise 13; L.C.M. $= 3 \times 5^2 \times 19 \times 43 = 61,275$.

29. See Exercise 15; L.C.M. $= 2^2 \times 3 \times 5 = 60$.

31. See Exercise 17; L.C.M. $= 2^2 \times 3^2 \times 5 = 180$.

33. $\frac{123}{215}$

35. $\frac{19}{24}$

37. $\frac{13}{30}$

39. $\frac{5}{12}$

41. $\dfrac{1,259}{26,445}$

*43. 1

6-4   *Equivalence and Order Relations*

1. (a), (b), (c), (e), (f), (i).

3. (a), (d), (e), (g), (h), (i).

5. **(a)** Yes; **(b)** yes; **(c)** yes.

7. **(a)** Yes; **(b)** no; **(c)** yes.

9. **(a)** No; **(b)** no; **(c)** yes.

11. **(a)** Yes; **(b)** no; **(c)** yes.

13. **(a)** $>$; **(b)** $<$; **(c)** $=$; **(d)** $>$; **(e)** $<$.

15. **(a)** As in logic p or $\sim p$; exactly one must hold. Here $a = b$ or $a \neq b$, exactly one must hold.
  (b) By the trichotomy law exactly one of the relations $a = b$, $a < b$, $a > b$ must hold. If the first of these relations does not hold ($a \neq b$), then exactly one of the other two must hold.

6-5   *The Set of Whole Numbers*

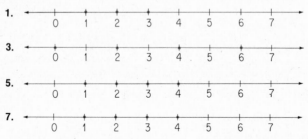

9. The graph is the empty set; there are no points in the graph.

11. $W = \{0, 1, 2, 3, 4\}$
  $\updownarrow \updownarrow \updownarrow \updownarrow \updownarrow$
  $C = \{1, 2, 3, 4, 5\}$

13. If $a \times b = 1$, then both $a = 1$ and $b = 1$. If $a \times b = 2$, then either $a = 1$ and $b = 2$ or $a = 2$ and $b = 1$. If $a \times b = 3$, then

either $a = 1$ and $b = 3$ or $a = 3$ and $b = 1$. If $a \times b = 4$, then $a = 1$ and $b = 4$, or $a = 2$ and $b = 2$, or $a = 4$ and $b = 1$.

15. None.

17. $W = \{0, 1, 2, 3, 4, \ldots, \quad n, \quad \ldots\}$
  $\updownarrow \updownarrow \updownarrow \updownarrow \updownarrow \qquad\quad \updownarrow$
  $C = \{1, 2, 3, 4, 5, \ldots, n + 1, \ldots\}$

6-6   *The Set of Integers*

1. $\{\ 1, \quad 2, \quad 3, \quad 4, \quad 5, \ldots, \quad n, \ldots\}$
  $\quad\ \ \updownarrow \quad\ \updownarrow \quad\ \updownarrow \quad\ \updownarrow \quad\ \updownarrow \qquad\ \updownarrow$
  $\{-1, -2, -3, -4, -5, \ldots, -n, \ldots\}$

3.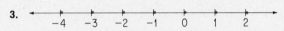

5.

7. True.

9. False.

11. True.

13. False; -0 is not a negative integer.

15. False; for example, $4 \div 3$ is not an integer.

17. No, zero is an integer but is neither positive nor negative and thus is not in the union of the set of positive integers and the set of negative integers.

19. $\{1,2, \quad 3,4, \quad 5,6, \quad 7, \ldots, 2n, 2n+1, \ldots\}$

$\{0,1,-1,2,-2,3,-3, \ldots, n, -n, \ldots\}$

21. Let $2k$ and $2m$ represent two even integers. Then

$$2k \times 2m = 2 \times 2 \times k \times m = 2(2km),$$

where $2km$ is an integer. Thus $2(2km)$ is an even integer.

23. Any integer is either even or odd. If an integer is even, then its square is even (as in Example 5). If an integer is odd, then its square is odd (as in Exercise 22). If the square of an integer is odd, the integer cannot be even and thus must be odd. If the square of an integer is even, the integer cannot be odd and thus must be even.

6-7 *The Set of Rational Numbers*

1. True.

3. True.

5. False.

7. $\frac{1}{3} + \frac{2}{3} = 1$

9. $\frac{4}{3} - \frac{1}{3} = 1$

19. No, the density property of the set of rational numbers states that between any two elements of the set there is always another element of the set.

11.   13.

Set	-5	$\frac{2}{3}$
Counting numbers	x	x
Whole numbers	x	x
Integers	✓	x
Rational numbers	✓	✓

15. Every rational number except 0 has a multiplicative inverse; the set of rational numbers is dense.

17. $\frac{1}{200}, \frac{1}{400}, \frac{1}{800}$. There are many other correct answers.

	Sentence	Counting numbers	Whole numbers	Integers	Rational numbers
21.	$n + 3 = 3$	none	0	0	0
23.	$3n = 5$	none	none	none	$\frac{5}{3}$

	Counting numbers	Whole numbers	Integers	Positive rationals	Rational numbers
25.	✓	✓	✓	✓	✓
27.	x	✓	✓	x	✓
29.	✓	✓	✓	✓	✓
31.	✓	✓	✓	✓	✓
33.	✓	✓	✓	✓	✓
35.	✓	✓	✓	✓	✓
37.	✓	✓	✓	✓	✓

6-8 *The Set of Real Numbers*

1. True.

3. True.

5. True.

7. 9.

Set	$-\frac{2}{3}$	0
Counting numbers	x	x
Whole numbers	x	✓
Integers	x	✓
Rational numbers	✓	✓
Real numbers	✓	✓

11. Terminating.

13. Repeating.

15. Terminating.

17. 0.8

19. 0.875

21. 1.8

23. $0.\overline{428571}$

25. $0.\overline{692307}$

27. $\frac{5}{11}$

29. $\frac{9}{11}$

31. $\frac{58}{111}$

33. $\frac{19}{9}$

35. $\frac{212}{99}$

37. $\frac{41}{333}$

39. (a) 0.7272727272;
(b) 0.3571571571;
(c) 0.8000000000;
(d) 0.4237373737.

41. $2.08, $2.37, $2.57, $2.64, $2.98.

43. $1.\overline{4}$, 1.45, $1.\overline{45}$, $1.4\overline{5}$, $1.\overline{5}$.

Among the many correct answers for Exercises 45 and 47 are:

*45. 0.2342   *49. (a) 5; (b) 120.

*47. 0.231

6-9   The Set of Complex Numbers

1. $2 + 3i$ 5. $5 + \sqrt{3}i$

3. $3 + 0i$ 7. True.

                                               9. True.

	11. -7	13. 0	15. $\sqrt{-9}$
Set	-7	0	$\sqrt{-9}$
Counting numbers	x	x	x
Whole numbers	x	✓	x
Integers	✓	✓	x
Rational numbers	✓	✓	x
Irrational numbers	x	x	x
Real numbers	✓	✓	x
Imaginary numbers	x	x	✓
Complex numbers	✓	✓	✓

	Sentence	Counting numbers	Integers	Rational numbers	Real numbers	Complex numbers
17.	$n + 5 = 2$	none	-3	-3	-3	-3
19.	$n^2 = 4$	2	$2,-2$	$2,-2$	$2,-2$	$2,-2$

	Counting numbers	Whole numbers	Integers	Rational numbers	Real numbers
21.	✓	✓	✓	✓	✓
23.	✓	✓	✓	✓	✓
25.	✓	✓	✓	✓	✓
27.	x	✓	✓	✓	✓
29.	x	x	✓	✓	✓
31.	✓	✓	✓	✓	✓
33.	x	x	x	✓	✓

7

An Introduction to ALGEBRA

7-1 *Sentences and Statements*

1. Identity.

3. Empty set.

5. Identity.

7. Empty set.

9. $\{3\}$

11. $\{0, 1, 2\}$

13. $\{0, 1, 2, 3, 4\}$

15. $\{\ldots, -2, -1, 0, 1, 2\}$

17. $\{\ldots, -2, -1, 0, 1, 2, 3, 4\}$

19. All real numbers less than 3.

21. All real numbers less than 5.

23. All real numbers except 3.

25. c

27. d

29. a

31. d

7-2 *Graphs of Sentences*

1. Point.

3. Ray.

5. Line segment.

7. Line.

9. Half-line.

11.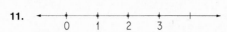

13.

15.

17.

19.

21.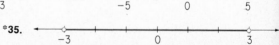

23.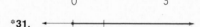

25.

27.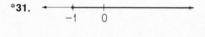

29.

*31.

*33.

*35.

7-3 *Sentences of The First Degree*

1. **(a)** Addition, $=$;
   (b) associative, $+$;
   (c) addition;
   (d) zero, $+$.

3. **(a)** Addition, $=$;
   (b) associative, $+$;
   (c) addition;
   (d) zero, $+$;
   (e) multiplication, $=$;
   (f) associative, $\times$;
   (g) multiplication;
   (h) one, $\times$.

5. $x = 3$

7. $x = -2$

9. $x = 14$

11. $x < 2$

13. $x > 2$

15. $x > -2$

17. $x > -6$

19. $x < -4$

21. Always true.

23. Not always true; $3 < 8$, $1 < 2$, $3 - 1 \not< 8 - 2$.

7-4 *Compound Sentences*

1. $\{1, 2, 3, 4\}$

3. $\{2\}$

5.

7.

9.

11.

13.

15.

17.

19.

21.

23.

*25.

*27.

7-5   *Linear Sentences In Two Variables*

1. $\{(1, 1)\}$

3.
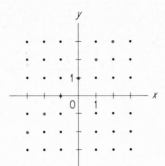

5.
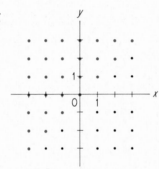

7.
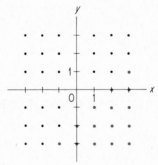

9.
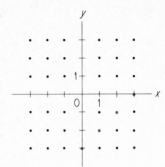

11.
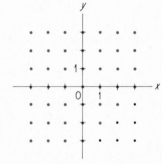

13. $\{(1, 2), (2, 1)\}$

15. $\{(1, 1), (1, 2), (1, 3), (2, 1), (2, 2), (3, 1)\}$

17. $\varnothing$

19. $\{(1, 4), (2, 3), (3, 2), (4, 1)\}$

21. $\{(1, 1), (1, 2), (2, 1)\}$

*23. $\{(1, 1, 1),\ (1, 1, 2),\ (1, 2, 1),\ (2, 1, 1),$
$(1, 2, 2), (2, 1, 2), (2, 2, 1), (2, 2, 2)\}$

7-6   *Relations and Functions*

1. **(a)** A function;
(b) not a function;
(c) a function;
(d) not a function;
(e) a function;
(f) not a function.

3.   See the diagram for Exercise 1 **(e)**.

5. **(a)** 5;   **(b)** 1;   **(c)** 26.

7. **(a)** 0;   **(b)** 0;   **(c)** 6.

9.

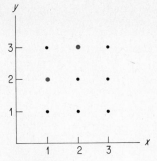

A function
Domain: $\{1, 2\}$;
Range: $\{2, 3\}$.

11.

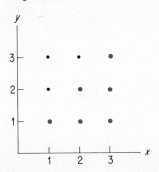

Not a function.

13.

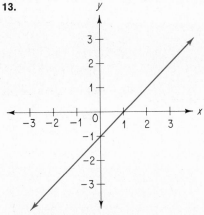

A function. The domain and the range are each the set of real numbers.

15.

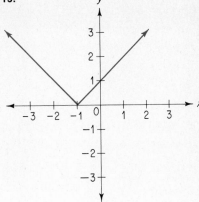

A function. The domain is the set of real numbers; the range is the set of non-negative real numbers.

17.

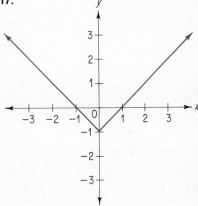

A function. The domain is the set of real numbers; the range is the set of real numbers greater than or equal to -1.

19.

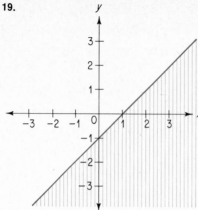

Not a function.

21.

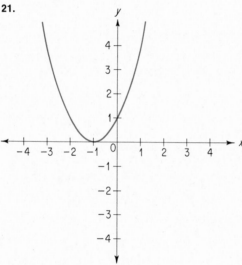

A function. The domain is the set of real numbers; the range is the set of non-negative real numbers.

7-7   Graphs on a Plane

1. (a) 5; (b) 5.

3. (a) 3; (b) -2.

5. (a) $-1\frac{1}{2}$; (b) 3.

7. (a) -2; (b) 6.

9. (a) $1\frac{4}{5}$; (b) $2\frac{1}{4}$.

11. $y = -2x + 7$

13. $y \geq -3x + 6$

15. $y \geq 2x - 4$

17.

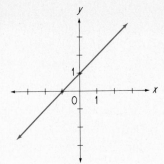

19.

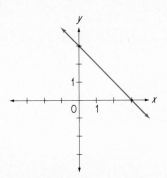

21.

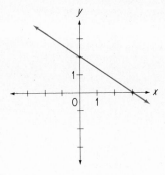

23.

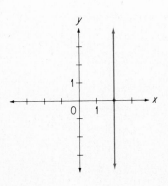

25.

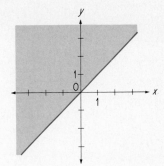

27.

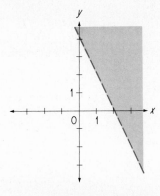

29.

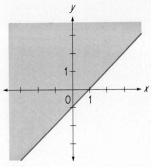

7-8   *Linear Systems*

1.

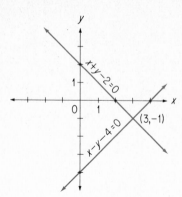

7.

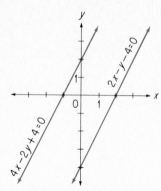

The lines are parallel; the solution set is the empty set.

3.

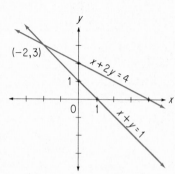

9.

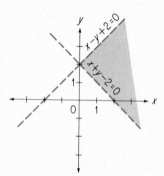

5.

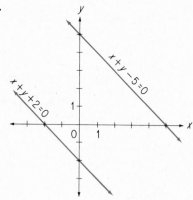

The graph consists of the union of the two lines.

11.

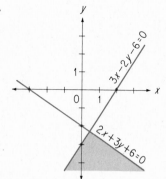

13.

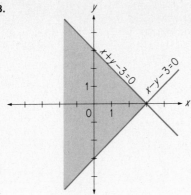

15.

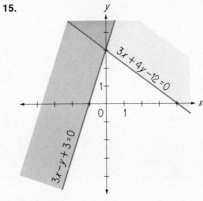

*17.

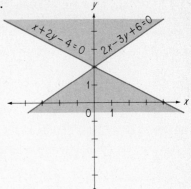

*19.

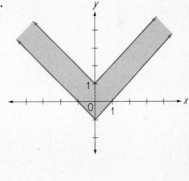

7-9 Linear Programming

1.

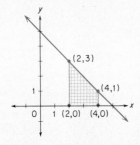

3. **(a)** Maximum value of 8 at $(2, 3)$;
   (b) minimum value of 2 at $(2, 0)$.

5. **(a)** Maximum value of 5 at $(3, 2)$;
   (b) minimum value of 0 at $(0, 0)$.

7. The conditions are $0 \le x$, $0 \le y$, $x + y \le 180$, and $5x + 3y \le 750$. The maximum of $3x + 2y$ occurs at $(105, 75)$; thus

under the assumptions of this exercise, 105 minutes of regular teaching and 75 minutes of television instruction would be best for the students.

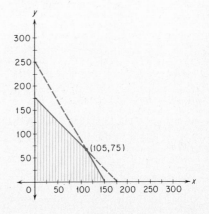

9. The maximum of $x + 2y$ occurs at (30, 150); thus under the assumptions of this exercise, 30 minutes of regular teaching and 150 minutes of television instruction would be best for the students.

8

An Introduction to
GEOMETRY

8-1   *Points, Lines, and Planes*

1.

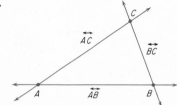

3. *ABC, ABD, ACD, BCD.*

5. **(a)** 1;
(b) 3;
(c) 6;
(d) 10;

(e) 15;
***(f)** 45;
***(g)** $n(n - 1)/2$.

7. True.

9. False.

11. False.

13. False.

15. False.

17. True.

8-2   *Plane Figures*

There are many correct answers for Exercises 1 through 7.

1.

5.

3.

7.

9. Empty set.

11. *M*

13. *P*

15. $\overset{\circ\text{---}\circ}{MN}$

17. $\triangle RMN$

19. $\overrightarrow{MP} \cup \overrightarrow{NV}$

21. *MN*

23. Interior $\triangle MRN$.

25. Interior of quadrilateral *STNM*.

27. Interior $\triangle RST$.

29. Empty set.

31. $\{B, D\}$

33. $\overline{BF}$

35. Interior $\triangle CBD$.

37. $\overrightarrow{FG}$

39. $\overset{\circ\text{---}\circ}{BD}$

8-3 *Space Figures*

1. (a) *M, N, O, P*;

   (b) $\overline{MN}, \overline{MO}, \overline{MP}, \overline{ON}, \overline{NP}, \overline{PO}$;

   (c) the triangular regions *MNO, PMN, PNO, PMO*.

3. (a) *A, B, C, D, E, F, G, H*;

   (b) $\overline{AB}, \overline{BC}, \overline{CD}, \overline{AD}, \overline{AE}, \overline{BF}, \overline{CG}, \overline{DH}$,
   $\overline{EF}, \overline{FG}, \overline{GH}, \overline{HE}$;

   (c) the square regions *ABCD, ABFE, BCGF, CDHG, DAEH, EFGH*.

5. (a) *J, K, L, M, N, O*;

   (b) $\overline{JK}, \overline{KL}, \overline{JL}, \overline{JM}, \overline{KN}, \overline{LO}, \overline{MN}, \overline{NO}$,
   $\overline{OM}$;

   (c) the triangular regions *JKL, MNO* and the rectangular regions *JKNM, KLON, LJMO*.

7. Pentagonal pyramid.

9. Hexagonal prism.

11. (Given figure.)

13. (Given figure.)

V	E	F
6	10	6
12	18	8
5	9	6
9	16	9

15. (b) and (c).

17. (a).

19.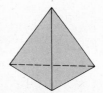

21.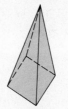

23.

25.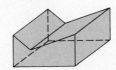

27.

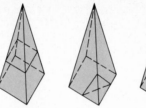

29.

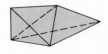

Number of painted faces	6	5	4	3	2	1	0
31. Number of cubes	0	0	0	8	12	6	1
*33. Number of cubes	0	0	0	8	36	54	27

8-4 *Plane Curves*

Exercises 1, 3, 5, and 7 may be done in many ways.

1.

7.

3.

5.

		V	*A*	*R*
9.	**(a)**	4	4	2
	(b)	4	5	3
11.	**(a)**	2	2	2
	(b)	2	3	3
13.	**(a)**	5	4	1
	(b)	16	24	10

15. $A = V + R - 2$

8-5 *Networks*

1. **(a)** 4;
    (b) 0;
    (c) traversable, *A, B, C, D.*

3. **(a)** 2;
    (b) 2;
    (c) traversable, *K,M.*

5. **(a)** 4;
    (b) 0;
    (c) traversable, *U, V, W, X.*

7. **(a)** 0;
    (b) 4;
    (c) not traversable.

9. The inspector can use a map for the highways involved as a network, determine the number of odd vertices, and know that each section can be traversed exactly once in a single trip if there are at most two odd vertices.

Think of the six regions as labeled and note the line segments that are needed as represented by arcs in the network. Since the network has four odd vertices (B, D, E, and F), the network is not traversable and the suggested broken line cannot be drawn.

11.

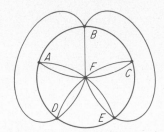

*13. Add one more bridge joining any two of the points A, B, C, D.

8-6 Linear Measures

1. 4

3. 3

5. 6

7. $m(\overline{P_2P_5}) + m(\overline{P_5P_7}) = m(\overline{P_2P_7})$

9. $m(\overline{P_1P_7}) = m(\overline{P_1P_4})$
$+ m(\overline{P_2P_7}) - m(\overline{P_1P_4} \cap \overline{P_2P_7})$

11. $m(\overline{P_{-3}P_4}) = m(\overline{P_1P_{-3}})$
$+ m(\overline{P_{-2}P_4}) - m(\overline{P_1P_{-3}} \cap \overline{P_{-2}P_4})$

13. $m(\overline{P_4P_{-1}}) = m(\overline{P_{-1}P_2}) + m(\overline{P_2P_4})$

15. $m(\overline{P_{-3}P_5}) = m(\overline{P_2P_{-3}})$
$+ m(\overline{P_0P_5}) - m(\overline{P_2P_{-3}} \cap \overline{P_0P_5})$

17. 14 feet.

19. 73 feet.

21. $3\frac{1}{2}$

23. (20) $\frac{1}{2}$ unit; (21) $\frac{1}{4}$ unit; (22) $\frac{1}{8}$ unit.

25. Approximate.

27. Exact.

8-7 Area and Volume Measures

1. 96 square inches.

3. 36 square inches.

5. 56 square inches.

7. 69 square inches.

9. $S = 6e^2$

11. The area measure of the square is multiplied by 4.

13. The area measure of the rectangle is multiplied by **(a)** 4; **(b)** 9; **(c)** $\frac{1}{4}$; **(d)** k^2.

15. The area measure of the polygon is multiplied by **(a)** 4; **(b)** k^2.

17. **(a)** 54 square yards;
   (b) 27 cubic yards.

19. **(a)** 150 square inches;
   (b) 125 cubic inches.

21. **(a)** 10.5 square feet;
   (b) 1.25 cubic feet.

23. 20 cubic inches.

25. The volume measure of the cube is multiplied by **(a)** 27; **(b)** $\frac{1}{8}$; **(c)** k^3.

8-8 Coordinates on a Plane

1. **(a)** 4, (2, 3); **(b)** 3, $(3\frac{1}{2}, 1)$;
   (c) 4, (2, 3); **(d)** 9, $(5, 2\frac{1}{2})$;
   (e) 5, $(-3, 4\frac{1}{2})$; **(f)** 4, $(-5, 2)$.

3. **(a)** $(4, -2)$; **(b)** $(2, 4)$; **(c)** $(-3, -6)$.

5. $(-5, 8)$

7. $C: (a, a)$

9. $S: (a - b, c)$

11.

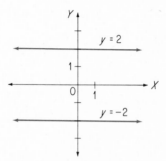

13.

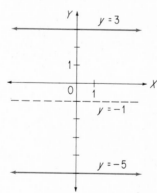

*15.

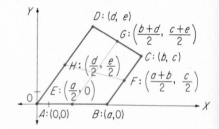

Consider $A: (0, 0)$, $B: (a, 0)$, $C: (b, c)$, and $D: (d, e)$. The midpoints of the sides are $E: \left(\frac{a}{2}, 0\right)$, $F: \left(\frac{a + b}{2}, \frac{c}{2}\right)$, $G: \left(\frac{b + d}{2}, \frac{c + e}{2}\right)$, and $H: \left(\frac{a}{2}, \frac{b}{2}\right)$.

The line segments $\overline{EG}$ and $\overline{HF}$ both have $\left(\frac{a + b + d}{2}, \frac{c + e}{2}\right)$ as midpoint and thus bisect each other.

8-9 Distances on a Plane

1. 5

3. 25

5. $\sqrt{18}$; that is, $3\sqrt{2}$.

7. The points of the circle with center $(0, 0)$ and radius 3.

9. The exterior points of the circle with center $(0, 0)$ and radius 8.

11. The points of the circular region with center $(1, 2)$ and radius 2.

13. The points of the circular region with center $(-1, -2)$ and radius 5.

15. $(x - 2)^2 + (y - 5)^2 = 9$

17. $(x + 3)^2 + (y + 1)^2 \leq 16$

19. $(x + 4)^2 + (y - 3)^2 > 16$

*21. The points of the solution set are the points of the circular region that are on or below the line.

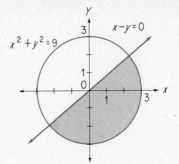

9
An Introduction to
PROBABILITY

9-1 Counting Problems

1. (a) 24; (b) 64.

3. 12

5. 24

7. 25; 20.

9. 1320

11. 27,216

13. (a) 12;
(b) 3.

*15. (a) 450;
(b) 180;
(c) 648;
(d) 200.

*17. $26 \times 25 \times 9 \times 9 \times 8$; that is, 421,200.

9-2 Definition of Probability

1. $\frac{1}{2}$

3. $\frac{1}{2}$

5. $\frac{5}{6}$

7. 0

9. $\frac{1}{13}$

11. $\frac{1}{4}$

13. 6

15. $\frac{1}{6}$

17. $\frac{1}{12}$

19. $\frac{7}{8}$

9-3 Sample Spaces

1. $\frac{1}{10}$

3. $\frac{2}{5}$

5. $\frac{3}{5}$

7. $\frac{1}{2}$

9.

First coin	Second coin	Third coin	Fourth coin
H	H	H	H
H	H	H	T
H	H	T	H
H	H	T	T
H	T	H	H
H	T	H	T
H	T	T	H
H	T	T	T
T	H	H	H
T	H	H	T
T	H	T	H
T	H	T	T
T	T	H	H
T	T	H	T
T	T	T	H
T	T	T	T

11. $\frac{5}{16}$

13.
$(1, 1), (1, 2), (1, 3), (1, 4), (1, 5), (1, 6),$
$(2, 1), (2, 2), (2, 3), (2, 4), (2, 5), (2, 6),$
$(3, 1), (3, 2), (3, 3), (3, 4), (3, 5), (3, 6),$
$(4, 1), (4, 2), (4, 3), (4, 4), (4, 5), (4, 6),$
$(5, 1), (5, 2), (5, 3), (5, 4), (5, 5), (5, 6),$
$(6, 1), (6, 2), (6, 3), (6, 4), (6, 5), (6, 6).$

15. $\frac{1}{36}$

17. $\frac{1}{18}$

19. $\frac{1}{6}$

21. $\frac{1}{6}$

23.
$R_1 R_2, R_2 R_1, W_1 R_1, W_2 R_1,$
$R_1 W_1, R_2 W_1, W_1 R_2, W_2 R_2,$
$R_1 W_2, R_2 W_2, W_1 W_2, W_2 W_1; \frac{1}{6}.$

25.
$R_1 R_2, R_2 R_1, R_3 R_1, W_1 R_1, W_2 R_1,$
$R_1 R_3, R_2 R_3, R_3 R_2, W_1 R_2, W_2 R_2,$
$R_1 W_1, R_2 W_1, R_3 W_1, W_1 R_3, W_2 R_3,$
$R_1 W_2, R_2 W_2, R_3 W_2, W_1 W_2, W_2 W_1; \frac{3}{10}.$

9-4 Computation of Probabilities

1. $\frac{2}{3}$

3. $\frac{2}{3}$

5. $\frac{1}{2}$

7. 1

9. $\frac{1}{2}$

11. $\frac{1}{52}$

13. $\frac{19}{52}$

15. (a) $\frac{1}{17}$; (b) 0; (c) $\frac{13}{204}$; (d) $\frac{1}{663}$; (e) $\frac{4}{17}$.

17. $\frac{1}{32}$

19. (a) $\frac{1}{12}$; (b) $\frac{1}{4}$; (c) $\frac{7}{12}$; (d) $\frac{3}{4}$.

Note: The following sample space is useful:

$H1, H2, H3, H4, H5, H6,$
$T1, T2, T3, T4, T5, T6.$

21. (a) $\dfrac{_{13}C_5}{_{52}C_5}$, that is, $\dfrac{33}{66,640}$;

(b) $\dfrac{_4C_4 \times _{48}C_1}{_{52}C_5}$, that is, $\dfrac{1}{54,145}$.

23. (a) $\frac{1}{64}$; (b) $\frac{5}{144}$; (c) $\frac{343}{1728}$; *(d) $\frac{1}{8}$.

25. (a) $\frac{1}{8}$; (b) $\frac{1}{8}$; (c) $\frac{1}{8}$; (d) $\frac{3}{8}$.

9-5 Odds and Mathematical Expectation

1. 1 to 3.

3. 1 to 3.

5. 3 to 1.

7. 11 to 2.

9. 2 to 7.

11. $4

13. $1.67

15. $2.00

17. The probability that both of the bills

drawn will be tens is $\frac{2}{5} \times \frac{1}{4} = \frac{1}{10}$. The probability that both will be fives is $\frac{3}{5} \times \frac{2}{4} = \frac{3}{10}$. The probability that one will be a five and one a ten is found as $(\frac{3}{5} \times \frac{2}{4}) + (\frac{2}{5} \times \frac{3}{4}) = \frac{3}{5}$. The mathematical expectation is then found to be

$$(\$20)(\tfrac{1}{10}) + (\$10)(\tfrac{3}{10}) + (\$15)(\tfrac{3}{5}) = \$14.$$

9-6 *Permutations*

1. 120

3. 56

5. 42

7. 10

9. 12!, that is, 479,001,600.

11. 30

13. 720

15. 6

17. 5

19. **(a)** 20;
       (b) 120.

21. 210

23. 9!; that is, 362,880.

25. $_nP_0 = 1$

9-7 *Combinations*

1. **(a)** 15;
       (b) 56.

3. **(a)** 21;
       (b) 35.

5. **(a)** 55;
       (b) 455.

7. *pqr, prq, qpr, qrp, rpq, rqp;*
       pqs, psq, qps, qsp, spq, sqp;
       prs, psr, rps, rsp, spr, srp;
       qrs, qsr, rqs, rsq, sqr, srq.
       Thus $_4C_3 = 4$.

9. $_nC_n = 1$

11. $_3C_0 + {_3C_1} + {_3C_2} + {_3C_3}$
       $= 1 + 3 + 3 + 1 = 8 = 2^3$

13. $_nC_0 + {_nC_1} + \cdots + {_nC_{n-1}} + {_nC_n} = 2^n$

15. $_5C_2$; that is, 10.

17. $_{52}C_{13}$; that is, 635,013,559,600.

19. Order is important.

21. $_{10}C_4$; that is, 210; $_7C_3 \times {_3C_1}$; that is, 105.

23. $\frac{1}{5}$

25. $\dfrac{_4C_4 \times {_{48}C_1}}{_{52}C_5}$, that is, $\dfrac{1}{54,145}$.

27. **(a)** $_{10}C_4$, that is, 210.
       (b) $_{10}P_4$, that is, 5040.

29. $_7P_7$, that is, 5040.

31. $_{15}C_9$, that is, 5005.

33. $_5P_5$, that is, 120.

35. $_8C_0 + {_8C_1} + {_8C_2} + \cdots + {_8C_8}$, that is, 256.

37. $_8C_1 + {_8C_2} + {_8C_3} + \frac{1}{2} \times {_8C_4}$, that is, 127.

10

An Introduction to
STATISTICS

10-1   *Uses and Misuses of Statistics*

1.   The fact that most accidents occur near home probably means that most of the miles driven are near the home of the driver. It does not necessarily mean that long trips are safer than short trips.

3.   The conclusion that mathematics is very popular is not justified by the previous statement even though the conclusion may be true. The fact that more students are studying mathematics reflects the fact that there are more students in college and that colleges are requiring more mathematics.

5.   The 100% sale probably indicates that the book was the required text in a course that all students had to elect, but the given evidence does not justify the given conclusion.

7.   The survey may not have been a representative sample of the voting population. Details of the sampling procedure would be needed to feel confident of the conclusion.

9.   One wonders how such a count could possibly have been made. Details of the sampling procedure would be needed to feel confident of the conclusion.

11.   Many people who have asthma go to Arizona because of the climate. Probably a larger part of the people in Arizona have asthma than for any other state. Thus the conclusion is not justified.

13.   People do not necessarily vote for the candidate of the party in which they are registered.

15.   It should be stated what average means in this statement. If average denotes the median, there would always be 50% at or below average.

17.   How short? How tall? What do you mean by aggressive? How do you measure aggressiveness?

19.   Effective for what conditions? How does brand A compare to other brands? What do you mean by effectiveness? How do you determine effectiveness?

21.   Fewer than whom? What age groups are under consideration? How much swimming is considered?

23.   What do you mean by success? What is implied by "student"?

25.   Are those who marry during their last year of college included? What is the source of this information?

27.   The small changes in temperature appear very large due to the fact that the temperature scale does not start at zero.

29.   No; a scale is needed before any such conclusions can be made.

10-2 *Collecting and Presenting Data*

7.

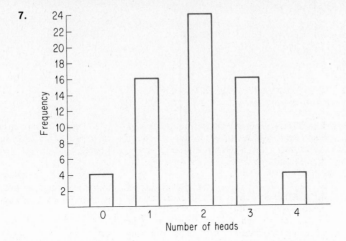

9.

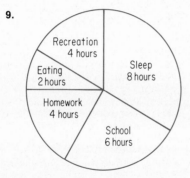

11. 47%

13. **(a)** Yes, 8 per cent; **(b)** no, per cents but not amounts are shown.

15. 169°

17. $80,000

19. 59 and 88.

21. The number of deaths each year decreases after age 75 because there are fewer people still alive then. However, on a percentage basis, the number of deaths per 1,000 increases. Compare the last two columns in the mortaility table after age 75.

23. Wyoming, South Dakota, North Dakota, West Virginia, and Mississippi.

25. California.

27. Total population is completely different from per cent change in population. For example, a state with 400,000 population could have a 25% increase with 100,000 people whereas for a state such as New York or California an increase of 1,000,000 people would still only be less than a 6% increase.

29. No region showed a decrease.

10-3 *Measures of Central Tendency*

1. **(a)** 83; **(b)** 82; **(c)** 85.

3. **(a)** 17; **(b)** 17; **(c)** none.

5. **(a)** 79; **(b)** 75; **(c)** 70.

7. Mean.

9. Mode is often used as related to style; that which is worn or done most often by most people.

11. 2460

13. 770

15. **(a)** 3400; **(b)** 77.

17. 125; she cannot obtain a 90 average if the tests are on a 100 point basis.

10-4 *Measures of Dispersion*

1. 43

3. $\{50, 60, 80, 85, 90, 92, 103\}$. There are many correct answers.

5. $\{57, 75, 80, 85, 90, 95, 100\}$. There are many correct answers.

7. Normal.

9. **(a)** 16%; **(b)** 2.5%; **(c)** 250; **(d)** 1600; **(e)** 8150.

11.
Scores:	d	d^2
7	-3	9
9	-1	1
10	0	0
11	1	1
13	3	9

$5\overline{)50}$ $\quad$ 20

$\overline{X} = 10$

$\sigma = \sqrt{\tfrac{20}{5}} = \sqrt{4} = 2$

10-5 *Binomial Distributions*

1. 120

3. 150

5. 100 to 140.

7. The mean is 32; $\sigma = 4$; $32 \pm 2\sigma$ provides a range of 24 to 40.

9. 45

11. 20 to 40.

INDEX

B

Babylonian numeration, 118
Base for an exponent, 129
Base:
  of a prism, 311
  of a pyramid, 311
Base of a system of numeration, 118, 133
BASIC, 194
Biconditional statement, 89
Biconditional symbol ($\leftrightarrow$), 89
Bimodal set, 410
Binary Boolean algebra, 81
Binary notation, 151
Binary operation, 159
Binary relation, 207
Binomial distribution, 421
Bones, Napier's, 125
Boole, George, 81
Boolean algebra, binary, 81

C

Cantor, Georg, 46
Cardinality of a set, 43
Cardinal number, 44, 48
  transfinite, 45
Cartesian coordinates, 276
Cartesian plane, 276
Cartesian product, 191, 266
Casting out nines, 8
Central tendency, measures of, 408
Chapter tests: *see* Tests
Circle, equation for a, 349
Circle graph, 403
Circuits, electric, 69, 74, 81
Circular permutations, 383
Circular region, 55, 350
Clock arithmetic, 170
Closed curve, simple, 316
Closed figure, 306
Closure property, 159
  for addition, 191, 214, 218, 224
  for multiplication, 191, 214, 219, 224
Coefficient of correlation, 419
Coincident lines, 300
Combinations, 384
Common factor, 202
Common fraction, 222
Common multiple, 204
Commutative group, 163, 218
Commutative property, 160
  for addition, 191, 214, 218, 224
  for multiplication, 192, 214, 219, 224

Complement of a set, 40
Complete number line, 230
Complex numbers, 237
Composite number, 196
Compound sentence, 260
Compound statement, 67
Computation:
  in base five, 140
  methods of, 123
  of probabilities, 369
Concave polygon, 306
Conclusion, 77, 96
Conditional statement ($p \to q$), 77
Conditions, necessary and sufficient, 89
Confidence limits, 421
Conformable matrices:
  for addition, 180
  for multiplication, 180
Congruent line segments, 327
Congruent modulo 5, 176
Conjecture, Goldbach's, 200
Conjunction ($p \wedge q$), 71
Connectives for statements, 67
Consistent statements, 103
Constant of proportionality, 271
Continuous data, 403
Contradictory statements, 104
Contrapositive of a statement, 84
Contrary statements, 103
Control chart, 422
Converse of a ststement, 84
Convex polygon, 306
Coordinate geometry, 297
Coordinates, 213
  Cartesian, 276
  on a line, 213, 326
  on a plane, 340
Correlation, coefficient of, 419
Correspondence, one-to-one, 34, 42
Counterexample, 160
Counting numbers, 34, 190
  properties of, 191
Counting problems, 355
  general principle for, 356
Cross-number puzzles, 215
Cube, 310
  model for a, 315
  volume measure of a, 336
Curve(s):
  arc of a, 318
  normal, 414
  plane, 316
  simple closed, 316
  vertex of a, 318

GLOSSARY

$\in$	is a member of
$\notin$	is not a member of
$\subseteq$	is included in (is a subset of)
$\subset$	is properly included in (is a proper subset of)
$\varnothing$	empty set
A'	the complement of set A
$A \leftrightarrow B$	set A is equivalent to set B
$n(B)$	the cardinal number of set B
$\aleph_0$	aleph-null
$A \cap B$	the intersection of sets A and B
$A \cup B$	the union of sets A and B
$\wedge$	and (conjunction)
$\vee$	or (disjunction)
$\sim p$	not p (negation)
$p \rightarrow q$	statement p implies statement q (conditional)
$p \leftrightarrow q$	p is equivalent to q (biconditional)
$\exists_x$	there exists x such that
$\forall_x$	for all x
$A \times B$	the Cartesian product of sets A and B